NETWORK STRATEGIES IN

Ashgate Economic Geography Series

Series Editors:
Michael Taylor, Peter Nijkamp, and Tom Leinbach

Innovative and stimulating, this quality series enlivens the field of economic geography and regional development, providing key volumes for academic use across a variety of disciplines. Exploring a broad range of interrelated topics, the series enhances our understanding of the dynamics of modern economies in developed and developing countries, as well as the dynamics of transition economies. It embraces both cutting edge research monographs and strongly themed edited volumes, thus offering significant added value to the field and to the individual topics addressed.

Other titles in the series:

Tourism and Regional Development
New Pathways
Edited by Maria Giaoutzi and Peter Nijkamp
ISBN: 978-0-7546-4746-1

Alternative Currency Movements as a Challenge to Globalisation?
A Case Study of Manchester's Local Currency Networks
Peter North
ISBN: 978-0-7546-4591-7

The New European Rurality
Strategies for Small Firms
Edited by Teresa de Noronha Vaz, Eleanor J. Morgan and Peter Nijkamp
ISBN: 978-0-7546-4536-8

Creativity and Space
Labour and the Restructuring of the German Advertising Industry
Joachim Thiel
ISBN: 978-0-7546-4328-9

Proximity, Distance and Diversity
Issues on Economic Interaction and Local Development
Edited by Arnoud Lagendijk and Päivi Oinas
ISBN: 978-0-7546-4074-5

Network Strategies in Europe
Developing the Future for Transport and ICT

Edited by

MARIA GIAOUTZI
National Technical University of Athens, Greece

and

PETER NIJKAMP
Free University, The Netherlands

LONDON AND NEW YORK

First published 2008 by Ashgate Publishing

2 Park Square, Milton Park, Abingdon, Oxon OX14 4RN
711 Third Avenue, New York, NY 10017, USA

Routledge is an imprint of the Taylor & Francis Group, an informa business

First issued in paperback 2016

British Library Cataloguing in Publication Data
Network strategies in Europe : developing the future for
transport and ICT. - (Ashgate economic geography series)
1. Data transmission systems
I. Nijkamp, Peter II. Giaoutzi, Maria, 1946-
621.3'82'094

Library of Congress Cataloging-in-Publication Data
Giaoutzi, Maria, 1946-
Network strategies in Europe : developing the future for transport and ICT / by Maria Giaoutzi and Peter Nijkamp.
p. cm. -- (Ashgate economic geography series)
Includes bibliographical references and index.
ISBN 978-0-7546-7330-9
1. Data transmission systems. I. Nijkamp, Peter. II. Title.

TK5105.G53 2008
388.068'4--dc22

2007046430

ISBN 13: 978-0-7546-7330-9 (hbk)
ISBN 13: 978-1-138-26759-6 (pbk)

Contents

List of Figures

List of Tables

Notes on Contributors

Ingeborg van Ansem
DHV BV
P.O. Box 1132
3800 BC Amersfoort (The Netherlands)
Ingeborg.vanansem@dhv.nl

Art Bleukx
Division Traffic and Infrastructure
Katholieke Universiteit Leuven
Kasteelpark Arenberg 40
B-3001 Heverlee (Belgium)
art.bleukx@bwk.kuleuven.be

Javier Campos
University of Las Palmas
Las Palmas G.C., Canary Islands (Spain)
jcampos@daea.ulpgc.es

Christos Dionelis
Department of Geography and Regional Planning
National Technical University of Athens
Athens (Greece)
cdion@survey.ntua.gr

Holmer Doornbos
Delft University of Technology,
Delft (The Netherlands)

Marina van Geenhuizen
Delft University of Technology,
Delft (The Netherlands)
m.s.vangeenhuizen@tbm.tudelft.nl

Maria Giaoutzi
Department of Geography and Regional Planning
National Technical University of Athens
Athens (Greece)
mgiaoutsi@central.ntua.gr

Moshe Givoni
Oxford University Centre for the Environment (OUCE)
South Parks Road,
Oxford, OX1 3QY (UK)
moshe.givoni@ouce.ox.ac.uk

Ekko van Ierland
Environmental Economics Group
Wageningen University
P.O. Box 8130
6700 EW Wageningen (The Netherlands)
ekko.vanierland@wur.nl

Ben Immers
Division Traffic and Infrastructure
Katholieke Universiteit Leuven
Kasteelpark Arenberg 40
B-3001 Heverlee (Belgium)
ben.immers@bwk.kuleuven.be

Dirk-Jan F. Kamann
Groningen Research Institute of Purchasing (GRIP)
University of Groningen
Lutkenieuwstraatje 4
9712 AX Groningen (The Netherlands)
d.j.f.kamann@rug.nl

Mark Koetse
Department of Spatial Economics
VU University Amsterdam
De Boelelaan 1105
1081 HV Amsterdam (The Netherlands)
mkoetse@feweb.vu.nl

Lars Lundqvist
KTH
Royal Institute of Technology
Stockholm (Sweden)
lars@infra.kth.se

John Mourmouris
International Economic Relations and Development Department
Demokritus University of Thrace
69100 Komotini (Greece)
jomour@eexi.gr

Peter Nijkamp
Department of Spatial Economics
VU University Amsterdam
De Boelelaan 1105
1081 HV Amsterdam (The Netherlands)
pnijkamp@feweb.vu.nl

George Patris
National Technical University of Athens
Athens (Greece)
gpatris@central.ntua.gr

Piet Rietveld
Department of Spatial Economics
VU University Amsterdam
De Boelelaan 1105
1081 HV Amsterdam (The Netherlands)
prietveld@feweb.vu.nl

Caroline Rodenburg
International Location Advisory Services (ILAS)
Ernst & Young
Euclideslaan 1
3584 BL Utrecht (The Netherlands)
Caroline.Rodenburg@nl.ey.com

Manuel Romero
University of Las Palmas
Las Palmas G.C., Canary Islands (Spain)
mromero@daea.ulpgc.es

Anastasia Stratigea
Department of Geography and Regional Planning
National Technical University of Athens
Athens (Greece)
stratige@central.ntua.gr

Luis Suarez-Villa
School of Social Ecology
University of California
Irvine, CA 92717-5150 (USA)
lsuarez@uci.edu

Vassilios Vescoukis
Department of Geography and Regional Planning
National Technical University of Athens
Athens (Greece)
v.vescoukis@cs.ntua.gr

Ron Vreeker
Department of Spatial Economics
VU University Amsterdam
De Boelelaan 1105
1081 HV Amsterdam (The Netherlands)
rvreeker@feweb.vu.nl

Preface

Since the early days of mankind, transportation and mobility have been one of the visible manifestations of progress, discovery and innovation. Transportation and mobility open up new horizons and create uniform opportunities, based on human and physical interaction. It is, therefore, no surprise that in building up the European Union the transportation and communication sector has become one of the foundation stones for an efficiently operating common market that would render Europe more competitive.

Clearly, transportation and communication are to be embedded in a broader societal, political and technological context. The embeddedness of this sector in a European unification strategy will induce the necessary socio-economic synergy, based on technological innovation and institutional reforms.

The present volume brings together various contributions on the economic, political, technological and institutional conditions for, and impacts of, an efficiently operating transportation and communication sector in Europe. It combines planning challenges with case-study experiences from various countries, and draws lessons from the multi-faceted fabric that makes up the European Union. In doing so, it presents also new research methodologies for tackling complex assessment issues in the European transportation and communication domain.

The ingredients for this volume were prepared during a NECTAR (Network on European Communication and Transportation Activity Research) workshop in Ikaria, on the pearls of the Aegean. A selection of various papers centring around the theme of transportation and communication was made and after a review process combined into the present edited volume. Our sincere thanks go to Roberto Patuelli who assisted us in the logistics of the review procedure, and to Mrs Patricia Ellman who made a formidable effort to turn the various contributions into beautiful English. And finally, the NECTAR organization has to be thanked for its continuous support for creating a high-level transportation and communication research expertise in Europe.

Maria Giaoutzi
Peter Nijkamp
Athens/Amsterdam, January 2008

Chapter 1

Spatial Interaction and Information: The Pillars of European Networks

Maria Giaoutzi and Peter Nijkamp

European Networks: Endless Roads

The 21st century will be the age of global interaction. The intensification of social, economic or information interaction is, however, not taking place randomly but will be based on organized patterns usually following network patterns and developments. Networks are systematically organized horizontal and vertical (multi-layer) interaction architectures that derive their existence from the embodied interaction efficiency in an open and accessible world. This holds for communication, for migration, and for physical transport. Given the strategic position of networks in transportation, we may, for example, argue that physical distances will largely lose their importance, and travelling on a world-wide scale will become a common phenomenon for the 'global citizen'. There is no natural limit to the mobility of individuals, while the transport of goods, services and information will experience accelerated growth as a result of competition at both global and local levels. The action radius of individuals and firms will be stretched out towards global levels, as a result of both technological progress and changes in lifestyle (see also Frandberg and Vilhelmson 2003). Hence, transportation systems and physical infrastructures will become the pillars for a new mobile society.

An important role will be played by the rapid development of ICT in general, and in relation to transport in particular. And, more specifically, information – as a virtual transportation means – will exert a great impact on the interaction and transportation patterns of people all over the world. Information will not only create a sense of global citizenship, but will also enhance the performance of transportation systems, so that these systems become more transparent and accessible (see also Urry 2003; Limtanakool et al. 2006).

The previous observations have a particular relevance for the emerging European space-economy. Until the 1990s, Europe used to be a historically-grown patchwork of fragmented spatial and national systems, characterized by national and regional identities and interests. The European network – in the sense of an open interactive system – hardly existed. It has taken almost 50 years – since the Treaty of Rome – to build up an open and accessible European system, where access, mobility and international travel can be enjoyed by almost all European cities.

The European network system is still in a state of emergence, and it will certainly take a few more decades before national and regional interests will have faded away to create an open and accessible European network. It is noteworthy that network concepts have prompted a rising interest in the past decades (see, for example, Cooke 2001). In general, networks are based on interactions between various nodes (for example, main actors or agencies) in an interactive system, and are driven by synergy effects emerging from an organized strategy by interaction partners. In other words, spatial and network externalities have a prominent place in network analysis (see, for example, Yilmaz et al. 2002; Russo et al. 2007).

The focus of the present volume is less on the modelling of networks and their synergy effects. We refer to studies by Wiberg (1993), Thorsen et al. (1999) or van der Laan (1998) for statistical and econometric studies on network topologies and structures. This volume aims to address in particular the strategic dimensions of networks, especially with regard to transportation and communication in the European policy space. This choice was instigated by the challenges put forward by European unification and accession efforts, by means of which a coherent and efficient European network system will have to be created. Hence, most studies in this book are chosen from the perspective of relevance for European network strategies in which many actors and different layers are involved. The topics addressed are: the emergence of new network structures and strategies, the implications of European integration policies for network operations and developments, and the assessment of network synergy effects.

Organization of the Book

In the light of the previous remarks, the present volume aims to offer a panoramic view of current analytical and policy issues regarding emerging European networks. Whenever appropriate, particular emphasis will be given to the interface between transportation and information systems and strategies, set against the background of recent developments in European transport and communication policy.

Part 1 begins with a novel contribution by Luis Suarez-Villa who introduces the concept of an experimental firm as an innovator in a dynamic network perspective. His argument is as follows. The rising importance of firms that are highly focused on research, where invention is a fundamental objective, poses new challenges to spatial development and policy. Such enterprises may be referred to as experimental firms, given their overarching emphasis on invention and innovation, their high reliance on external networks and communication systems, fluid internal organization, multidisciplinary needs, and the fundamental importance of maintaining high outputs of new discoveries. Networks can play an important role in the development of firm-level innovative capacity. In many experimental firms, establishing network relations with other firms can provide the vital resources needed to come up with new inventions and innovations. Suarez-Villa's contribution considers the relationship between experimental firms, their networks and innovative capacity, along with their implications for spatial development and policy. He also considers how networks influence innovative capacity at the level of firms. The role of collaboration in networks and its importance for enhancing firm-level innovative capacity is explored as well. Next, the case of experimental firms in the biotechnology sector is given special attention, in order to consider the factors that promote collaboration and their effect on innovative capacity, along with a summary of some empirical evidence. Finally, he discusses how spatial development theory and policy might address experimental firms, their innovative capacity and dynamic networks, given their importance for regions and locales.

The second contribution in Part 1 is offered by Ben Immers and Art Bleukx, who focus their attention on the robustness of emerging network structures, in particular from the viewpoint of reliability in the usage of networks. Reliability of travel times is a subject that is rapidly gaining in importance in many countries and many policy plans are currently focusing on this issue. Several definitions and indicators of reliability can be found in the current literature. The problem is that these definitions and indicators are not always as clear as we would like them to be. The authors then propose a number of way in which the definitions and indicators should be improved in order to make them directly applicable. Furthermore, little is known about the factors on which reliability depends, in particular, the relationship between the robustness of networks and the reliability of travel time. The authors then identify the factors that determine the reliability of travel time. Robustness is one of them. Their contribution addresses in more detail the following aspects of robustness: redundancy, interdependency, resilience, and flexibility. The meaning of these terms, as well as their relevance for the robustness of networks is explained. Finally, several results of a series of model calculations on a test network are used to show the impact of various (robustness) aspects on the reliability of travel time.

Next, Dirk-Jan Kamann deals with the new challenges imposed by electronic B2B contacts and patterns. In his study, he combines various scientific fields to assess the potential impact of E-contacts on locational behaviour. The purchasing literature is used to select those goods and services that are E-Contact sensitive. From this, a differentiation is made between suppliers and types of *relations.* In the next step, regional science literature is used to evaluate the E-Contact sensitivity of relations, based on the degree of routinization of activities and the nature of contacts required to perform these activities. Extrapolating the results, new opportunities for actors are identified, in new roles. This results in speculative new criteria on which actors will be evaluated. This has effects on the required characteristics of logistic flows and the optimal location for establishing activities. Using this, a dynamic location model is formulated for logistic activities. It predicts a continuous process of preferential shifts between locations. Nowadays, new locations with time hazard-free access and a high perceived connectivity to central areas become places of attraction. When, as a result of their popularity, congestion rises, the trend will be reversed, and another location becomes the new place after which the cycle continues again.

Information systems are apparently critical for spatial dynamics. As current network technology developments rapidly penetrate citizens' life in modern cities, more ICT technology-based services are made available, initially as luxuries and eventually as necessities. Social interaction patterns are drastically changing, and so are many aspects of urban organization and governance. The urge to reduce the gap between the purely technological aspects of new ICT technologies and their commercialization results in a grid of several to some extent converging but largely independent technologies and services. As the prospect for upgrading urban services increases, driven by social demand, the integration of technologies may provide the means for a more efficient governance of urban services. The focus of this chapter, written by Maria Giaoutzi and Vassilios Vescoukis, is on the impacts of ICT on urban policy making and applications, with emphasis on the attributes introduced by the integration of mobile information technologies. After a brief review of the theoretical findings on the likely impacts of ICT on urban patterns, a reference framework describing the functioning of the urban cyber context, on the one hand, and the operation patterns of ICT services, on the other, is presented. Next, ICT applications will be highlighted with emphasis on the layers of operation of each particular application, after which the ICT service supply-demand interaction with the urban policy framework is also discussed, followed by a description of urban information location-based services (LBS), with an application to car parking management, where integration and convergence issues are incorporated.

Spatial and transportation networks are often vehicles for competition, and hence may increase economic efficiency. An interesting example is offered by airline networks. An assessment of competition in network industries – with particular emphasis on air transport systems – is next offered by Javier Campos and Manuel Romero. They argue that any company operating in a network industry may use the structure of its network – the location of the nodes and the links between them – as a means by itself to exercise market power. To provide competition agencies with adequate instruments to evaluate the relevance of this problem, their study proposes a new concentration measurement tool based on the use of origin-destination matrices. The European air transport sector provides a suitable example to test this tool, not only because of its network structure, but also as a result of the liberalization process it was involved in from 1987 to 1997. Their empirical results suggest that competition has increased, but a number of market imperfections still persist, possibly preventing an economically optimal resource utilization from a pan-European viewpoint. In fact, the hub-and-spoke network structure appears to dominate the industry, giving advantages to some carriers and possibly preventing a larger degree of competition.

The final contribution in Part 1, written by Vassilios Vescoukis, offers an analysis of the role of enabling technologies like ICT for spatial interactions, especially on the job markets. The speed of technological evolution, especially in the domain of network technology and infrastructure, is so fast that traditional vocational training often fails to provide students with up-to-date knowledge and practical skills. This results in a gap between the classroom and the job practice, which is even wider in non-urban regions, often not favoured by central administrations. Several actions have been taken in order to fill this gap, ranging from continuing professional education to long-term solutions dealing with the structure, the content, and tools used in vocational training. Network technologies are *enabling technologies* that support new knowledge delivery approaches using independent-of-time-and-space learning paradigms. Their independence from space enables education actions to take place in any geographic location and thus, the inherent handicaps of non-urban areas (for example, rural and border areas) in this field may be treated. One field where new network-based educational paradigms can be effectively applied is on-the-job training, which has been recognized as a very important educational activity but often carried out without the necessary efficiency. The authors discuss an alternative to traditional on-the-job training activities in technical vocational training, based on distributed network technologies, which favours the remote areas. Their approach is based on business simulation activities, conducted asynchronously and independently of physical location, over a computer network – usually the Internet. Spatial interactions and network technologies are apparently mutually related.

The next part of the present volume (Part 2) focuses attention on the dynamics of the European Union, in particular its extension and the resulting network developments. The first chapter in Part 2 deals with the enlargement of the European Union and the consequent evolution of transport linkage and patterns. It is written by Christos Dionelis and Maria Giaoutzi. The authors argue that the recent enlargement is a unique historic opportunity to further promote the integration of the European continent, and to extend a zone of stability and prosperity over the whole of Europe. The enlargement will have a major impact on EU transport policy for both old and new Member States and particularly for the new Members. In order to ensure the development of sustainable transport in its neighbouring countries, the EU should take account of the legislative and democratic framework in the acceding countries of the EU, trans-boundary connection and cooperation, the need for modernization and optimization of the use of the existing infrastructure, and the promotion of environmentally-friendly transport modes. As demand for transport keeps increasing, the Community's answer cannot be just to build new infrastructure and open up markets. The transport system needs to be optimized to meet the demands of enlargement and sustainable development. Currently, transport questions in the Union are handled by a coherent set of rules and provisions, which form the Common Transport Policy (CTP). Traditionally, as transport problems directly reflect the segmentation of Europe's transport systems and markets themselves, the primary role of the CTP was to remove the existing technical and institutional barriers between the Member States. The enlargement will be a new major challenge for the successful implementation of the CTP. At the heart of this policy, sustainable mobility has been set as the main objective; this will continue to rule future actions towards the integration of the new regions into the 'old European core'. The authors highlight the main characteristics of the transport scene in Europe, and the main expressions of the CTP in terms of infrastructure development, that is, the transport networks and corridors. Some additional anticipated implications of the enlargement for the Union's transport infrastructure are also tackled in this contribution.

The next chapter in Part 2 is produced by Lars Lundqvist. He deals mainly with cumbersome issues related to the evaluation of model results along key dimensions of European spatial policy. Polycentric development has emerged as a key concept during the European Spatial Development Perspective process. One of the motivations is to avoid further excessive economic and demographic concentration in the core area of the EU. A methodology for analysing the impacts of EU transport policies on polycentric development is explained here. Building on the earlier tradition of analysing key conditions for regional development, the author combines indicators for the following dimensions: mass, competitiveness, connectivity, and development trend, into a composite indicator of development potential. This composite indicator is used to compare the impacts of transport policy scenarios on the development potential at a detailed territorial level. By analysing the spatial pattern of these impacts, the effects on the potential for polycentric development are illustrated. The results indicate that transport policy (investments and/or pricing) can potentially be used to encourage various forms of polycentric development. The author then offers explicit analyses of the potential trade-offs between the aims for efficiency (competitiveness) and equity (cohesion). The analysis focuses on efficiency in terms of total GDP or development potential and the corresponding inequities measured in both absolute and relative terms. The results indicate that there are trade-offs between efficiency and the indicator of absolute inequity. Transport investments appear to increase both efficiency and inequity in comparison with the reference scenario.

Networks may manifest themselves in various ways, ranging from fully integrated networks, at one extreme, to corridors at the other. In their contribution on network versus corridor concepts, Christos Dionelis, Maria Giaoutzi and John Mourmouris highlight that recent developments in transport research and policy making call for sound tools and operational concepts in order to cope with the complexity of the existing problems in infrastructure patterns. The competition and cooperation between the various modes of transport, the conflicting interests, and the allocation of limited funds for transport infrastructure developments are issues requesting rational decision making on European networks. Their study focuses on the evolving nature of such problems, dealing with the shift in emphasis, from the network to the corridor concept, in the EU Common Transport Policy (CTP), in order to enable a broader range of specific attributes to be incorporated into the analysis and evaluation phase of the planning process. Their contribution presents the framework of the CTP, its objectives and some of its main achievements. The emphasis is placed on the Trans-European Transport Network (TEN), the Pan-European Transport Corridors, and the Transport Infrastructure Needs Assessment (TINA) Network in order to understand the developments in the EU context that have created the need for the evolution of these concepts. Next, their study presents a set of indicators used in practice for the analysis and evaluation of a transport corridor or network, while it also examines the appraisal framework for the evaluation of the emerging EU transport corridors and networks.

Network integration is a particularly relevant issue in European transport networks, in particular in the TEN system. In the next chapter, Anastasia Stratigea, Maria Giaoutzi and Constantin Koutsopoulos observe, that during recent decades, there has been a remarkable change in mobility patterns at a European level as a result of various social, economic, political as well as technological driving forces. This change has significantly affected the nature and structure of European networks. The EU, while recognizing the role of infrastructure networks as strategic resources, has put emphasis on policies promoting infrastructure network development and integration, in order to ensure the provision of the necessary infrastructure, a prerequisite for achieving the twin goals of the smooth functioning of the internal market and of social and economic cohesion. Emerging in this context are the Trans-European and Pan-European Transport Networks (TEN-T and PEN-T, respectively), which are considered to play a crucial role as the heart of the interaction between European regions. As such, they form the core of the EU's CTP. A key feature of TEN-T and PEN-T is their emphasis on multimodal, intermodal and interoperable networks. These are considered as key aspects of network integration at the European level, and represent the core of transport policy and network integration at a European level.

The final contribution to Part 2 of this volume is written by Vassilios Vescoukis and Maria Giaoutzi. Their contribution is devoted to European telecommunication networks. The Trans-European Networks for Telecommunication (TEN-Telecom) aim to stimulate the use of global telecommunications networks in areas of high socio-economic value, by promoting across Europe new multimedia applications and generic services of common interest made possible by such infrastructures. TEN-Telecom adopts the multi-network approach, addressing not single-carrier telecommunications networks, but, instead, services which seamlessly integrate fixed and mobile network components. In one sense, TEN-Telecom helps participating parties to bridge the gap between technical evolution and market operations, accelerating the deployment of new services to the market, and encouraging investments in services of common interest that are not likely to obtain immediate funding. The authors outline the opportunities and barriers involved.

The final part of this volume is devoted to the evaluation of network integration policies, at various institutional, geographical and modal levels. The first contribution in this part is provided by Moshe Givoni, who analyses competition and substitution effects between two different transport models that are critical for European integration, viz. airlines and railways. Under airline and railway integration the High-Speed Train (HST) may be a substitute for the aircraft, but not the service provider, on one leg of the journey. Although the airline remains the provider of the HST service this service is operated by the railway company. His chapter analyses the network effect of such an operation, termed 'airline and railway integration', at one major international airport, London Heathrow. The analysis is based on a current definition in the literature for mode substitution, but eventually presents a new criterion. In this criterion, attention is drawn to the fact that on some routes the HST is much more time-consuming than the aircraft route and this must be taken into consideration. The analysis also shows the potential to free up runway capacity at Heathrow following mode substitution; and this was found to be about 20 per cent of the airport current capacity. Other network effects arising from connecting the airport to the (HST) rail network are also explored. His analysis demonstrates the dynamic – direct and indirect – effects of network integration.

Another interesting contribution is offered by Marina van Geenhuizen and Holmer Doornbos, who look into the role of airports as nodes in knowledge networks, in relation to the development of the urban space-economy. In studies on the agglomeration advantages of urban areas, much attention is paid to benefits from local knowledge spillovers. However, much less attention is given to the global nature of some urban knowledge, the concomitant needs of knowledge workers to travel to distant places by air, and the benefits of having to travel only a short distance to an international airport to serve these needs. In this chapter, the role of international airports in providing access to global knowledge networks is explored by taking Amsterdam Schiphol Airport (AMS) in the Netherlands as an example. On the basis of notions from resource dependence theory and rough set analysis of a small, selected, sample of young, innovative companies, factors are identified that determine the spatial layout of knowledge networks and the importance of AMS as a node in employing global knowledge networks. Furthermore, the use of ICT is explored to identify whether there is substitution between long-distance physical contacts and electronic communication. This is followed by a profile description of young high-technology companies for which access to an international airport is vital.

Networks cover many modalities, and one of them is waterways. In a subsequent chapter of Part 3, Ron Vreeker, Peter Nijkamp, Ingeborg van Ansem and Ekko van Ierland pay attention to the role of seaports in international waterway networks. They address one of the most ambitious plans for port expansion, viz. the expansion of the Port of Rotterdam into a location in the North Sea. As a maritime nation, The Netherlands has a strong historical position in the transport of goods. It is thus no wonder that port expansion has always been looked on favourably. This has, for example, led to the major offshore expansion of the Port of Rotterdam in the 1970s, by reclaiming a new port area from the North Sea, known as Maasvlakte. This area is gradually reaching its capacity limits, and consequently an intensive debate has started on the question whether another large area might be reclaimed from the sea. The present study maps out the strategic issues involved, the various choice alternatives and the relevant policy criteria for evaluating this large-scale project. Then, a multicriteria analysis is deployed to identify the best possible choice possibility for port expansion.

Another example of increasingly important networks in a complex spatial constellation is offered by urban transport networks. Cities are nuclei of economic forces, and hence urban networks provide a platform of action for the many actors involved. Effectively operating transport networks in cities are therefore of critical importance. Cities in Europe offer a surprising variety of urban network developments, ranging from advanced mass transport systems to an upgrading of existing systems. An interesting illustration of the latter strategy is offered by the Athens Tramway system, as part of an integrated restructuring of the urban transport network. The authors, George Patris, Maria Giaoutzi and John Mourmouris, focus in particular on decision tools developed for the implementation study of the Athens Tramway system. The aim of their contribution is to present the methodology developed for the evaluation of the alternatives proposed for the selection of the best-case scenario. First, their study elaborates on the rationale underlying the choice of tools utilized in the evaluation process for the problem under study. And next, they pay attention to the evaluation procedures as such, that have been undertaken in successive stages using a set of indicators – both qualitative and quantitative – as an input to the multicriteria approach that aims to identify a compromise best-case scenario. They conclude their study by pointing out the strong and weak points of the implemented approach.

And finally, attention is paid to the institutional coordination of urban network policy, from the viewpoint of ideas from fiscal federalism applied to parking policy in urban transport systems. The authors, Caroline Rodenburg, Mark Koetse and Piet Rietveld, observe that parking is an issue that is increasingly calling for attention in policy making and in scientific research, mainly because of its great many externalities. Many local authorities have therefore introduced parking tariffs or put restrictions on parking duration and supply in order to internalize some of the externalities. As well as efficiency arguments, other objectives of parking policy may exist, such as the generation of revenues. To what extent economic efficiency and equity play a role in actual policy making depends on, amongst other things, the goals of decision makers at various levels of government. It is therefore not surprising that substantial variations may exist between municipalities, and that the interests of local, regional and central governments may diverge. In this respect, the governmental structure underlying the parking policies may have a great impact on the outcomes of the policies implemented. Indeed, from an institutional standpoint there has been an interest in the influence of governmental structures on the effectiveness and outcomes of policy making. One of the reasons for this is the potential heterogeneity of the relationship. In their contribution, the authors look at the institutional background of Dutch parking policies in order to analyse the influence of federal structures on parking policy and the functioning of the Dutch parking market. The authors illustrate their arguments by presenting results from three case studies on different forms of Dutch parking policies.

This volume offers a diversity of novel insights into the underlying driving forces of network developments in Europe. Cohesion and unification need, as foundation stones, the presence of advanced and accessible spatial networks. The functioning of these networks is not only decided by the technological sophistication of these systems, but also by their usage. A critical success factor for efficient usage is given by modern ICT developments. But again, ICT application and client-oriented usage are necessary conditions, but by no means sufficient success factors. In the first place, the physical geography may be a limiting factor for the high performance of networks, and institutional impediments have to be removed as well. It is thus clear that there is an overwhelming need for appropriate assessment techniques and evaluation methods that may inform public policy actors. The present volume provides not only a wealth of insights into the diversity in the emerging European network, but also interesting examples of operational decision-support approaches for balanced European network strategies.

References

Cooke, P. (2001), 'Regional Innovation Systems, Clusters, and the Knowledge Economy', *Industrial and Corporate Change* 10:14, 945–74.

Frandberg, L. and Vilhelmson, B. (2003), 'Personal Mobility', *Environment and Planning A* 35:10, 1751–68.

van der Laan, L. (1998), 'Changing Urban Systems', *Regional Studies* 32:3, 235–47.

Limtanakool, N., Dijst, M. and Schwanen, T. (2006), 'The Influence of Socioeconomic Characteristics, Land Use and Travel Time Considerations on Mode Choice for Medium- and Longer-Distance Trips', *Journal of Transport Geography* 14, 327–41.

Russo, G., Reggiani, A. and Nijkamp, P. (2007), 'Spatial Activity and Labour Market Patterns', *Annals of Regional Science*, forthcoming.

Thorsen, I., Uboe, J. and Naevdal, G. (1999), 'A Network Approach to Commuting', *Journal of Regional Science* 39:1, 73–101.

Urry, J. (2003), 'Social Networks: Travel and Talk', *British Journal of Sociology* 54:2, 155–75.

Wiberg, U. (1993), 'Medium-Sized Cities and Renewal Strategies', *Papers in Regional Science* 72:2, 253–69.

Yilmaz, S., Haynes, K.E. and Dinc, M. (2002), 'Geographic and Network Neighbours', *Journal of Regional Science* 42:2, 339–60.

PART 1
Analysis of Network Structures and Strategies

Chapter 2

Networks, Innovative Capacity and the Experimental Firm: Implications for Regional Development Policy

Luis Suarez-Villa

Introduction

The rising importance of firms that are highly focused on research, where invention is a fundamental objective, poses new challenges to regional development policy. Such enterprises may be referred to as *experimental firms*, given their overarching emphasis on invention and innovation, their high reliance on external networks, their fluid internal organization, their need for multidisciplinary research talents, and the fundamental importance of sustaining a continuous stream of new discoveries. The best examples of these organizations can be found in the most research-intensive activities of our time, in such fields as biotechnology, software design, bioinformatics, nanotechnology, or biorobotics.

The survival of most experimental firms depends on what may be referred to as their *innovative capacity*, or the possibility of coming up with new inventions. Innovative capacity was originally introduced as an indicator of macro-level inventive output (see www.innovativecapacity.com and Suarez-Villa 1990, 1993, 1996, 1997, 2000, 2001a, 2001b, 2002c; Suarez-Villa and Hasnath 1993). However, it can also be applied to individual firms and to networks of firms, in order to determine their research intensity and technological capability. The case of experimental firms seems very suitable in this regard, given the overarching importance of invention and of inventive productivity.

Networks can play an important role in the development of firm-level innovative capacity. In many experimental firms, establishing network relations with other firms can provide the vital resources needed to come up with new inventions. Relational links involving inter-firm collaboration have therefore become common in research-intensive sectors where many small- and medium-size firms are found, with limited resources of their own and strong pressures to sustain or increase the flow of new discoveries.

Innovative Capacity in Networks

The main objective of developing a firm's innovative capacity is to come up with discoveries that can be turned into innovations. Networks support this process by allowing firms to secure advantages that they would not be able to obtain on their own. There are various ways in which networks enable firms to enhance their innovative capacity. All of them involve collaboration built on relationships with other firms, where trust is of fundamental importance.

Access to Resources

Network relations can provide access to new knowledge and other intangible resources needed to support a firm's innovative capacity. This is often accomplished when firms with diverse talents and specialties are part of a network and pool their resources in a collaborative way. In fields such as biotechnology, bioinformatics, software design, robotics and nanotechnology, for example, multidisciplinary talents are essential for research and can seldom be found together in a single firm. Collaboration through networks can therefore be vital for firms in those sectors, since it is often the only way they can obtain access to all the talents needed to invent and innovate in a cost-effective way.

In software specification and design, for example, linguists, computer engineers, graphic artists and marketing experts often collaborate with software specialists to come up with new products (see, for example, Clark and Edwards 1999; Tsang 2000). In biotechnology, biologists, physicians, pharmacologists, software engineers and chemists often have to be brought together to find the kinds of discoveries that are at the heart of biotech firms' innovative capacity (see Liebeskind et al. 1996; Robbins-Roth 2000; Baldi and Brunak 2001; Orsenigo et al. 2001). In robotics, mechanical engineers, zoologists, electrical engineers, orthopaedic specialists and software engineers often have to pool their talents to come up with new devices (see, for example, Taubes 2000; Menzel and D'Aluisio 2000; Webb and Consi 2001).

Similarly, networks facilitate access to the kinds of tangible resources which are needed to support innovative capacity, such as investment capital. Developing a firm's innovative capacity often requires substantial financial resources that are simply not available through the conventional financial institutions. The latter are often too risk-averse to back research projects for which there is significant uncertainty, either because it is impossible to predict end results or to determine precisely the returns on investment. Financially stressed firms are common in many new sectors and activities. In biotechnology, for example, the 'cash-burn' rates of firms are such that about one-third of all firms typically have less than 12 months' capital on hand to support their operations (see Suarez-Villa and Walrod 2003). For those firms, networks are vital in making contact with potential partners who can provide the financial resources to undertake or continue a research project.

Access to supportive resource relationships with other firms involving outsourcing or subcontracting arrangements that can enhance a firm's innovative capacity are also often found through networks. Such relations support innovative capacity by making it possible to save internal resources which can be redeployed to support research projects. Activities or tasks that are outsourced allow firms to dedicate more resources to research when the outsourced activities can be more effectively carried out by the subcontracted firms (see, for example, Suarez-Villa and Fischer 1995; Suarez-Villa and Karlsson 1996; Suarez-Villa and Rama 1996; Suarez-Villa and Walrod 1997). In that way, therefore, firms' internal resources may not only be preserved and reallocated strategically towards research, but they may also be enhanced if the outsourcing arrangement can add value at a lower cost than could be done internally.

As firm specialization deepens within a sector or activity, networks become vital for securing such inter-firm relationships. Indeed, many firms' outsourcing arrangements today resemble a vast web or net of links to other firms, where resource savings often translate into greater material support for research. Concentrating on an activity that a firm can do best and letting others take over ancillary operations is often a key to success for firms doing cutting-edge research. Developing a strong innovative capacity can often depend on such strategic decisions. As a new technology sector develops and specialization deepens, strategic considerations are bound to make inter-firm networks more important for securing the expertise needed to invent and innovate.

Innovative capacity can also be enhanced through the creation of *networked research alliances* (see Suarez-Villa 1998, 2002a). Such alliances can be vital in acquiring the basic resources needed to sustain any programme involving invention and innovation. In biotechnology, for example, strategic research alliances commonly develop between large firms in related sectors, such as pharmaceuticals, and small but highly innovative biotech enterprises. The larger firms often provide the financial resources that their small partners do not have, while the small firms provide the new knowledge and research capabilities that their large partners cannot deploy in-house. Increasingly, such alliances are occurring within networks, as partners are needed to support such areas as production, marketing and distribution. Thus, multiple alliances are becoming a common feature of the networked research alliances that are occurring in such fields as biotechnology, software design, pharmaceuticals, advanced electronics, and nanotechnology.

Networks also often provide the resources needed to incubate new firms or their inventions. Many new firms involved in cutting-edge research do not have the resources to subsist or develop much innovative capacity on their own. Becoming part of a network is often the only way they can have access to the expensive resources and equipment needed to develop any innovative capacity. In general, the more resource-intensive a firm's research is, the more likely it will have to rely on resources provided by others. This dependence on a network often evolves into interdependence, as complementary participants come into contact and develop collaborative relationships based on mutual needs. Past research on innovation networks has provided evidence of this dynamic in various locales over the years (see, for example, Håkansson 1987; Camagni 1991; Axelsson and Easton 1992; Ratti et al. 1997).

Incubation through a network often involves small gains and little growth. However, once a 'tipping point' is reached and rapid growth begins, if the network has incubated a firm that reaches that stage of development, this may result in many benefits for its network partners. In biotechnology, for example, the incubation of new firms often involves providing the expertise needed to see their products through the various and lengthy clinical trials needed to secure approval for a product. Without such help most biotech firms in existence today would have been unlikely to survive their start-up phase.

Flexibility

Diluting hierarchies and the barriers to entry and association that they create is an important attribute of networks. This means that firms high up in a hierarchy are less able to control access to new sectors and activities. Those companies often erect barriers, which can be embedded in oligopolistic contracting arrangements or pricing schemes, that impede new or smaller firms from operating independently of their power. Diluting those oligopolistic hierarchies often results in greater flexibility for firms that join a network, in terms of their ability to craft research relationships that enhance their innovative capacity.

Allowing easier entry to new research areas and their markets can therefore result from inter-firm networks where hierarchies or the dominance of a few firms are diminished. Older, larger and more established firms may find their control over any given sector eroded, as newer and more innovative enterprises with higher innovative capacity become competitors. Increased competition within those new sectors and research activities, between the larger, established firms and the new entrants (and between the new firms) may therefore result. Freer interactions and transactions between firms can also help expand the size of a network, thereby increasing the opportunities available to start and develop new research projects, and to select partners for them. As a result, the innovative capacity of firms involved may be increased if these relations lead to new inventions and innovations.

Supporting innovative capacity through the flexibility obtained from network relations may lead to temporary arrangements targeted at solving a specific research problem, or it may foster wider-ranging collaborative relationships dealing with diverse research needs. An example of the former is the development of 'grid computing' networks involving many different firms, aimed at generating large amounts of computing power for research projects that require it, such as those found in bioinformatics and gene-decoding research (see Baldi and Brunak 2001; Waldrop 2002). Instead of spending vast sums on supercomputers to perform the necessary analyses on their own, firms involved in such networks tap the underutilized computing power found in most organizations, by linking up many personal computers in a network that generates a level of computing power similar to that provided by supercomputers. Examples of the wide-ranging variety of collaboration that enhances innovative capacity are common in biotechnology, where collaboration can provide much needed support for research and accounts for significant spillovers (see Forrest and Martin 1992; Liebeskind et al. 1996; Audretsch and Stephan 1999; Suarez-Villa and Walrod 2003).

Internally, the greater flexibility obtained through the network may allow participants to structure their operations to support their innovative capacity, so that they can respond quickly to changing conditions. In software design, for example, open source collaboration through networks has allowed participants to work in parallel, creating new applications that multiply the potential uses of a new design at a fraction of the time it would take to do so sequentially (see Raymond 1999; Moody 2001; Holtgrewe and Werle 2001). Such flexibility has allowed open source innovators to bypass or leapfrog the dominance of the largest software firms, such as Microsoft, by taking advantage of the research opportunities provided by networks (see, for example, Wayner 2000; Mustonen 2003).

Another example of how inter-firm networks provide flexibility that can be turned into support for innovative capacity is found in modular industrial systems (see, for example, Sanchez and Mahoney 1996; Thomke and Reinertsen 1998; Schilling 2000). Their rising importance has signaled a shift away from vertical integration towards a kind of horizontal specialization that is supported by network relations, where firms with high levels of innovative capacity tend to target research. Modular systems have become common in the electronics industry, although they are also being introduced in traditional sectors, such as automobile manufacturing. The networks supporting modular systems allow some firms to focus on invention and innovation, subcontracting manufacturing and other activities to other firms. Some of the best examples of rapid growth of companies linked to modular system networks can be found in firms such as Solectron, a provider of engineering services, including research, for the electronics industry. In less than 12 years, Solectron grew from a single Silicon Valley plant to 50 establishments in three continents, while its employment and revenues grew 2000- and 8000-fold, respectively (Sturgeon 2003). Other firms in the electronics sector, such as Flextronics, Sanmina/SCI and Jabil Circuit, have also experienced spectacular growth by finding a network niche that can increase and sustain high levels of innovative capacity.

Gaining flexibility through networks to enhance innovative capacity can also involve the 'spinning-in' of firms with high levels of innovative capacity. An acquiring firm may thus be able to internalize a higher level of innovative capacity by taking in talents and resources of firms that can produce disruptive new technologies, while at the same time reducing competitive pressures. This strategy is pursued in sectors where many small, highly innovative firms can complement or sustain the acquiring firm's own innovative capacity. One of the best examples of the spin-in strategy is Cisco System's practice of acquiring small firms with promising capabilities and patents before they grow large or become significant competitors to itself (see, for example, Bunnell 2000).

Scale Advantages

Networks can generate increasing returns to scale both to themselves and to the firms that participate in them. The advantages and resources gained through larger scale can then be used to increase and sustain firm-level innovative capacity. Substantial increases in resources to support innovative capacity can occur even when rising network scale results in greater specialization for firms. In fact, specializing in a certain niche as a network expands can be a very rewarding strategy, as shown by the earlier example of electronics firms involved in modular system networks.

A network's value (as an entity) can increase exponentially as it expands by improving access or including new members. In many networks targeting research, the more extensive a network becomes, the more value it may acquire as a repository of the collective innovative capacity of its members. This phenomenon, which entails increasing value with abundance of access, is diametrically opposed to one of the most fundamental principles of economics. That principle, which has become part of virtually all assumptions and teaching in the field of economics, is that value results from scarcity. In many research-oriented networks, by contrast, abundance of access (rather than its scarcity) actually produces greater value, for the innovative capacity of the network as an entity and for its members as individual firms.

The expansion of a network's scale often translates into growth for the firms which are involved. Moreover, firm-level growth is often a product of external collaborative relationships developed through the network (see Freeman 1991; Powell et al. 1999; Riccaboni and Pammolli 2002). Collaboration is therefore a fundamental factor in the expansion of network scale, and it is also a major ingredient for boosting innovative capacity at both the firm- and network-wide levels.

The rise of collaborative relationships as a factor in the growth of networks and of the firms involved in them can be vital in the entry of new participants. The resulting expansion then usually leads to greater specialization for many firms, and eventually to a deepening division of labour within the network.

Larger scale for a network can increase opportunities for cross-fertilizing ideas that may lead to major discoveries, thereby compounding the innovative capacity of some (or all) participants. This is probably more relevant for complex research undertakings, where many resources must be brought into play in order to have any chance of making new discoveries. In biotechnology, for example, research often involves parallel tasking, where researchers in various labs, in different organizations or firms around the world, test different approaches and combinations to try to come up with a new discovery. Complex expertise is needed in such projects, and specialists from many disciplines such as pharmacology,, biology, genetics, medicine, software engineering and chemistry, from different firms and diverse cultural, ethnic and national backgrounds, must be brought together through the network in order to gather a critical mass of innovative capacity. Testing to gain regulatory approval often compounds the complexity of such projects, where the clinical testing process can take as long as eight years, requiring much coordinated testing and evaluation to overcome all the hurdles along the way. Even so, typically one out of 10,000 compounds tested as potential new biotech-generated medications actually gain regulatory approval, conferring on biotech research what is perhaps the worst odds of coming up with a successful new product of any economic activity in existence (see Suarez-Villa and Walrod 2003). In such cases, network scale can make a substantial difference, if parallel testing and knowledge sharing by various members can enhance the collective innovative capacity of participants, making it possible to find new discoveries and reduce the amount of time needed for regulatory approval.

Related to network scale is the need of many research-intensive firms to capture increasing returns to volume. Much innovative capacity and substantial cost may be needed to come up with the first sample or prototype of a new product, while subsequent units cost very little or nothing to turn out. The field of software design offers many examples of this case. Microsoft's Windows software, for example, required substantial resources and millions of dollars in costs to be designed. By contrast, however, subsequent copies of the software cost virtually nothing, or at least no more than a diskette's manufacturing cost. Similarly, the design of financial risk management software can take up substantial resources and monetary cost, but once it is completed it can be applied to an infinite number of transactions at little or no marginal cost. In virtually all cases involving increasing returns to volume, the network-supported innovative capacity is vital for coming up with a new discovery or product. In this regard, having the necessary innovative capacity to be first to come up with the product is crucial. Once a product is innovated, then network scale effects take over its marketing and distribution.

Finally, networked research alliances often rely on network scale to a substantial extent in order to generate the innovative capacity needed to make inter-firm relations effective. Such alliances often involve a division of labour involving different firms, with some becoming highly specialized while others retain or acquire a broader scope (see Suarez-Villa 1998, 2002a). Some sets of firms, for example, may become important nodes in the network (the lead firms) while others take up a position as providers or suppliers, focusing on certain research activities. This phenomenon has become obvious in the electronics industry where, for example, lead firms such as Ericsson and Hewlett Packard have sold off much of their non-research operations, such as manufacturing and distribution, in order to concentrate on high-end research activities. However, in order to sustain their research objectives, such firms often have to rely on a constellation of small firms that have highly specialized expertise on aspects of a research project. In such cases, network scale can be very important for generating the necessary innovative capacity, at the level of the entire network (collectively), in order to come up with new discoveries in a timely way.

Decentralization

Networks can also encourage a fragmentation of decision making, as control of a sector or activity by a few firms is loosened. This can translate into greater autonomy for firms to decide how to bolster their innovative capacity. A result may be to boost the independent initiative of firms with respect to invention and innovation, to reach out to other firms, select research partners, and structure collaborative relationships in ways that can raise their innovative capacity.

One of the important effects of network-induced decentralization is to spread risk and responsibilities on research projects. Considering that invention is one of the riskiest activities that can be undertaken, the sharing of risk can be an important factor for many small and financially stressed firms. Some of the best examples on how network-induced decentralization can lead to better risk management are found in biotechnology. Biotech has perhaps the highest risk of any economic activity in existence today. As noted earlier, the chance of a new biotech-generated medical compound achieving regulatory approval in the United States is about one in 10,000. And this extremely lopsided ratio does not even take into account whether a new product will be profitable if it happens to be marketed. Many biotech products that do manage to gain regulatory approval are not successful once marketed, thereby increasing considerably the odds against both being approved and marketed successfully.

Participating in a networked research alliance with diverse R2R (research unit-to-research unit) relationships helps reduce risk by distributing responsibility amongst the various partners (see Suarez-Villa 2002a). In this way, the failure of a given research effort need not have a catastrophic effect on the fortunes of a firm, or on the collective fortunes of the firms in the network. Managing risk through the network also affords many opportunities for learning, particularly with the easy accessibility to new knowledge provided by information technology (see, for example, Forrest and Martin 1992; Barley et al. 1992; Powell et al. 1996; Teece 1998; Wellman 1999; Schwartz et al. 2000). Moreover, networks also afford the possibility of diluting risk further once a research project is underway, by drawing in partners with expertise that may not already be available to the alliance.

Another effect of network-induced decentralization is the possibility of raising intra-firm creativity in unusual ways so as to boost innovative capacity. In some firms, peripheral groups are established that operate outside the established lines of control and authority, informally becoming autonomous subunits in their own right. More often than not, such groups seek to generate tacit knowledge that can be kept within the firm, or perhaps even within the group, leading to new ways of thinking and discoveries. A key to the groups' success is typically network relations with other groups or firms that would be difficult or impossible to link up with through the conventional channels originating within the firm at large. The result of this strategy has often been significant technological breakthroughs and a boost to the firm's innovative capacity.

Examples of this kind of decentralized, network-induced effort to boost innovative capacity can be found in many research-intensive firms. At Sun Microsystems, for example, the Java software was created when a group of researchers was allowed to set up its own autonomous group outside the firm's direct control, thus being left free to experiment on their own and network with other firms (see Buderi 2000a). The biopharmaceutical firm GlaxoSmithKline divided its research unit into eight groups or centres to come up with new medications, leaving each one to compete for resources within and outside the firm and make its own contacts (see Pilling and Guerrera 2000). As a result, the firm's innovative capacity was boosted considerably, leading to a new series of biotech-based medications that revolutionized its segment of the biopharmaceuticals market. At Intel, its long line of microprocessor inventions and innovations started in the 1980s when a group of researchers was given free rein to deviate from the firm's narrow core business of producing custom microchips for calculator manufacturers, leaving it free to network with other researchers outside the firm (Buderi 2000a; Takahashi 2000). IBM in the 1990s established experimental 'Horizon Three' groups within the company to autonomously pursue 'pervasive computing' discoveries. The groups' objective involves finding ways to extend computing power from conventional appliances, such as laptops and personal computers, into cell phones, furniture, grooming devices, clothing, and other personal goods (Waters 2001). Earlier on, IBM had established a modular research system involving various autonomous groups that came up with some of its most important computer products (see Buderi 2000a; Baldwin and Clark 2000).

Network-induced decentralization can also involve the spinning-off of internal units as independent firms in order to boost innovative capacity. Such spin-offs can take over specific research projects that would be more capably run outside the larger firm. This can occur because management and resources in the spun-off firms can be more effectively targeted at the research tasks that need to be accomplished. The spun-off firms may also have more autonomy to attract talent that the larger firm would not be able to capture, or to establish links with other firms that it would not otherwise be able to do, because of divided loyalties, conflicts of interest or competitive factors. Spinning off units as new firms also often provides a way to get venture capitalists to invest in a project, which they might never consider if the project was housed inside the larger firm. Examples of spun-off units can be found in AT&T's creation of Lucent Technologies to boost its innovative capabilities. Lucent, in turn, later spun off several units as new firms, such as Flarion to come up with wireless data inventions, CyberlQ to find new ways to manage Internet traffic and content, and SyChip to invent new microchip designs for wireless Internet appliances, leaving them free to network with other firms according to their research needs (Buderi 2000b). In the 1990s Intel spun off Vivonic as a new firm, with the aim of inventing new fitness-planning software for handheld communication devices, also releasing the spun-off enterprise to network with other firms on its own (Buderi 2000b).

The kind of decentralization fostered by networks can also help firms reorganize their research operations to boost innovative capacity. To do so, research activities may need to be restructured so that they become more integrated with other components of the firm, such as production and customer service. Through such integration with in-house operations, new knowledge may be gained that can help turn out new inventions and innovations and raise innovative capacity, for example, by getting customers or production personnel involved in research efforts. This might require flattening or inverting the division of authority between units within the firm. The result of a flattening or inversion of the internal hierarchy may be a more fluid organization, capable of networking more effectively with other firms and establishing external research relationships (see, for example, Ashkenas et al. 1995; Clarke and Clegg 1998; Evans and Wurster 2000).

The four ways discussed in this section are by no means the only vehicles through which networks can influence innovative capacity. However, they seem to be the ones with the strongest relationship to research intensity and effectiveness. The following section will explore how these network-induced effects are shaping the emergence of a new organizational form in some of the most research-intensive activities of our time.

Collaboration, Networked Innovative Capacity and the Experimental Firm: The Case of Biotechnology

Collaboration is the most important factor involving the growth of innovative capacity through networks. Collaboration here refers to relationships involving research projects that engage at least several firms, where all of them share some of the new knowledge gained and add research value through their expertise. Collaboration through networks increases innovative capacity by making it possible to experiment and perform the necessary research tasks more effectively. The effectiveness gained must, in the end, result in significant quantitative or qualitative improvements of invention and innovation, over and above what each individual firm could achieve on its own.

The term *networked innovative capacity* refers to the innovative capacity which a firm acquires through collaborative relationships grounded in inter-firm research networks. Conceptually, at least, it is therefore necessary to differentiate between the innovative capacity that a firm could achieve on its own, and that which is gained through membership in the research network. Institutionalized forms of intellectual property, such as patenting, can take into account either individual ownership (by a person or a company) or shared ownership. Although all patents are awarded to persons, they are often reassigned by the latter to corporations or other entities. Employment and research contracts are often the key to such reassignments. The same usually applies to licensing. In terms of collaborative relationships grounded in inter-firm research networks, intellectual property rights would therefore be determined and allocated on the basis of individual contributions, taking into account that each individual participant is ultimately linked to a firm in the network.

The network's contribution to the innovative capacity of participating firms would likely be derived from providing easier access to both tangible and intangible resources, such as talents, expertise and laboratory hardware. Network-induced innovative capacity may also be supported by organizational intangibles, such as greater flexibility in structuring internal and external activities and relations, increased scale or specialization which lead to greater returns, and a decentralization of decision making that leads to greater creativity both internally within a firm and for the network's members as a whole.

Experimental Firms: Biotechnology

The biotechnology sector provides some of the best examples of what may be referred to as the *experimental firm.* Biotechnology is the most research-intensive activity in existence today. Perhaps no other sector depends on research for its survival and development as biotechnology does. In the United States, for example, biotech research spending amounts to approximately 80 per cent of the sector's total revenues (Suarez-Villa and Walrod 2003). This reality is in marked contrast with those of other sectors. In pharmaceuticals, for example, research spending amounts to only 20 per cent of revenues, while in the aerospace, electronics and defence sectors it is no more than about 5 per cent of total revenues (see, for example, McKelvey 1995; Acharya 1999; Gaisford 2001; Achilladelis and Antonakis 2001; Orsenigo et al. 2001). Such a high emphasis on research spending means that invention and innovation are extremely important for the survival of most biotech firms.

Networking and collaboration are very important for most biotech firms. Because of the complexity of most biotech research and the limited resources of most firms, network-based collaboration has become a major organizational characteristic of this sector. Inter-firm collaboration in biotechnology research is possibly the highest it has ever been in any economic activity. This allows most firms to gain access to resources that they would not be able to obtain on their own. Vast networks of collaborative arrangements permeate the biotech sector in the United States, leading to relationships that provide access to new knowledge, talents, finance and hardware, without which most firms would be unable to survive.

The biotech sector's experimental firms and the networked innovative capacity that supports them depend greatly on a culture of creativity in research, which is nurtured by inter-firm relations. To a great extent, the internal culture of creativity generated in biotech's experimental firms seems to be conditioned by their external relations. That culture of creativity is one which relies on allowing individual and group autonomous decisions on research projects, with respect to external relations and internally within each firm (see Suarez-Villa 2002b). Emphasizing process over outcomes is also part of that culture, despite the resource limitations, high risk and uncertainty faced by most biotech firms. In this very challenging environment, the survival of most biotech firms depends on a constructive convergence of the external and internal decisions that shape how research projects are structured, who participates in them, whether they are curiosity driven or expediency oriented, and the level of trust that can be deployed.

The culture of creativity in biotechnology's experimental firms is also nurtured by a diversity of disciplines or fields. Genetics, biology, pharmacology, medicine, computer science and software design all contribute substantially to biotech research. Much of the expertise from these fields is obtained through network relations, involving time, personal knowledge of individual and group dynamics, and reciprocity. By and large, therefore, biotech's experimental firms rely on intangible resources obtained from their network relations and the kind of collaboration in which they engage.

Collaboration and Networked Innovative Capacity

Network-induced collaboration in biotech's experimental firms can involve two main modes. Both of these modes can contribute directly to improve a firm's networked innovative capacity. One of them involves research alliances. Alliances can be wide-ranging, or they can involve one or several research projects (see, for example, Arora and Gambardella 1994; Gerybadze 1995; Gomes-Casseres 1996; Child 1998; Doz and Hamel 1998; Orsenigo et al. 2001). Wide-ranging alliances are often long term and involve relationships based on personal knowledge of groups and individuals, where much professional trust is invested. Project-based alliances tend to be more limited in scope and time, and often target a specific project and related activities. In general, wide-ranging alliances may be preferred for complex projects that require much time, diverse knowledge, and resources from numerous parties.

Research alliances ultimately improve a firm's networked innovative capacity by raising the output and quality of inventions and innovations. First, such alliances typically involve establishing long-lasting relations with other firms. The relations involved can lead to the acquisition of new knowledge and skills over and above what a firm could obtain on its own. Second, the inter-firm relations developed through alliances can allow access to research lab equipment, facilities and raw materials (such as test samples, organisms) needed to undertake research projects, which a firm would not be able to afford by its own means. Third, such relations can allow participation in research projects which a firm cannot undertake on its own, and which may eventually lead to profitable products. Fourth, the inter-firm relationships developed through alliances can save internal resources (human, financial, equipment, space or facilities), so that they can be redeployed to support internal research capabilities more strongly.

Outsourcing arrangements are the second mode of collaboration in networks. Outsourcing would involve subcontracting research activities with other firms (see, for example, Grabher 1993; Bragg 1998; Domberger 1998; Heywood 2001). Subcontracting can occur in two ways: subcontracting out research work to other firms, or taking in work from (or being subcontracted by) other firms. A third possibility is to do both: to subcontract out and to take in work at the same time. The latter is a fairly common practice amongst firms that emphasize relational links with other firms, where reciprocity is important (see Suarez-Villa and Fischer 1995; Suarez-Villa and Karlsson 1996; Suarez-Villa and Walrod 2003).

Outsourcing arrangements differ from research alliances in the quality of the relationships involved, which tend to be less of a relational kind and more contractual (but can actually involve both). Outsourcing arrangements also tend to have a more limited time window and are usually short term. The scope of the links or relations in outsourcing arrangements tends to be narrower than with alliances, and it is usually limited to specific research tasks or activities. For example, a research outsourcing arrangement may involve a clinical testing regimen required to move an invention from Phase I to Phase II in the regulatory testing process, and no more than that. Or, outsourcing might only involve decoding the genetic information of a specific organism by the subcontracted firm, with no further link or relationship beyond that task.

Outsourcing arrangements eventually lead to increases in the output or quality of inventions and innovations, thereby improving a firm's networked innovative capacity. However, they can be expected to have a much more limited impact on innovative capacity than the alliances. Outsourcing arrangements usually improve a firm's networked innovative capacity by having a subcontractor add research value (through expertise, facilities or equipment), which could not be gained by the firm on its own, or which can be obtained at lower cost than the firm could achieve on its own. A second way in which outsourcing contributes to raise innovative capacity is by allowing the firm to save internal resources which can then be ploughed back or redeployed to support other internal research activities. This can lead to a stronger research intensity and it can become part of a strategic research plan.

Alliances and outsourcing relations often involve highly complementary relationships between small but highly innovative biotechnology enterprises and larger firms from other sectors, such as pharmaceuticals, chemicals or agriculture. In most such cases, the larger firms provide resources that are urgently needed by their small biotech partners. The biotech firms, in turn, provide the larger partners with state-of-the-art research capabilities, knowledge, discoveries or patents that can eventually be turned into new products. Alliances and outsourcing relationships also occur between biotech firms, most commonly between specialized enterprises that focus on different aspects of research and can undertake certain activities or tasks more effectively.

Factors Promoting Networked Collaboration in Experimental Firms

The most important factors promoting networked collaboration in biotechnology's experimental firms are a diverse lot. As they condition collaboration, these factors also become determinants of networked innovative capacity. The first factor promoting collaboration is the high and increasing complexity of biotech research. Complexity in biotech is developing at various levels. One level involves complex knowledge and research tasks. This is part of a long-term process of change which became obvious, for example, as the field of cell biology evolved to spawn the new field of genomics, and as the latter in turn provided the basis from which the new areas of proteomics and physiomics have been emerging (see Acharya 1999; Ernst and Young LLP 2000; Gaisford 2001; Suarez-Villa and Walrod 2003). A second level of increasing complexity involves links between biotech research and other disciplines, such as the link between information technology and genetics (which is at the core of the new field of bioinformatics). The third level of increasing complexity involves hardware, or the highly sophisticated and expensive laboratory equipment (and in some cases supercomputers) needed to undertake commercial biotech research.

Increasing complexity at all of these levels makes it very difficult for biotech firms to undertake research alone, by their own means or in a self-sufficient way. This situation is in contrast with the case of other high technology industries, such as electronics, where many inventions and innovations were developed in virtual isolation or by stand-alone tinkering. For biotechnology, more than for any other economic activity, the time of the lone inventor or isolated group making a significant discovery or innovation has never really existed. From the start, the biotech sector has relied on substantial external collaboration insofar as nearly every significant invention or innovation is concerned.

The second factor promoting collaboration in biotech's experimental firms is the very high risk and uncertainty involving this sector. Research in biotechnology is the riskiest of any economic activity in existence today. For most new discoveries it is very difficult to predict their commercial viability, because of the unforeseen and complex effects that many biotech compounds have, and the very lengthy and unpredictable regulatory testing process. Regulatory testing, on its own, introduces a high element of risk not found in any other high technology industry. In the United States, six to ten years can pass for a new discovery to receive regulatory approval from the federal Food and Drug Administration (the agency charged with certifying the safety and effectiveness of any potential biotech product). The testing process employed by the FDA comprises several stages that act as 'filters', with a very high number of potential new products being eliminated as testing advances from one stage to the next. As an example, it has been estimated that one out of 10,000 compounds initially submitted survives the regulatory testing process and gains approval to be marketed to the public (Suarez-Villa and Walrod 2003). This extremely high rejection rate virtually forces many biotech firms to become 'factories' of invention and innovation, where networked innovative capacity becomes a fundamental necessity.

The third major factor inducing networked collaboration in biotechnology is the convergence of various disciplines and expertise in biotech research. In biotech, a very diverse combination of disciplines and expertise must be engaged in order to come up with nearly every discovery. The multidisciplinary needs of biotech research, referred to earlier, are thus a major incentive to engage in network-driven, inter-firm collaboration. Bringing together specialists from the fields of biology, pharmacology, genetics, medicine, software design, computer science and chemistry has become increasingly common in most biotech research projects. These combinations of specialists and fields may eventually produce entirely new disciplines. The new field of bioinformatics, for example, combines knowledge from software design, biology, computer engineering and medicine, and has become essential for research in genetics, proteomics and physiomics. Another example is the new biotech-based chemical industry, which is replacing traditional industrial chemical processes, as enzymes created through biotechnology replace highly noxious chemicals that were long used in many heavy manufacturing activities. Creating those new enzymes has required a convergence of knowledge in biology, chemical engineering, genetics, computer simulation and environmental science. This difficult convergence of fields that is so typical of biotech research provides a fundamental incentive for collaboration.

The very high cost of research is the fourth factor promoting collaboration in biotechnology. Biotech research and development costs are amongst the highest of any economic sector in existence. Moreover, the costs of research are often not recovered by new products. Only about one-third of all new biotechnology products that are marketed tend to earn back their research and development costs. However,

Table 2.1 Factors promoting collaboration in biotechnology research

Factor	Features
High complexity of biotech research	Increasingly complex knowledge and research tasks, links between biotech research and other disciplines, such as computer science (bioinformatics), pharmaceuticals (biopharmaceuticals), medicine (biomedicine), nanotechnology (bionanotech)
High risk and uncertainty	Great difficulty in predicting commercial viability of any discovery, very lengthy regulatory/testing process and outcomes
Convergence of various disciplines and expertise	Diverse multidisciplinary, specialized talents needed in research (biology, pharmacology, proteomics, medicine, genetics, software design, chemistry, computer science, for example)

many of those that do manage to cover such costs are not ultimately profitable (see Suarez-Villa and Walrod 2003). Thus, raising most biotech firms' innovative capacity depends greatly on external relations that can spread the cost of research amongst partner firms in a network.

The weight of high research costs may explain why biotechnology, as a sector, has never made any profits in the United States (Ernst and Young LLP 2000). The net annual losses endured by the biotech sector in the US typically amounted to about one-half to two-thirds of all research and development spending through most of the 1990s. Only a few biotech companies, individually, tend to be profitable and they are usually the larger and well-established firms. Many small- and medium-sized firms therefore barely end up covering their research expenses. Their best hope to reach profitability is to find partner firms which can share costs or provide research financing.

Adding to these pressures is the continuing detachment of commercial biotechnology from academic research in the United States. Initially, biotechnology was strongly linked with academic research. Major universities were the most important initial sources of commercial biotechnology in the 1980s (see, for example, Kenney 1986). Costs were thus to a great extent carried by institutions. However, the very different objectives of academic and commercial research have driven many firms away from this kind of collaboration, to seek arrangements with other firms instead. An important reason for this ongoing detachment is the high competitive pressure experienced by many biotech firms to immediately appropriate any new discoveries and patent them, which is in marked contrast with the open diffusion objectives of academic research. In the diminishing number of cases where biotech firms sustain links with academia, their primary objective seems to be to recruit personnel away from universities to do commercial research.

The very limited financial resources that most biotech firms have to support research is the fifth factor inducing research collaboration. Biotech firms 'burn' capital at a very fast rate, mainly because of the lengthy testing regimens required to gain regulatory approval and the high costs of research. Throughout the 1990s, for example, one-third of all biotech firms in the United States had less than 12 months to survive with the capital they had on hand (Ernst and Young LLP 2000). One-half to two-thirds of all firms had less than 24 months to survive based on their capital 'burn rate'. This situation is not attractive to venture capitalists, who typically seek lower risks and shorter turnaround times for their investments. It should therefore come as no surprise that in 1999 the amount of venture capital raised for biotechnology in the U.S. was less than 0.5 per cent of the total market capitalization of the sector. This is a very small proportion by any standard, particularly in a year that established records for venture capital investment in nearly every high technology activity. Despite this dire situation, it is surprising that no more than 10 per cent of all biotech firms failed in any given year since the early 1980s. The best possible explanation that can be offered for firm survival in these circumstances is the very important role played by networked collaboration and its effect on raising firms' innovative capacity.

Empirical Evidence on Collaboration

The empirical evidence available on collaboration in biotech, and its effect on firms' invention and innovation, is meagre. Obtaining establishment data on research is extremely difficult in the biotechnology sector because of the high privacy with which firms guard their research activities, given their importance for survival. Beyond this aspect, many firms' research activities also have public health and security implications, all of which creates much reticence about providing internal information on nearly every aspect of research and development, relationships with other firms, and internal economic data. Despite these obstacles, however, some empirical evidence can be provided on the effect of networked collaboration on firm-level innovative capacity.

Details on the empirical evidence discussed below can be found in Suarez-Villa and Walrod (2003) and will not be replicated in this contribution. Surveys of US biotechnology establishments were undertaken, based on Dun and Bradstreet Information Services' universe (808 establishments total) in three 4-digit SIC (Standard Industrial Classification) categories which fit the most commonly accepted definition of biotechnology used by trade associations and industry analysts (SIC 2835: In Vitro or In Vivo Diagnostic Substances, 2836: Biological Products, 8731: Commercial Biological Research). Over 300 statistical tests were performed on the database to determine the impact of collaboration on research intensity, involving various sets of internal economic data. Control groups were used to gauge differences in collaboration through alliances and different outsourcing modes, and their impact on research intensity and innovation.

The empirical evidence showed that alliances, which are typically structured through network-type arrangements, are the most important means of collaboration in biotechnology. Two-thirds of all biotech establishments were engaged in alliances with other establishments or with other firms. Virtually all alliances were relational, involving a great deal of trust and familiarity with partners. Using various measures that can provide some indication of establishment-level innovative capacity, such as patents awarded during previous years, number of new products in the research discovery phase, number of new products being tested in preclinical trials, research expenditures and assets, showed that collaboration through alliances with other firms or establishments provided the strongest indicator of innovative capacity. Establishments that were not engaged in alliances showed far lower measures of innovative capacity, compared with those that were so engaged.

Tests of outsourcing arrangements, which are also commonly structured through networks, involved: those where establishments were primarily subcontracted to perform research tasks for others; those where establishments were primarily subcontracting out research work to others; and those where both types of situations occurred (two-way subcontracting). Most outsourcing arrangements were also relational in scope and quality. Compared with establishments that were not engaged in any outsourcing arrangements, the ones that subcontracted out or were subcontracted had stronger measures of innovative capacity. However, establishments engaged in alliances had by far the strongest positive influence on innovative capacity.

Establishments that were not involved in collaborative arrangements either through alliances or outsourcing relationships were in the worst possible situation. In general, their measures of innovative capacity were far lower than those engaged in alliances or outsourcing arrangements. Their research intensities were also substantially lower than those found for the other establishments. Also telling about their prospects for survival was the fact that their level of assets and financial resources were significantly lower than those of establishments engaged in network-related collaboration through alliances or outsourcing.

Implications for Regional Development and Policy

The rise of experimental firms, their unrelenting pursuit of innovative capacity, and the networks they rely on, have important implications for regional development and policy. Since its beginnings, the field of regional development (and its policy scope) has focused its attention either on macro-level phenomena, such as trade flows, income, employment and investment, or on micro-aspects related to individual firms (in isolation) and the factors affecting their location. Theoretical overviews coupled with methodological tools have developed an analytical apparatus that invariably targets a macro- or micro-level aspect, often mirroring similar treatments in the larger field of economics, but adding a spatial perspective.

Taking into account the experimental firm and its operational context may, however, require a radical deviation from traditional regional development theories and methods. Invention and innovation, as the riskiest and most uncertain economic activities that can be undertaken, defy the standard treatment provided by mainstream economic theory and its regional progeny, where total certainty, perfect foresight, complete knowledge, and an all-encompassing emphasis on perfect maximization are gospel. When these central tenets are not adopted (explicitly or implicitly), mainstream models and theories tend to fall apart. Developing new approaches that can account as fully as possible for the high risk and uncertainty of invention and innovation should be a priority, if we are to move regional development concepts and methods into closer contact with the reality of the experimental firm.

The networks that support the experimental firm also defy what is perhaps the most important and long-standing tenet of standard economic theory, which assumes value to increase with scarcity. Contrary to standard economic theory, the value of networks increases with abundance of access or membership. Qualitative factors also play an important role, where relational transactions or interactions can have a major effect on value, over and above purely contractual links. Regional development theory and methods are greatly limited by not being able to account for the dynamics of networks and the qualitative aspects of relations that occur in them. As such, our current repertory of regional development concepts and methods is mostly inadequate to take full account of the networks in which experimental firms thrive, and the relationships that occur in them.

The obvious (and uneasy) coexistence of collaboration and competition in the experimental firm is another aspect that needs to be taken into account in regional development theory and policy. Contrary to another major tenet of standard economic theory, which assumes perfect competition to provide the greatest benefit to end-users or consumers, collaboration can provide the greatest benefit (over competition) when it becomes the only means through which new inventions or innovations can be found and marketed. In biotechnology, for example, such inventions can save numerous lives, or they can introduce major increases in efficiency that might never occur if competition were forced on firms where collaboration is the key to survival and technological discovery.

To be effective, regional policies targeting innovative activities and their sectors must provide a platform to support the innovative capacity of experimental firms and the networks they depend on. A platform of incentives must necessarily be based on a diagnostic indicator that can provide a measure of local and regional invention and innovation, at the level of firms, networks and sectors. The indicator should be *innovative capacity*, which is based on a measure of network- and firm-level invention and innovation to determine changes over time, between sectors, and also between firms. Interregional and national measures of innovative capacity may also be developed, to take into account changes over time and space (see Suarez-Villa 1990, 1993, 2001a, 2002c, and www.innovativecapacity.com). This kind of indicator can be compiled and monitored on a regular basis, much as income, employment or investment data currently are, and it could become a standard component of economic and business statistics (see the Real World Applications page in www.innovativecapacity.com).

Regional policies must also take into account the importance of networks for innovative activities and the experimental firm. Policy incentives must be built on a solid foundation of information. Qualitative and quantitative measures of various aspects of networks would be essential. An indicator of network intensity, for example, could provide information on the extent to which firms in any given region or locale depend on network relations to research, invent and innovate. Similarly, tracking the tangible and intangible interactions that occur in networks, and the extent to which they are relational or contractual, could provide much needed perspective on the significance of network-based transactions. As with the previously mentioned indicator of innovative capacity, such diagnostic information could be made a standard component of regularly compiled business and economic statistics.

Compiling data on the networked innovative capacity, or that which is gained primarily through network effects (and which individual firms would not obtain by their own means or without the networks) should also be considered. This measure would further refine the innovative capacity indicator mentioned previously, in order to gain additional information on the value of networks for regional and local development. A measure of the networked innovative capacity could then be related to conventional economic and social statistics, such as employment, income and investment, that are vital for determining the economic health of regions and localities.

Finally, the relationship between localities and experimental firms seems to be worthy of attention as we strive to develop our existing knowledge of regional development. In some cases, localities may become extensions of the networks of experimental firms they comprise. In such situations, we may face employment markets related to research and innovation that are significantly different from those posited by traditional notions of labour supply and demand. Employment relations may, for example, be deconstructed (as we currently know them), such that research talents become not specific to firms (as currently assumed) but to networks of firms and the localities that host them. As extensions of networks of experimental firms, localities might then become more like fluid employment networks for research activities, where temporary alliances between individuals and firms occur in much the same way as they happen between experimental firms.

References

Acharya, R. (1999), *The Emergence and Growth of Biotechnology* (Cheltenham: Edward Elgar).

Achilladelis, B. and Antonakis, N. (2001), 'The Dynamics of Technological Innovation: The Case of the Pharmaceutical Industry', *Research Policy* 30, 535–88.

Arora, A. and Gambardella, A. (1994), 'Evaluating Technological Information and Utilizing It: Scientific Knowledge, Technological Capability, and External Linkages in Biotechnology', *Journal of Economic Behavior and Organization* 24, 91–114.

Ashkenas, R., Ulrich, D., Jick, T. and Kerr, S. (1995), *The Boundaryless Organization* (San Francisco: Jossey-Bass).

Audretsch, D. and Stephan, P.E. (1999), 'Knowledge Spillovers in Biotechnology: sources and incentives', *Journal of Evolutionary Economics*, 9, No. 1/February.

Axelsson, B. and Easton, G. (eds) (1992), *Industrial Networks: A New View of Reality* (London: Routledge).

Baldi, P. and Brunak, S. (2001), *Bioinformatics* (Cambridge: MIT Press).

Baldwin, C.Y. and Clark, K.B. (2000), *Design Rules. Volume I: The Power of Modularity* (Cambridge: MIT Press).

Barley, S.R., Freeman, J. and Hybels, R.C. (1992), 'Strategic Alliances in Commercial Biotechnology', in N. Nohria and R.G. Eccles (eds), *Networks and Organizations* (Boston: Harvard Business School Press), pp. 311–47.

Bragg, S.M. (1998), *Outsourcing* (New York: Wiley).

Buderi, R. (2000a), *Engines of Tomorrow: How the World's Best Companies Are Using their Research Labs to Win the Future* (New York: Simon and Schuster).

—— (2000b), 'Lucent Ventures into the Future', *Technology Review* November/December, 95–106.

Bunnell, D. (2000), *Making the Cisco Connection: The Story Behind the Real Internet Superpower* (New York: Wiley).

Camagni, R. (ed.) (1991), *Innovation Networks: Spatial Perspectives* (London: Belhaven).

Child, J. (1998), *Strategies of Cooperation: Managing Alliances, Networks and Joint Ventures* (Oxford: Oxford University Press).

Clark, J. and Edwards, O. (1999), *Netscape Time: The Making of the Billion-Dollar Start-Up that Took On Microsoft* (New York: St. Martin's).

Clarke, T. and Clegg, S. (1998), *Changing Paradigms: The Transformation of Management Knowledge for the 21st Century* (London: Harper Collins).

Domberger, S. (1998), *The Contracting Organization: A Strategic Guide to Outsourcing* (Oxford: Oxford University Press).

Doz, Y. L. and Hamel, G. (1998), *Alliance Advantage: The Art of Creating Value through Partnering* (Boston: Harvard Business School Press).

Ernst and Young LLP (2000), *Convergence: The Biotechnology Industry Report* (Palo Alto: Ernst and Young LLP).

Evans, P. and Wurster, T.S. (2000), *Blown to Bits: How the New Economics of Information Transforms Strategy* (Boston: Harvard Business School Press).

Forrest, J.E. and Martin, M.J.C. (1992), 'Strategic Alliances between Large and Small Research Intensive Organizations: Experiences in the Biotechnology Industry', *R&D Management* 22, 41–53.

Freeman, C. (1991), 'Networks of Innovators: A Synthesis of Research', *Research Policy* 20, 499–514.

Gaisford, J.D. (2001), *The Economics of Biotechnology* (Cheltenham: Edward Elgar).

Gerybadze, A. (1995), *Strategic Alliances and Process Redesign: Effective Management and Restructuring of Cooperative Projects and Networks* (Berlin: de Gruyter).

Gomes-Casseres, B. (1996), *The Alliance Revolution: The New Shape of Business Rivalry* (Cambridge: Harvard University Press).

Grabher, G. (ed.) (1993), *The Embedded Firm: On the Socioeconomics of Industrial Networks* (London: Routledge).

Håkansson, H. (ed.) (1987), *Industrial Technological Development: A Network Approach* (London: Croom Helm).

Heywood, J.B. (2001), *The Outsourcing Dilemma: The Search for Competitiveness* (London: Financial Times Prentice Hall).

Holtgrewe, U. and Werle, R. (2001), 'De-commodifying Software? Open Source Software between Business Strategy and Social Movement', *Science Studies* 14, 43–65.

Kenney, M. (1986), *Biotechnology: The University-Industrial Complex* (New Haven: Yale University Press).

Liebeskind, J., Oliver, A., Zucker, L. and Brewer, M. (1996), 'Social Networks, Learning and Flexibility: Sourcing Scientific Knowledge in New Biotechnology Firms', *Organization Science* 3, 783–831.

McKelvey, M. (1995), *Evolutionary Innovation: The Business of Biotechnology* (Oxford: Oxford University Press).

Menzel, P. and D'Aluisio, F. (2000), *Robo Sapiens: Evolution of a New Species* (Cambridge: MIT Press).

Moody, G. (2001), *Rebel Code: The Inside Story of Linux and the Open Source Revolution* (Cambridge: Perseus).

Mustonen, M. (2003), 'Copyleft: The Economics of Linux and Other Open Source Software', *Information Economics and Policy* 15, 99–121.

Orsenigo, L., Pammolli, F. and Riccaboni, M. (2001), 'Technological Change and Network Dynamics: Lessons from the Pharmaceutical Industry', *Research Policy* 30, 485–508.

Pilling, D. and Guerrera, F. (2000), 'Drug Giant Plans Radical Research Move', *Financial Times* 11 November: 16.

Powell, W.W., Doput, K.W. and Smith-Doerr, L. (1996), 'Interorganizational Collaboration and the Locus of Innovation: Networks of Learning in Biotechnology', *Administrative Science Quarterly* 41, 116–45.

Powell, W.W., Doput, K.W., Smith-Doerr, L. and Owen-Smith, J. (1999), 'Network Position and Firm Performance: Organizational Returns to Collaboration in the Biotechnology Industry', in S. Andrews and D. Knocke (eds), *Research in the Sociology of Organizations* (Greenwich: JAI Press), pp. 129–59.

Ratti, R., Bramanti, A. and Gordon, R. (eds) (1997), *The Dynamics of Innovative Regions: The GREMI Approach* (Aldershot: Ashgate).

Raymond, E.S. (1999), *The Cathedral and the Bazaar: Musings on Linux and Open Source by an Accidental Revolutionary* (Cambridge: O'Reilly).

Riccaboni, M. and Pammolli, F. (2002), 'On Firm Growth in Networks', *Research Policy* 31, 1405–16.

Robbins-Roth, C. (2000), *From Alchemy to IPO: The Business of Biotechnology* (Cambridge: Perseus).

Sanchez, R.R. and Mahoney, J.T. (1996), 'Modularity, Flexibility, and Knowledge Management in Product and Organization Design', *Strategic Management Journal* 17, 63–76.

Schilling, M. (2000), 'Towards a General Modular Systems Theory and its Application to Inter-firm Product Modularity', *Academy of Management Review* 25, 312–34.

Schwartz, D., Divitini, M. and Brasethvik, T. (2000), 'Knowledge Management in the Internet Age', in D. Schwartz, M. Divitini and T. Brasethvik (eds), *Internet-based Knowledge Management and Organizational Memory* (London: Idea Group Publishing), pp. 1–13.

Sturgeon, T.J. (2003), 'What Really Goes on in Silicon Valley? Spatial Clustering and Dispersal in Modular Production Networks', *Journal of Economic Geography* 3, 199–225.

Suarez-Villa, L. (1990), 'Invention, Inventive Learning, and Innovative Capacity', *Behavioral Science* 35, 290–310.

—— (1993), 'The Dynamics of Regional Invention and Innovation: Innovative Capacity and Regional Change in the Twentieth Century', *Geographical Analysis* 25, 147–64.

—— (1996), 'Innovative Capacity, Infrastructure and Regional Policy', in D.F. Batten and C. Karlsson (eds), *Infrastructure and the Complexity of Economic Development* (Berlin: Springer-Verlag), pp. 251–69.

—— (1997), 'Innovative Capacity, Infrastructure and Regional Inversion: Is There a Long-term Dynamic?', in C.S. Bertuglia, S. Lombardo and P. Nijkamp (eds), *Innovative Behaviour in Space and Time* (Berlin: Springer-Verlag), pp. 291–305.

—— (1998), 'The Structures of Cooperation: Downscaling, Outsourcing and the Networked Alliance', *Small Business Economics* 10, 5–16.

—— (2000), *Invention and the Rise of Technocapitalism* (Lanham, New York and Oxford: Rowman and Littlefield).

— (2001a), 'Inventive Knowledge and the Sources of New Technology: Regional Changes in Innovative Capacity in the United States', in M.M. Fischer and J. Fröhlich (eds), *Knowledge, Complexity and Innovation Systems* (Berlin: Springer-Verlag), pp. 165–80.

— (2001b), 'The Rise of Technocapitalism', *Science Studies* 14, 4–20.

— (2002a), 'Networked Alliances and Innovation', in Z.J. Acs, H.L.F. de Groot and P. Nijkamp (eds), *The Emergence of the Knowledge Economy: A Regional Perspective* (Berlin and New York: Springer-Verlag), pp. 65–80.

— (2002b), 'Technocapitalism and the New Ecology of Entrepreneurship'. Paper presented at the International Workshop on Entrepreneurship in the Modern Space Economy: Evolutionary and Policy Perspectives, Tinbergen Institute, Amsterdam, June.

— (2002c), 'Policies or Market Incentives? Major Changes in the Geographical Sources of Technology in the United States, 1945-1995', in B. Johansson, C. Karlsson and R. Stough (eds), *Regional Policies and Comparative Advantage* (Cheltenham: Edward Elgar), pp. 127–50.

Suarez-Villa, L. and Fischer, M.M. (1995), 'Technology, Organization and Export-driven Research and Development in Austria's Electronics Industry', *Regional Studies* 29, 19–42.

Suarez-Villa, L. and Hasnath, S.A. (1993), 'The Effect of Infrastructure on Invention: Innovative Capacity and the Dynamics of Public Construction Investment', *Technological Forecasting and Social Change* 44, 333–58.

Suarez-Villa, L. and Karlsson, C. (1996), 'The Development of Sweden's R&D-intensive Electronics Industries: Exports, Outsourcing and Territorial Distribution', *Environment and Planning A* 28, 783–818.

Suarez-Villa, L. and Rama, R. (1996), 'Outsourcing, R&D and the Pattern of Intra-metropolitan Location: The Electronics Industries of Madrid', *Urban Studies* 33, 1155–97.

Suarez-Villa, L. and Walrod, W. (1997), 'Operational Strategy, R&D, and Intra-metropolitan Clustering in a Polycentric Structure: The Advanced Electronics Industries of the Los Angeles Basin', *Urban Studies* 34, 1343–80.

— (2003), 'The Collaborative Economy of Biotechnology: Alliances, Outsourcing, and R&D', *International Journal of Technology Management* 26, forthcoming.

Takahashi, D. (2000), 'Reinventing the Intrapreneur', *Red Herring* September, 189–96.

Taubes, G. (2000), 'Biologists and Engineers Create a New Generation of Robots that Imitate Life', *Science* 288:April 7, 80–83.

Teece, D. (1998), 'Capturing Value from Knowledge Assets: The New Economy, Markets for Know-how, and Intangible Assets', *California Management Review* 40, 55–79.

Thomke, S. and Reinertsen, D. (1998), 'Agile Product Development: Managing Development Flexibility in Uncertain Environments', *California Management Review* 41, 8–30.

Tsang, C.D. (2000), *Microsoft First Generation: The Success Secrets of the Visionaries Who Launched a Technology Empire* (New York: Wiley).

Waldrop, M.M. (2002), 'Grid Computing', *Technology Review* May, 31–7.

Waters, R. (2001), 'Never Forget to Nurture the Next Big Idea', *Financial Times* 15 May: 8.

Wayner, P. (2000), *Free for All: How Linux and the Free Software Movement Undercut the High-Tech Titans* (New York: Harper Business).

Webb, B. and Consi, T.R. (eds) (2001) *Biorobotics* (Cambridge: MIT Press) <http://www.innovativecapacity.com> (including 9 sections).

Wellman, B. (1999), *Networks in the Global Village* (Boulder: Westview Press).

Chapter 3

On the Robustness of Transportation Network Structures

Ben Immers and Art Bleukx

Introduction

There is a growing awareness that, in the past, insufficient consideration has been given to what is called the 'robustness' of the road network. It is only during the last decade that considerable research interest has started to emerge for this important aspect of the transportation system. Current research interests, see, for example, Bell and Cassir (2000) and Bell and Iida (2003), include network design, travel behaviour under conditions of uncertainty and the definition of reliability measures for networks.

There may be some confusion around the concepts of 'robustness' and 'reliability'. We consider *reliability* of travel time to be a user-oriented quality of the transportation system. *Robustness*, on the other hand, is a characteristic of the transportation network itself. Offering a high degree of reliability to the user often requires a robust system, especially when dealing with overloaded network links.

A very clear review of the research on the vulnerability (the opposite of robustness) of transportation networks is given in Berdica (2002). Definitions are given of vulnerability and related terms and the paper contains a useful survey of ongoing research and outstanding issues. Some authors (for example, Chen et al. 2002) distinguish between different types of reliability. *Connectivity reliability* is only concerned with the probability that network nodes remain connected. *Capacity reliability* assumes a fixed demand and considers the probability that the capacity of one or more network arcs is insufficient to handle this demand. Finally, *travel time reliability* considers the probability that a trip can be made within a specified interval of time.

Assessing the probability that a certain trip can be completed is, however, not sufficient. A complicating issue is that travellers may have varying perceptions of travel times and variations in travel time, as is emphasized by Hilbers et al. (2004), who distinguish between objective and subjective travel time reliability. Furthermore the consequences of a disruption in the transportation network are largely dependent on a traveller's knowledge or information about the system (Nicholson et al. 2003).

Indices for Reliability

An important issue is the measurement of reliability. Several indices for reliability have been proposed. Hilbers et al. (2004) give an overview and distinguish three broad groups of indices. The first group comprises measures based on variations in experienced travel time; the second group takes delays in travel time as a point of departure; and the third group consists of measures based on the probability of arriving at the destination on time, that is, a possible delay should not exceed a certain accepted limit. In all of these indices it is necessary to be specific about the expected travel time. This could be some average of experienced travel time, or it could be the free-flow travel time.

The concepts of reliability and robustness may be approached from different angles. In our research efforts, we mainly focus on the importance of good transportation network design. An adequate network is first of all aimed at minimizing the risk of network overloading. On the other hand, if overloading does occur, efforts should be directed at minimizing the consequences.

In the first part of our chapter, we survey a number of important aspects of reliability and network robustness; in the second part, we illustrate our findings by some model calculations on a test network.

Reliability and Robustness: Problem Survey

Reliability

We define reliability of travel time to be the 'degree of certainty with which a traveller is able to estimate his own travel time'. When making a trip, a traveller expects a certain travel time, based on his knowledge of the route and on past experience. Slight variations in actual travel time are natural to the system and completely acceptable to the users of the system. The system is considered to be reliable if expected and actually experienced travel times closely agree. If a traveller frequently faces unexpected delays, the system is regarded as unreliable.

It is especially the delays that cause inconvenience and costs, the early arrivals are of lesser concern. Figure 3.1 shows some typical distributions of travel times. From top to bottom the situation becomes progressively more unreliable. In order not to be late at his destination, the traveller will initially prefer to stay on the safe side and take plenty of time for his journey (that is, the 90 per cent point). But, after arriving much too early for a number of times, he might start moving back to the middle of the travel time distribution, only to be surprised when some unexpected delay hits him again.

The degree of certainty with which a traveller is able to estimate his own travel time depends on the probability distribution of travel times, the stability of travel time, the information available to the traveller, and the alternative travel options open to him. The importance of the distribution of travel times has been illustrated above. By 'stability', we mean the rate of change in travel time in the case of changes in trip demand or network capacity. High instability complicates the prediction of travel

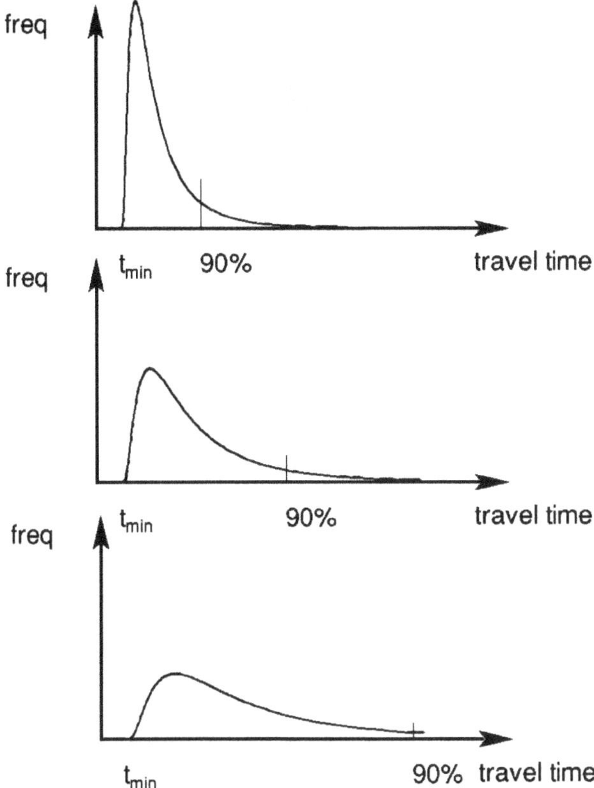

Figure 3.1 Distributions of travel time

times. The availability of information and the existence of alternative travel options are important because they influence the degree of inconvenience experienced by a traveller in the case of sudden disruptions.

Finally we note that not all travellers perceive a delay in the same way. The valuation of lost time depends on the general characteristics of the traveller, such as occupation and age, and on the characteristics of the trip itself, such as the purpose of the journey (see de Jong et al. 2004).

Figure 3.2 schematically shows the important factors affecting the reliability of travel time. Travel time and degree of uncertainty depend on traffic disruptions brought about by changes in traffic demand and supply. These variations in the demand and supply pattern may be grouped along several dimensions. An important subdivision is the distinction between expected and unexpected situations. These, in their turn, may be subdivided into frequently occurring circumstances and exceptional events. Table 3.1 shows examples of the different possibilities. As the examples in the table show, overloading is sometimes caused by variations in demand (rush hour traffic, public events), but more often it involves a change in the supply of road capacity (normal and extreme weather conditions, incidents ranging from minor accidents to catastrophic events, road works ranging from the routine to major overhauls).

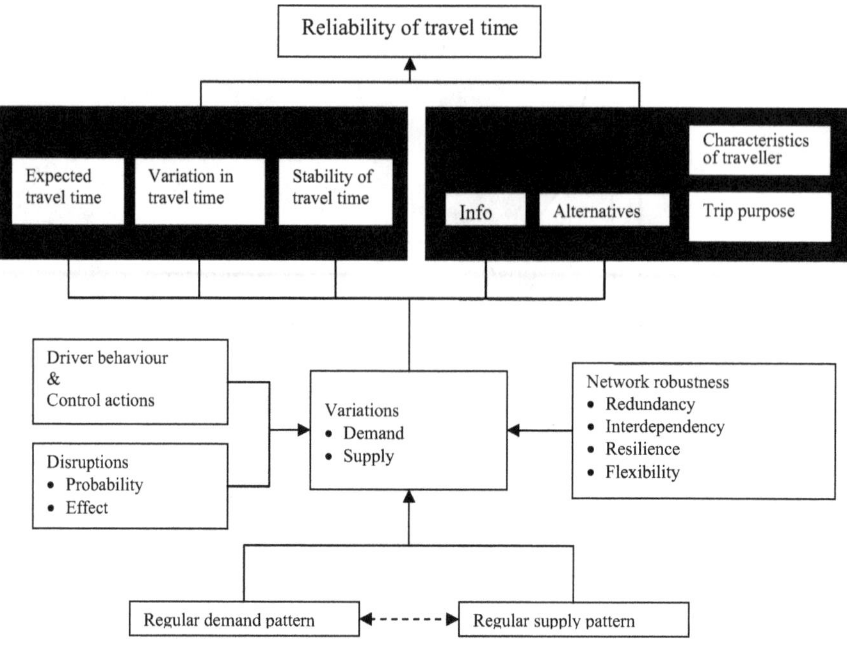

Figure 3.2 Factors affecting the reliability of travel time

Table 3.1 Causes of variation in demand and supply

	Everyday occurrences	**Exceptional conditions**
Expected situations	Rush hour Bad weather Minor road works	Public events Major road works
Unexpected situations	Minor accidents	Calamities Extreme weather conditions

The robustness of the network, together with driver behaviour and control actions undertaken by network operators, ultimately determines the effects of the traffic disruptions on travel time.

Besides network robustness, which we will discuss in the next section, the behaviour of drivers and network operators is of importance. Assuming, for example, a blockage caused by an incident, the use of alternative routes depends on timely information issued by network operators and the willingness of drivers to deviate from their chosen routes.

Network Robustness

'Robustness' is defined as the degree to which a system is capable of functioning according to its design specifications (capacity, speed and so on) in the case of serious disruptions.

Taking a number of corrective measures may enhance the robustness of the transportation system. These measures include the introduction of a certain redundancy or spare capacity into the system and minimizing the interdependency of system components to prevent a local disturbance from propagating throughout the entire system. In our opinion, the related notions of resilience and flexibility also have a bearing on the robustness of a system.

Redundancy The robustness of a system may be improved by introducing a certain amount of redundancy or spare capacity into the system. Strictly speaking, redundancy means the existence of more than one means to accomplish a given function. There are two types of redundancy: active and passive (or stand-by) redundancy. In the case of active redundancy, both main system and spare system operate together in normal conditions, but each system is capable of handling the complete task on its own in case of failure of the other. Passive redundancy means that the back-up system is activated only upon failure of the main system.

A lack of redundancy may have catastrophic consequences. Examples are the recent major disruptions in the electricity supply in New York, England and Italy. In the Netherlands the service quality on the national railway network has suffered considerably from a lack of redundancy.

On the road network, insufficient spare capacity may also lead to degradation in the quality of service. This could potentially have grave consequences in situations necessitating a rapid evacuation of the population. On a smaller scale, even relatively minor incidents may cause sizeable congestion on the road network that not only causes delays but also may interfere with the emergency services reaching the location of the incident. Incidental situations such as major roadworks, extreme weather conditions and large-scale public events also require some redundancy in road capacity.

Minimizing interdependency The location of a link or a node is important, in the sense that in certain cases congestion and associated unreliability are confined to the link concerned or to a small part of the network. In other cases, congestion at a centrally located link or node may cause a series of cascading failures, disrupting traffic on large parts of the network. These cascading failures are enhanced by the presence of all types of interdependencies between system components (Alderson 2002).

Possible options to minimize interdependency in infrastructure networks are:

- maintaining a hierarchy of essentially independent but well-connected functional road subsystems;
- reducing the vulnerability of main network nodes: for example, by limiting the number of branches at an intersection, and optimizing the distance between nodes.

Resilience and flexibility Organic, biological systems are characterized by a high degree of robustness, which they achieve mainly by possessing substantial redundancy and sometimes by a spreading of functionality throughout the organism. Also, organic systems turn out to be resistant to adverse environmental conditions because they are capable of rapidly recovering from temporary strain, and because they are able to gradually adjust to changing conditions in the long term.

It is interesting to apply these notions to the transportation system because the transportation system itself shows signs of a living self-organizing system (Immers and van Koningsbruggen 2001).

Resilience is the capability of the transport system to repeatedly recover, preferably within a short period of time, from a temporary overload. The resilience of the transportation system is enhanced by the availability of fast professional emergency services.

Flexibility: The robustness of the transportation system may also be measured by the extent to which the system is able to carry out more and other functions than those for which it was originally designed. Flexibility is, therefore, the property which enables a system to evolve with new requirements. As an example we could mention the flexible layout of network components enabling the network to adjust capacity according to demand or the ability to use the existing motorway network for road trains or double-stack containers.

Reliability and Robustness: An Example

Network Structure and Travel Time Availability

By means of simulation using a Macroscopic Dynamic Assignment Model (MaDAM), which is part of the OMNITRANS suite (2007), we examined what type of road network structure is most capable of adjusting in a flexible way to fluctuations in demand and supply. Therefore, we used the network presented in Figure 3.3 and a demand pattern represented by the origin-destination (O-D) table in Table 3.2. We briefly present the main conclusions of that research.

In principle, one can distinguish two broad types of network structure:

- a backbone network with a number of subsidiary road networks;
- a hierarchy of essentially independent functional road subsystems.

In a backbone network, each subsidiary network functions as a feeder system to the next higher system. This type of network structure provides for some robustness because of the generic nature of the backbone network, but it also results in a certain vulnerability because a single blockage of an important node may disrupt the entire functioning of the system.

In the second type of network, each subsystem can act as a back-up system to the next higher level, in the case of an emergency. This type of network is aimed at a separation of functions. A malfunction of one of the subsystems causes no disturbances in any of the other subsystems. For example, regional traffic cannot

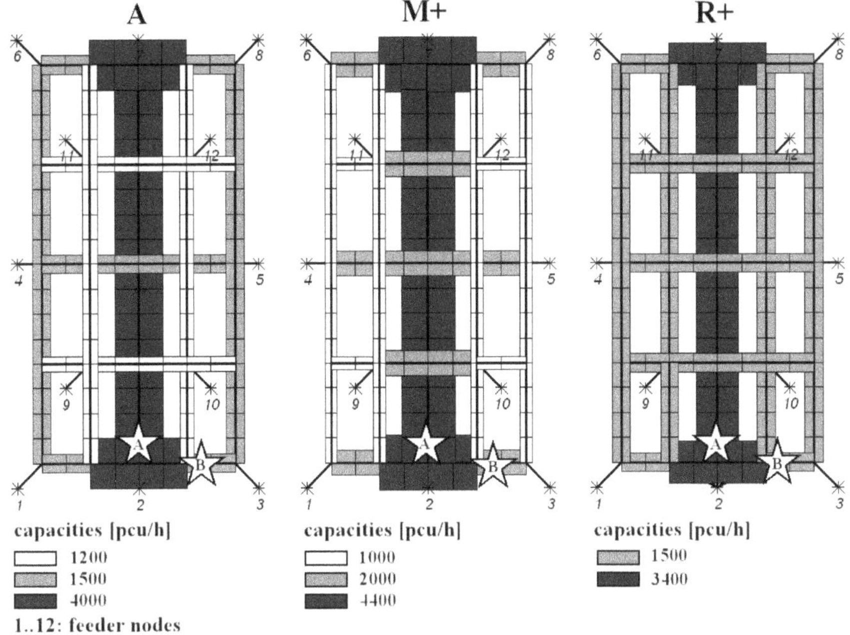

Figure 3.3 Alternative road networks considered in simulation

Table 3.2 Traffic assigned to the networks; base load in Step 1

From node 1-12 to node 2	Number of trips per hour
1	125
2	0
3	125
4	200
5	200
6	100
7	300
8	100
9	50
10	50
11	75
12	75

cause congestion on long-distance national roads. Additional advantages, besides robustness, of a hierarchical structure are that each system does not have to be constructed to the highest design standards, and that the infrastructure is geared to the user's requirements.

In order to be able to assess the impacts of the network structure on travel time variability, we designed the following setting:

Variations in the structure of the network are represented by three alternative networks (see Figure 3.3).

- *Reference network (A)*: The reference network is loosely inspired by the road network east of Brussels, covering an area stretching to about 20 km from the capital. It is characterized by a main motorway heading to Brussels and connecting to the beltway around the city, indicated by node 2 in Figure 3.3. A system of subsidiary roads, mainly feeding onto the motorway, provides access to and from the region.
- *Motorway-plus network (M+)*: By concentrating more flow capacity on the motorways, as compared with the reference network, we obtain the M+ network. In this network, the backbone function of the motorway is emphasized. The predominant function of the underlying network is to provide access to the motorway.
- *Regional-plus network (R+)*: This network tends towards a hierarchy of independent road subsystems as described above. Flow capacity in this network is more evenly distributed between motorways and regional roads. In this way, many regional, short-distance, trips can be accommodated by the regional network, thus relieving the motorway links.

The networks are comparable in that total network capacity (throughput in personcars per hour (pcu/h)) is the same for each of the three networks.

Variations in demand and supply are modelled in the following way:

- The traffic load (O-D table) will be increased in eight consecutive steps starting with free flow conditions (base load, see Table 3.1) and ending with serious congestion in the network (eight times the base load). At a traffic load factor of about 4, the networks start to show some signs overloading at certain locations. All flow is directed from the 11 origin nodes towards the only destination, node 2 in Figure 3.3, representing the access to the beltway (commuter traffic, strong orientation towards zone 2). The O-D table for the base load in Step 1 is given in Table 3.2.
- Local variations in supply are modelled assuming an incident on one (heavily loaded) link, which is either a motorway link or a regional road link. The incident locations are indicated by the stars marked A and B in Figure 3.3. We assume that the incident reduces link capacity to 25 per cent of the original capacity. As to the route choice behaviour of travellers, we assume that 50 per cent of all travellers persevere in their once chosen route as if no incident had happened, for the other 50 per cent of travellers we assume that they are susceptible to the changes in travel time due to the incident. They divert to other routes, thus avoiding the incident.

Traffic was assigned to each of the three networks using a dynamic traffic loading algorithm based on an equilibrium assignment. As an indicator for the performance of the networks in terms of traffic time sensitivity, we computed the average network speed, defined as the 'total distance covered by all travellers divided by the total travel time spent by all travellers'. Figure 3.4 shows how the average network speed develops under changing load conditions.

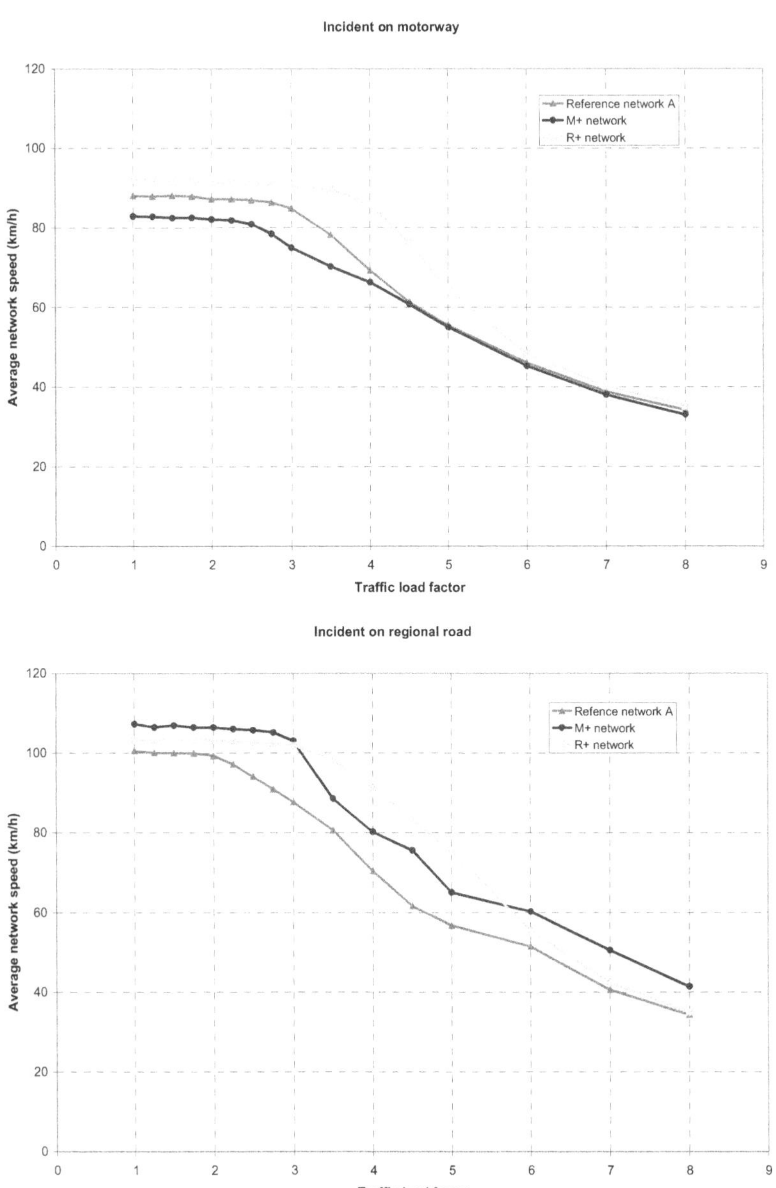

Figure 3.4 Performance curves for the three considered network structures

The following interesting observations can be made:

- The M+ network performs slightly better than the other networks under normal conditions (normal flows, no incidents).
- High traffic loads provoke a serious deterioration of the network quality for the M+ network. The reference network A, and in particular the R+ network, are less sensitive to a serious increase in demand.
- Incidents on the motorway have a serious impact on network performance, even when we are dealing with low-demand conditions. The impact of an incident on the regional network is limited as long as demand is low, but, as soon as total demand increases, the impact of an incident on the regional network becomes significant.
- The M+ network is more sensitive to incidents and changes in demand than the reference network and the R+ network. The performance of the R+ network is the best under all prevailing conditions.

Figure 3.5 shows the reduction in average network speed as a result of an incident on the motorway (upper panel) and on a regional road (lower panel). When there is an incident on the motorway, there is a serious decline in network speed on the M+ network, even under moderate traffic-load conditions. In contrast, an incident on the motorway has less serious repercussions for the reference network, and in particular for the R+ network. The reductions in network speed are smaller and occur at higher loads in the latter networks (R+ and reference network).

When dealing with an incident on a regional road, again the R+ network performs better under moderate traffic-load conditions. At higher traffic loads the reductions in speed increase in the R+ network, which is predominantly caused by the superior performance of the R+ network under incident free conditions. Note that the changes in average network speed are significantly smaller if an incident happens on a regional road.

Summarizing, we conclude that the performance of the R+ network is superior under various conditions. In particular, the stability of network performance at increasing network loads is an interesting characteristic. This characteristic is often referred to as the property of 'graceful degradation'. It makes the R+ network structure the most appropriate when dealing with heavily loaded network conditions.

The sensitivity of network structure to increases in network load is clearly demonstrated in Figure 3.6, which shows the gradients of the average network speed curves, given earlier in Figure 3.4. The M+ network shows the highest gradient, and this already occurs under moderate traffic load conditions. The R+ network is quite insensitive to increases in demand (up to load factor 3), and the gradient is quite moderate.

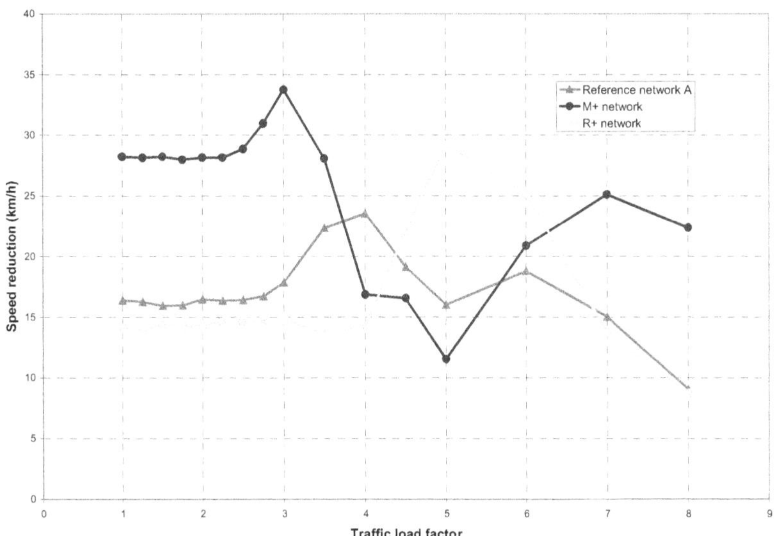

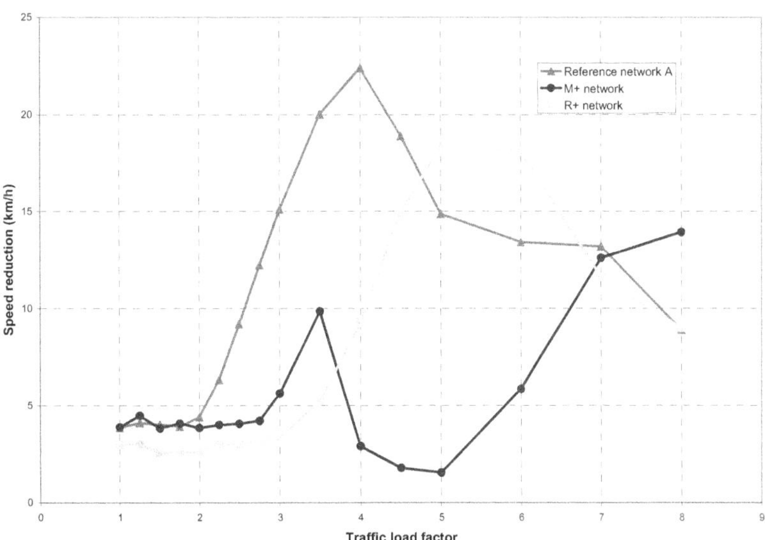

Figure 3.5 Reductions in average network speeds due to an incident

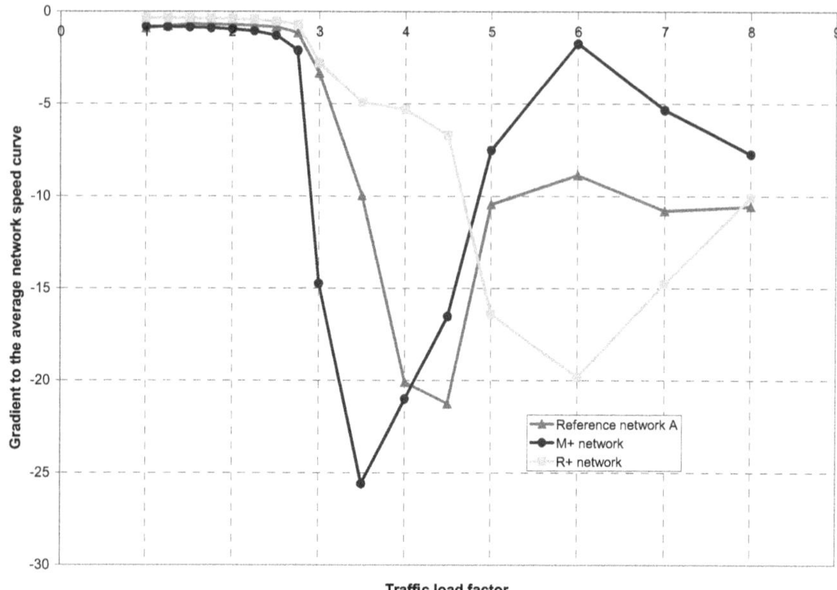

Figure 3.6 Sensitivity to demand variation of the various networks

Interpretation of the Simulation Results and Policy Implications

The results discussed above clearly indicate that, when dealing with growing demand, the R+ network has two interesting properties:

- it exhibits the characteristic of 'graceful degradation';
- it is less sensitive to incidents than other network structures considered.

We believe that the above characteristics are highly relevant for present-day traffic conditions. Three major issues can be mentioned:

- During recent decades investment in additional network capacity has not kept pace with growth in traffic demand. As a consequence, relatively little capacity is available to cater for increasing traffic flows. This leads to heavily loaded networks during peak periods and, as people change their behaviour to avoid congestion, the length of these peak periods continues to increase.
- To mitigate the impact of increasing traffic loads on network performance, various dynamic traffic management measures are being implemented. These measures may have a positive impact on traffic flow conditions, in that higher traffic volumes can be accommodated, without higher congestion levels. However, the opportunities to apply dynamic traffic management measures

strongly depend on the network structure, such as the availability of alternative routes, and so on.

- The higher traffic loads make the network more susceptible to small variations in demand or supply. In countries like the USA, Germany and the Netherlands, 40 to 50 per cent of congestion is caused by incidents and other unexpected occurrences, with an accompanying serious impact on the reliability of travel times.

Conclusions and Discussion

Our study leads to the following (early) conclusions and observations.

- Network robustness plays a vital role in providing travel time reliability for travellers. The robustness of networks depends on several factors: redundancy, network structure which influences the interdependency of network components, resilience, and flexibility.
- A hierarchical network structure encompassing independent mutually connected road subsystems may considerably improve the robustness of a transport system if compared with a backbone network with a number of subsidiary networks.

The analysis given in this chapter is obviously only exploratory. In order to describe the traffic performance correctly, we should be able to incorporate spill-back effects and (if possible) hysteresis effects (capacity drop). This means that we actually need a very sophisticated traffic assignment model which allows us to model the traffic performance of a network very accurately.

The O-D table used in the example is simple, as all trips are directed towards node 2. We would like to extend the analysis by compiling other O-D tables, thereby varying not only the size but also the distribution of demand. Moreover, a robustness check of a network should imply quite a bit of variation in demand and supply, thus representing real-world conditions. Therefore, in the next step of our analysis, stochastic characteristics of demand and supply will be dealt with using Monte Carlo techniques (see Meeuwissen et al. 2004).

As indicated in Section 3.1, various factors determine the robustness of a network against fluctuations in demand and supply. Future analysis will therefore focus on calculating the contribution (impact) of a series of network characteristics, such as, for example: network topology (grid structure, radial structure, and so on), hierarchy in the network, interconnectivity of various networks, node density, and so on. Furthermore the robustness of the network will be described in greater detail using a series of assessment criteria: for example, travel time distribution, median of travel time, average network speed, and so on.

References

Alderson, D. (2002), 'Cascading Failures in Infrastructure Networks'. Paper presented at the Workshop on Large-Scale Engineering Networks, Institute for Pure and Applied Mathematics, UCLA.

Bell, M.G.H. and Cassir, C. (2000), *Reliability of Transport Networks* (Baldock Philadelphia: Research Studies Press Ltd).

Bell, M.G.H. and Iida, Y. (eds) (2003), *The Network Reliability of Transport, Proceedings of the 1st International Symposium on Transport Network Reliability (INSTR)*. (Oxford: Pergamon).

Berdica, K. (2002), 'An Introduction to Road Vulnerability: What Has Been Done, Is Done and Should Be Done', *Transport Policy* 9:2, 117–27.

Chen, A., Yang, H., Lo, H.K. and Tang, W.H. (2002), 'Capacity Reliability of a Road Network: An Assessment Methodology and Numerical Results', *Transportation Research Part B: Methodological* 36:3, 225–52.

Hilbers, H., van Eck, J.R.R. and Snellen, D. (2004), *Behalve de Dagelijkse Files: Over Betrouwbaarheid van Reistijd* (*Except for the Daily Queues: On the Reliability of Travel Times*) (in Dutch) (The Hague: NAi Uitgevers).

Immers, L.H. and van Koningsbruggen, P. (2001), 'Self-Organisation in the Transport Sector', in *Proceedings of the ECCON 2001 Annual Meeting on Chaos and Complexity in Organisations*, Lage Vuursche.

de Jong, G., Kroes, E., Plasmeijer, R., Sanders, P. and Warffemius, P. (2004), 'The Value of Reliability'. Rand Europe, ITS Leeds and Transport Research Centre (AVV), Dutch Ministry of Transport, Public Works and Water Management.

Meeuwissen, A.M.H., Snelder, M. and Schrijver, J.M. (2004), 'Statistische Analyse Variabiliteit Reistijden voor SMARA' (Statistcal Analysis of Travel Time Variability Using SMARA). Report Inro Verkeer & Vervoer 2004-31, TNO Inro, Delft.

Nicholson, A.J., Schmöker, J. and Bell, M.G.H. (2003), 'Assessing Transport Reliability: Malevolence and User Knowledge', in M.G.H. Bell and Y. Iida (eds), *The Network Reliability of Transport. Proceedings of the 1st International Symposium on Transportation Network Reliability (INSTR)* (Oxford: Pergamon).

OMNITRANS International (2007) <http://www.omnitrans-international.com>.

Chapter 4

Electronic B2B, Impact and Experiences so Far

Dirk-Jan F. Kamann

E-Contacts: High Expectations and Disillusions

Electronic Data Interchange (EDI), but especially web-based electronic data communication, has raised high expectations. The leading multinationals started large, expensive and prestigious E-Commerce and Electronic Procurement projects. However, the bursting dot.com bubble plus the problems encountered when trying to implement all the web-based dreams, resulted in changed attitudes. It now transpires that many projects that were started were ill-founded or prompted by greed or fashion rather than sound business or scientific arguments. Now both companies and scientists have a more realistic view about reality, the following questions may be raised: What can the impact of the 'E-thing' *potentially* be? What is the effect on the way companies will deal with each other in the future? What will be the effect on flows of goods and information between firms? What will be the effect on transportation of the goods involved? These questions are the basis for this contribution.

With the term 'E-Contacts' we refer to all information, passing between actors in a network by electronic means. This includes EDI, E-Commerce, E-Business, E-Procurement and E-Mail. According to Forrester Research (2000), the consumer-oriented E-Commerce sales value (B2C) in Europe is assumed to grow from $3 million in 1999 to $232 million in the near future. The value of business-to-business transactions (B2B) will be six times as high. For that reason, we will focus on the B2B part, in combination with other E-Contacts that are complementary to these B2B contacts, such as E-mail and EDI.

As stated, the purpose of this contribution is to analyse *(1)* the impact of E-Contacts on the way companies deal with each other in networks; *(2)* the resulting effects on logistical flows; and *(3)* the associated effect on the locational requirements of companies. The structure of the chapter is as follows. To establish the possible effects on goods and services, we start with the purchasing literature. This literature enables us to differentiate between various types of goods and services and the type of business relations, associated with them. Using the characteristics of these different types of relations, we will analyse how 'sensitive' they are to the impacts of E-Contacts. For this analysis, we will draw on the regional science literature, and will use the degree of standardized routines in the activities related to these contacts as a basis. As a result of this analysis, we find new possibilities for actors in the network, with new roles to be played. Associated with these new roles, we will discuss the requirements that the various actors have to fulfil in order to play these roles successfully. These demands also include logistical aspects. Therefore, we will formulate the expected changes in logistical demands, and what these will mean for the locational behaviour of companies in a number of hypotheses. These hypotheses result in a dynamic location model. Having described the model, we will discuss the result of an empirical test of our theoretically based hypotheses and model. This empirical test was based on a series of interviews with different types of actors. We will finish with our conclusions and recommendations for further research.

Differentiation in Supplier Relations: Purchasing Literature

The Purchasing Cube

To enable us to differentiate between the various types of *relations* that companies – or, actors, in more general terms – have in a network, we use a concept from the purchasing literature: portfolio analysis (Fisher 1970; Kraljic 1983; Hughes et al. 1998; Olsen and Ellram 1997). The most recent form of portfolio analysis is what is called the *purchasing cube* (Kamann 1999, 2000), which in fact is the assembled result of three matrices (Figure 4.1).

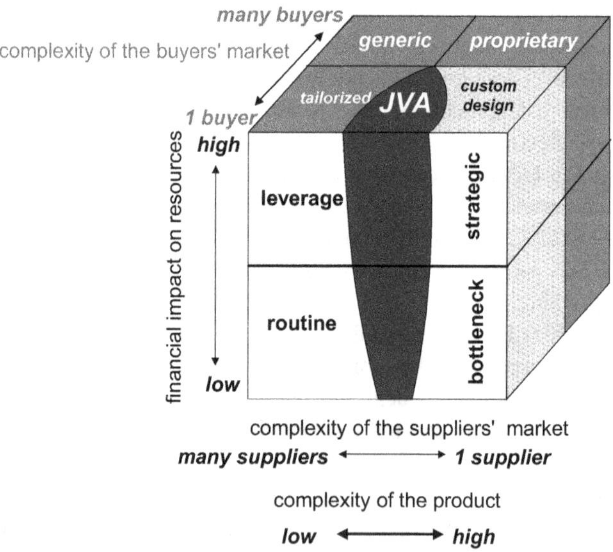

Figure 4.1 The purchasing cube

Here, the role of portfolio analysis is its ability to differentiate between different types of products, the associated suppliers and relations with those suppliers, including the relevant supplier strategies. Of course, there are many types of actors in a network and, as a result, many types of configurations of actors and many types of relationships (see Hughes et al. 1998). However, the three-dimensional cube has already turned out to be rather sophisticated in practical use. Therefore, although we are aware of the simplification, we will have to be satisfied on this point.

By analysing these relations between actors – derived from the cube – more closely, predictions can be formulated about the effects of E-Contacts and the way actors will deal and interact in networks in the future.

Differentiated Products, Suppliers and Relations

The front end of the cube contains the Kraljic-matrix. This matrix has four product categories, characterized by two axes:

1. the complexity of the supply market, in practice: Is it easy to find another supplier for this product? Are there many suppliers?
2. the financial impact: How much do we spend on this product category?

In theory, the top two matrices contain 80 per cent of the total expenditure in purchasing and 20 per cent of the suppliers. The division line usually hovers at around 2–4 per cent of the total spent. *Product* groups with more than 3 per cent of the total spent are above the line. The resulting four quadrants and their categories are:

- *routine* products: stationery, catering, 'small fry' (many suppliers, low purchasing value);
- *leverage* products: steel, oil, energy, standard raw materials/commodities (many suppliers; high value);
- *bottleneck* products: screws, nuts and bolts of special alloys, special parts of Original Equipment Manufacturers (OEM) origin, special products (few suppliers; low value);
- *strategic* products: more expensive special products; results of special product developments by suppliers and co-designers in cooperative projects ('few suppliers; high value').

Using the product categories, the associated *suppliers* are categorized along the same lines. Each category is associated with a different supplier strategy, leading to different supplier relations. These are:

- routine: no 'messing around with things'; logistics important; complete dependence of a single supplier to optimize Vendor-Managed Inventory and other services in the field of ordering (catalogue systems) and paying (purchasing cards, one invoice a month, and so on); supplier can be easily substituted; avoidance of organizational costs like handling, ordering, administration receives much attention;

- leverage: optimize contract, supply chain and overheads: supplier visits for audits; supplier has to come up with new ideas to reduce total costs; logistical reliability key factor, next to qualifiers such as price and (standard) quality;
- bottleneck: try to replace with routine product; try to let leverage supplier deal with it, especially the handling;
- strategic: partnering, durability in relations, trust, high switching costs, small number of suppliers involved, contact intensive, logistics in development stage less important, but in the production stage and supply it is more important.

For a number of categories we find that a number of services are added to the products: logistics, supply chain coordination, payments, way of ordering. Increasingly, the potential to combine physical flows of goods with digitized information to create an attractive package for the buyer is pursued.

The Supplier's Point of View

The second matrix reflects the point of view of the supplier (Hughes et al. 1998). Again, the horizontal axis gives the number of suppliers (many <> few). The vertical axis this time gives the complexity of the buyer's market: many or few buyers. When we flip the matrix and put it on top of the cube, we get four columns:

- *tailorized* products (Dell computers); suppliers combine through mass customization and flexible production technology a larger number of market segments;
- *generic* products (oil); standard products, usually with industry norms, traded as commodities; potential spot-markets;
- the *custom design* situation: one-to-one: one supplier – one buyer;
- *proprietary* products (Microsoft): unique (branded) products with a de facto monopolistic supplier (selling through agents, distributors, dealers) and many atomistic buyers.

With suppliers of generic and proprietary products, specifications are fixed: no discussion is possible. With generic products, logistical services can be added to the product to differentiate it from the 'plain' commodities and spot markets. With tailorized products, specifications vary over a defined range within certain limits. Custom-designed products are unique in their buyer specifications.

Joint Value Analysis (JVA)

The Fisher-matrix (1970) is the third matrix. It resembles the Kraljic-matrix but this time the horizontal axis reflects the complexity of the *product*. For more complex products, more intensive cooperation is likely, especially for high values involved. Cooperation focuses on a trajectory of joint value analysis (JVA) and joint value engineering. All stakeholders in the value chain should be represented in the analysis and re-engineering: how to redesign the product, produce it most efficiently, package and transport it, and, finally, how it fits best on the display, shelf or freezer. The aim is to reduce total costs of the entire trajectory through the supply chain, from initial producer to final user (and recycler), while maximizing user value. JVA is assumed to take place in the transcending area between leverage and strategic products. This is because, with very simple products, it usually makes less sense to go through this type of analysis to change specifications – only the logistics surrounding them. The really very complicated products, on the other hand, will fit into the category 'co-design' or 'co-development'.

JVA has two stages: in the first stage, many contacts and discussions are required between the stakeholders. Once the optimal configuration and specifications are established, the second stage can start. From then onwards, JVA actually blends in with the category of tailorized products.

Above, we found differences between supplier relations. In order to trace the impact potential of E-Contacts, we first have to make a short detour to see *which aspects* of these relations can be used as criteria.

E-Contact Potential: The Regional Science Literature

In order to establish the possibilities of E-Contacts, the results of research of the regional sciences are used (Goddard 1973). The more routinized activities are, the easier it is to substitute the contacts related to them with telecommunications. The more knowledge-intensive – 'paradigm transcending' – activities are, the more the contacts associated with them have to be face-to-face contacts. These are more difficult to replace with telecommunications. Telecommunication does have a complementary effect, but does not substitute. Here, it is assumed that E-Contacts in nature and influence do not differ in a significant way from telecommunication contacts when it comes to contact requirements. That means that, in the case of increasing knowledge and intensive face-to-face contacts, substitution with E-Contacts is less likely. This has the following implications for the *relations* and the associated *contacts* with suppliers:

- A large block of *tailorized/generic/proprietary products:* after the routine of comparing similar supplier offers, a standard contract is closed, coming from the database of the buyer/supplier; orders are generated by users in the organization (using catalogue systems or not), by end-users, by scanners (in check-outs or bins/tanks) and/or by vendor-managed inventory systems or material requirements planning (MRP) systems; relations are relatively impersonal and routine-based; exceptions can be found in the first stage (negotiation and contracting) and the vendor evaluation stage, where more face-to-face contacts can be found;
- The first stage of JVA contains a high number of face-to-face contacts, supplemented with E-mail, transfer of drawings and calculations by web-based technologies; the *second stage* falls into the previous category described in terms of ordering and coordination;

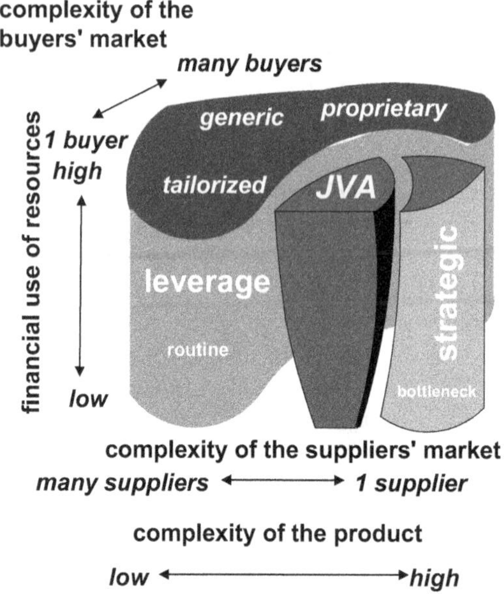

Figure 4.2 Clustering inputs and suppliers based on E-contact potential

- For the category of custom design, we expect little substitution to take place; less routine activities and relatively many face-to-face contacts discussing new applications, techniques, (im)possibilities and brainstorming sessions have to take place; the production stage moves towards the large block on Figure 4.2 above in terms of E-Contacts and possible effects.

Figure 4.2 represents the result.

New Roles for New Players

Brokers, Capacity Suppliers and Co-developers

On the basis of the previous section and further extrapolating the impacts, three new roles for suppliers can be identified. These roles differ in the type of contacts taking place, the type of activities taking place, the knowledge level, and the competence profile. We are aware that, again, these roles – and the actors who play them – are described as archetypes and a simplification of the Real World; most companies will have features of various roles and very few 'pure' types can be found. Still, for the sake of purity, we will describe the 'pure' role characteristics and actor features.

The first role is played by the *broker*. Brokers do not themselves produce in the traditional sense of the word, they *organize* supply and demand. In their most archetypal form, they are virtual organizations. Some brokers may have a *collecting* role: they collect a complete 'shopping basket' of *products* as specified by the buyer from various suppliers. They supply the buyer through orders or through scanners. Brokers also can collect *buyers* in order to obtain higher volume and, because of that, discount (leverage effect). In fact, this is the modern version of the purchasing cooperative. Producers can do the same and ask a broker to act as a sales cooperative. For both cases, it applies that if a large market share of producers or buyers is involved, cooperative action is subject to cartel legislation.

The second role is played by the *capacity supplier.* This actor owns and uses production facilities. Buyers reserve a '*time slot*' in the production schedule for a certain time period. Products can be generic, but also tailorized. They can be components, modules or final products. To assemble various components and modules into a final product can be a capacity supplier's job.

The third role is performed by the *co-developer.* The buyer will have an intensive and long-term relation with this type of actor. Sometimes, a *team* of co-developers (Praat and Alders 1998) will work together, maintaining frequent contacts. When the prototype is ready, decisions have to be taken about which capacity supplier will produce certain components and modules or whether he will perform the entire assembly.

The various players only can perform their roles properly, when the respective business processes are well linked up. From the increased emphasis on Supply Chain Management, and Materials Management in the literature, we derive the view that logistics can be seen as the *bonding agent* between the business processes of all actors involved.

Actually, within the group of all logistical activities, we again differentiate between *brokers* (who arrange and coordinate transport, distribution, and so on), *capacity suppliers* (who serve as the 'wheels': the 'mamma-daddy' companies but also the specialists in co-packing, order picking, boxing and other logistical services), and, finally, the *co-developers* (who design the entire supply chain and logistics).

The New Triangle of Players

The three new types of players are ordered using two bi-polar scales:

- reflecting the nature of *relations* and *contacts:* 'high extent of (possible) use of E-Contacts' versus 'high level of non-substitutable face-to-face contacts'; the more routinized an activity associated with the contact, the more likely E-Contacts can be used;
- reflecting the nature of *activities* of the player: 'virtual organization' (without physical production) versus 'in-house production'.

Logistics is located in the middle as the bonding agent (Figure 4.3).

From Figure 4.3, we come to '*the network of the future*': a focal company, surrounded by brokers, co-designers (sometimes), capacity suppliers and brokers, surrounded by capacity suppliers, and so on (Figure 4.4).

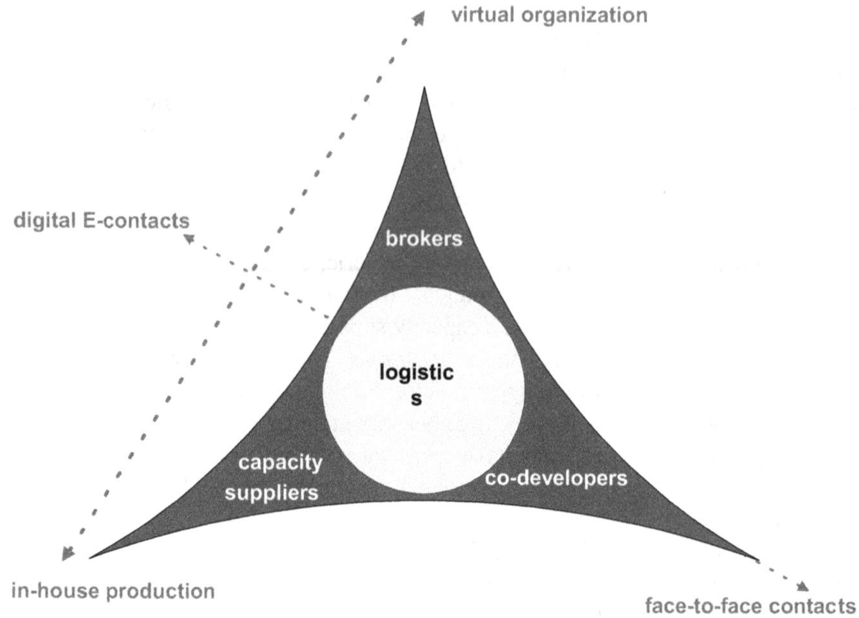

Figure 4.3 New players in new roles

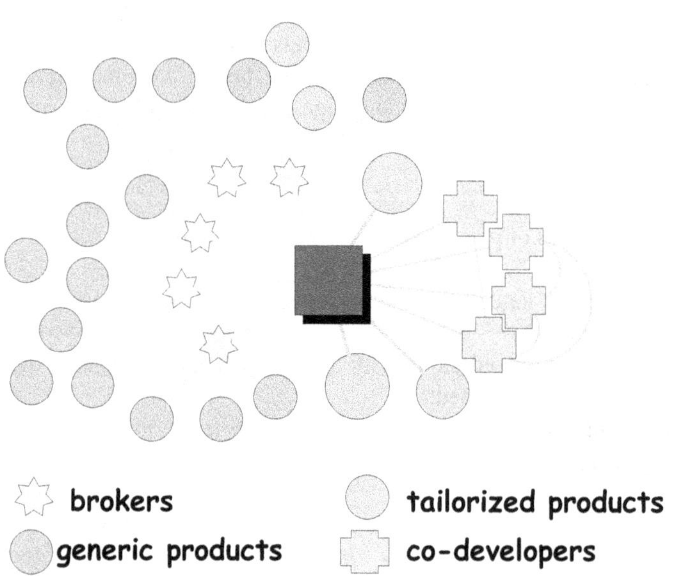

Figure 4.4 The network of the future

Towards Our Model: Empirical Support

Our model describes three archetypes of actors, but real life rarely consists of archetypes. For this reason, we looked at different sectors to see what trends could be derived (Kamann 1988, 2003). These trends should either support our predictions or contradict them. Figure 4.5 represents some of our findings. We specially focused on the questions related to the scales used in Figure 4.3.

In Figure 4.5, we find that the companies and industries have different strategies. The first type consists of typical *'true' virtual companies* like Internet auctions that are just marketplaces. Transport of the goods purchased is provided and arranged by the seller (for example, eBay). As long as they stick to their original concept, they remain true virtual, based on digital contacts. The second type is the *virtual company that moves into more in-house activities:* keeping stock, distribution houses, and so on: the Amazon.com example. A third category is the *'direct selling' company* that uses web-based selling: easyJet. We will find that, for instance, mail order companies will also come in this category. For them, the Internet is just a substitute for telephone orders: a new channel. Airlines have already outsourced their 'digital contacts' on the sales side to a shared reservation system: Galileo or Amadeus for example. We find these days that airlines sell their shares in these reservation systems and move into direct selling through the Internet. As a company, they therefore move towards the left in the matrix. We have included tour operators as a type of company with a *mixture of strategies:* through vertical integration, the

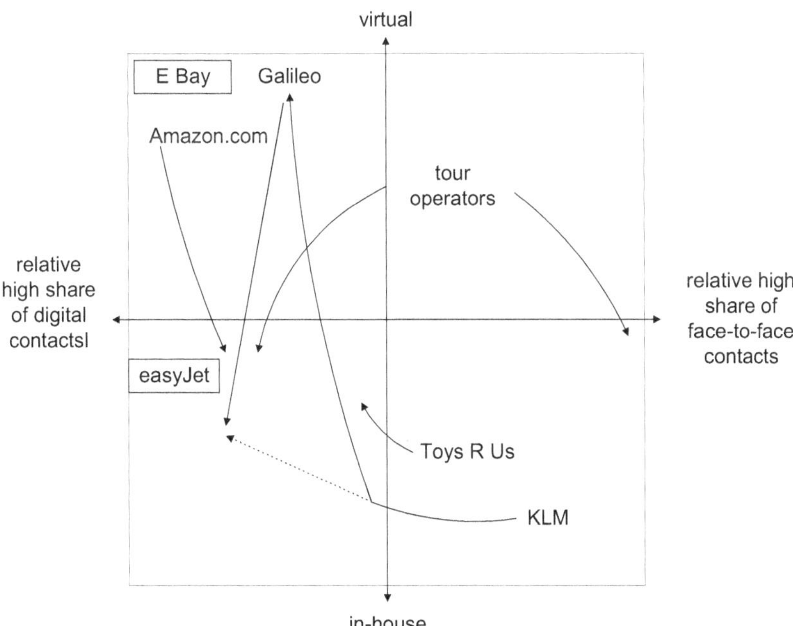

Figure 4.5 Some types of companies and their trends

large multinational tour operators move towards more in-house production. On the sales side, the face-to-face contacts increase through the ownership of travel agents. On the other hand, they also moved into web-based direct selling. Contacts with suppliers – the hotels, airlines, catering, and so on – are usually electronic: EDI, E-mail or through electronic reservation systems. Or, by fax or telex.

What we see is two different movements:

1. some true virtual web-based companies may move towards more in-house activities;
2. some true in-house-based companies (the 'brick-and-mortar' companies) start using the web as a new sales channel which makes them move towards companies with a higher share of digital standard contacts.

However, the moment the true virtual companies start moving towards more in-house production, they have to compete with specialized capacity suppliers: for example, in distribution, stock-keeping, order-picking, and so on. They have to possess the proper *capabilities* (Kay 1993). However, these do not belong to their core capabilities that were the very reason for their existence (the creative marketing concept, for instance). This makes them vulnerable, since they may not have the managerial skills – and routines – to fulfil those activities in an appropriate way. A rather different situation exists for companies that start from the opposite side – the in-house producers that just add web-based selling to their sales activity as a new channel. They already have the appropriate skills – routines – to run the 'physical' parts of the business. All they have to do is to buy the expertise – skills, routines – that are required to set up and run a web-based sales activity. Which actually is more difficult than just starting a home page. For these companies, the challenge lies in making the company processes fit the new sales techniques. Companies that fail to do so will fail to reap the benefits from their web-sales activity and may end up with the same fate as the virtual companies that moved away from virtuality towards more in-house activities. Finally, a trend we did not mention here is a rather general trend to outsource activities to specialized suppliers. In less inventive companies, these are leverage suppliers, in more innovative companies, these are more strategic partners. Figure 4.6 above gives a summary.

Location Demands

The Players in Their New Roles

Having checked our model with real life, we will return to our main issue: What is the impact of this all on logistics and the specific locational demands of the various actors in their new roles? Of course, these actors can be new *and* existing actors, developing new roles. In fact, what we see happening at the moment is that many existing companies are moving towards new roles. From the existing network literature with a spatial dimension (Kamann 1988, 2003), we can derive conclusions as to whether actors are:

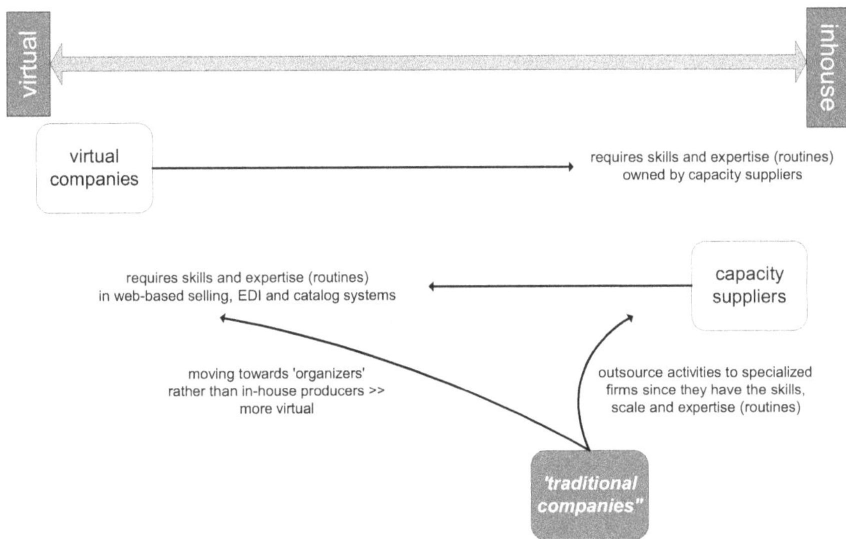

Figure 4.6 Trends among companies

- relatively footloose or not: that means location *indifferent*;
- focused on *industrial agglomeration effects*: that means they want to be located close to their buyers and/or suppliers;
- focused on *urban agglomeration effects*: that means they want to be located near large urban areas (see central place theory, Christaller 1933).

The questions that come up from the previous sections are related to the nature of *(1)* the demand that logistics is facing in order to fulfil the function of 'bonding agent'; and *(2)* the demand in terms of locational requirements and preferences that are the result of that: What will be the locational preferences of the logistics firms?

The Logistics Service Providers: Time Hazard-Free Access

In order to fulfil the requirements of the various players in the network in its function as bonding agent, the predictability and reliability of supply times are becoming more important. Given this increased need for the fine-tuning of the logistical processes, for the transport company, reliable *access time* – to suppliers and buyers – becomes an important factor. This implies that, as part of their locational behaviour, both suppliers (the capacity suppliers) *and* buyers (divided over other capacity suppliers, retailers and final customers) will look for proper infrastructural facilities. Distance per se plays a less important role. This is because distance is translated into cost price, while time uncertainty is reflected in the reliability of delivery. To arrive too late can result in a production stop of a plant for one hour, or the *time slot* in which the particular batch of raw materials had to be used, has expired. The financial consequences of such an occurrence can be higher than the additional (predictable)

Table 4.1 Location requirements of players in their new roles

Category player	Sub-group	Footloose or agglomeration-oriented
Brokers	General	*Footloose;* in 'purest virtual' form, brokers can be located anywhere
	Commodity specialists	In case of clear *(inter)national centres* in the trade of certain products, these centres are likely locations because of the presence of (informal) contacts (see the focus concept: Kamann, 1988, 2003)
	Hybrid brokers	Broker activities originating from existing wholesale activities will follow the locational preferences of these wholesalers, close to large urban areas; oriented on urban agglomerations
Capacity suppliers		Will follow 'normal' industrial location preferences: varying from relatively footloose to focused on *industrial agglomerations;* access to good physical infrastructure is precondition for increased importance of just-in-time, zero-stock and vendor-managed inventory
Co-developer		Because of high content of face-to-face contacts predominant focus on *industrial agglomerations;* in practice, buyers are internationally dispersed and industrial agglomeration is in economic space rather than in geographical space

costs of distance per se. Logistic service providers will therefore have a preference for locations with risk or *time hazard-free access* to: *(1) industrial* agglomerations for collecting goods and delivering goods; and *(2)* large *urban agglomerations* for deliveries to buyers in those areas (retailers, wholesalers and urban-oriented firms on industrial estates around the cities).

Logistic service providers live in a paradoxical situation: on the one hand, they want to be as close as possible to the market (for inputs and delivery points). On the other hand, they want to reach scale effects which favour centralized locations. However, centralization may result in larger distances (and therefore longer travel time); an optimization model may help to make the right choice.

Summarizing: brokers can be relatively footloose, while, on the contrary, capacity suppliers and logistical service providers prefer locations with hazard-free access times. For capacity suppliers, these locations can usually be found on industrial estates with proper infrastructural connectivity within the context of industrial agglomerations. Logistics providers will prefer central locations with hazard-free access times to suppliers and delivery points. In the case of co-developers, it all depends on the role and importance of the actual production of the products developed. The larger that part becomes, the more they move into the direction of capacity suppliers.

A Dynamic Model for Locational Behaviour for Logistics Providers

Given the purpose of this contribution, the locational preference of logistics providers will now be discussed in more detail. These companies aim for:

- *centrality* in their locational behaviour: that is, close to industrial and urban agglomerations;
- *certainty* in access time; or, in other words, minimizing time hazards.

Because of this, the preferred location should have high *connectivity,* together with a low access time hazard. This results in the following dynamic model (Figure 4.7).

A high perceived certainty related to access means a location has a good perceived connectivity. That results in improved perceived effective centrality. Because of that, the location will become more attractive for logistics providers. This will have a positive impact on locational preferences. However, that results in more companies moving there, which results in increased traffic flows to and from that particular location. That causes congestion, which has a *negative* impact on the certainty of access time. Then, a negative spiral movement sets in, reducing the locational preference of logistics providers for that location and shifting it towards another location. In other words: the perception that a location is attractive results in the establishment of companies, but a *collectively shared perception* turns it *un*attractive again, causing logistics providers to move out again to a location that they find more attractive. Or, new investments will occur in other locations.

As the reviewer of this contribution states: 'There are costs associated with relocations. At some point (and in some locations), the cost of relocation might exceed the cost of congestion (borne by the logistics providers, their contacts, or the

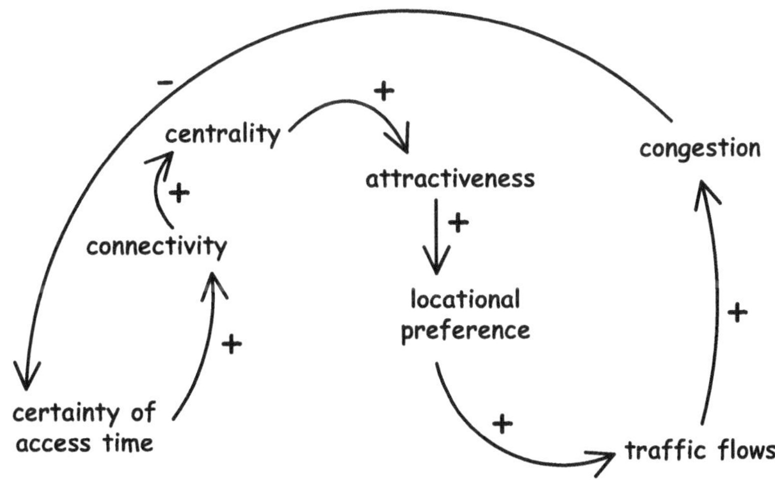

Figure 4.7 Dynamic locational preference model logistics providers

public at large). What influence might a polycentric urban structure have on those relocation costs, versus (say) a monocentric urban structure? The latter may have a more limited choice of locations compared with the former, and (because of the more limited number of potential new locations) the costs might be higher.' I agree that these points are certainly worth considering. However, I assume that, although these issues may result in slightly deviating results, the model per se remains valid. It would nevertheless be interesting to test the model in real-life situations and refine the general model described here along the lines of the comments of the reviewer.

We conclude from this that a continuous reorientation takes place by logistics providers. Again and again, new locations are declared to be 'the' favourite location, and, as a result, they suffer from their own popularity. A continuous process of 'reorientation' and 'further moving' collection/distribution centres with a perceived optimal mix of centrality and hazard-free access follows from our theoretical model. But, what can be done if there is nowhere to go to anymore? Or, as our reviewer puts it: 'if potential new locations are exhausted?' My prediction would be that pressure will be put on infrastructure-providing authorities to create these new locations, or to solve bottlenecks in order to regenerate old locations into new ones.

Results of Interviews with 26 Firms

Our reflections are theoretical. To see whether companies recognize the trends described and act accordingly, we interviewed representatives from 26 firms belonging to various types of companies: transport companies, food producers, and suppliers of stationery in the typical 'routine quadrant' where logistics and Vendor Managed Inventory (VMI) are assumed to be important. The results (Table 4.2) show that companies in general do become aware of the role of time slots and time hazards. Nevertheless, few companies had actually decided to *relocate* because of it.

Table 4.2 Interview results from 26 firms

Transport companies	Capacity supplier food	Wholesalers stationery
Increased importance of lead-time and time-slots	Stricter time windows for buyers	Time slots of importance
Reliability key element	Strict requirements from suppliers	Vmi makes the difference
Locational preferences according to model	Location indifferent	Follows locational preference model
More outsourced logistics	Externalizes problem (outsourced logistics)	Client comes or delivery service?
Increased awareness of traffic-related time hazards	No congestion problems (periphery)	Becoming more traffic risk aware
'Wheels' versus 'full service'		Large leaders <> small laggers

We did find that more and more companies solve the problem by outsourcing transport and distribution. This way, they externalize the problem to specialists. Transport companies that operate in the relatively congestion-free North did agree with the trend and emphasized that for trips to the West, they certainly take it into account. In the North, they only experience the problem in the typical sugar beet season. Companies providing 'routine' products and acting on vendor-managed inventory systems – stationery – seem to be most aware of the trend. These typically were the larger companies: they were acting accordingly.

Conclusion

Using a differentiation between various product categories, supplier relations and the contacts associated with them can be differentiated as well. The level of routinization of activities and the related contacts are used to predict the potential impact of E-Contacts. On the basis of these predictions, new roles can be described for brokers, capacity suppliers and co-developers. Logistics would act as a bonding agent of the business processes of the actors in the network. To fulfil this function properly, logistics providers and routine goods suppliers (modern wholesalers) acting on vendor-managed inventory will prefer locations with time hazard-free access to (and from) industrial and urban agglomerations. This results in a dynamic process of continuous reorientation where again and again new locations become 'the' favourite location for distribution centres. However, these favourites end up becoming unpopular because of the resulting congestion.

References

Christaller, W. (1933), *Die Zentralen Orte in Süddeutschland*, (Jena: Gustav Fisher).

Fisher, L. (1970), *Industrial Marketing: An Analytical Approach to Planning and Execution* (Princeton: Brandon/Systems).

Forrester Research Inc. (2000), <http://www.forrester.com>.

Goddard, J. (1973), *Office Employment, Urban Development and Regional Policy* (Dublin: An Foras Forbartha).

Hughes, J., Ralf, M. and Michel, B. (1998), *Transform Your Supply Chain: Releasing Value in Business* (London: International Thomson).

Kamann, D.-J.F. (1988, 2003), *Externe Organisatie vanuit een Netwerkperspectief (Industrial Organization from a Network Perspective)* (Groningen: Charlotte Heymanns).

—— (1999), *Inkoop vanuit een Netwerkperspectief* (Groningen: Charlotte Heymanns).

—— (2000), 'Kraljic Krijgt Extra Dimensie (Kraljic Gets Additional Dimension)', *Tijdschrift voor Inkoop en Logistiek* 4, 8–12.

Kay, J. (1993), *Foundations of Corporate Success* (Oxford: Oxford University Press).

Kraljic, P. (1983), 'Purchasing Must Become Supply Management', *Harvard Business Review* 61:5, 109–17.

Olsen, R.F. and Ellram, L.M. (1997), A Portfolio Approach to Supplier Relationships, *Industrial Marketing Management* 26, 101–113.

Praat, H. and Alders, B. (1998), *Toeleveranciers en Productontwikkeling (Subcontractors and Product Development)*, (Zoetermeer: NEVAT).

Chapter 5

Integrated Mobile Spatial Information Systems and Urban Services

Maria Giaoutzi and Vassilios Vescoukis

Telecommunications and Urban Structure

Information and telecommunication technologies are in the process of completely transforming economic and social life. Their influence is felt at all levels of society, and in all contexts where communication takes place.

Telecommunication systems, in the policy context of today, very often appear as a substitute for the physical movement of people and services. However, the growing use of telecommunications goes far beyond the influence these may have on people's everyday lives, since these technologies drastically increase the complexity in urban patterns by increasing the number and type of interactions between individuals, firms, and technical systems, as well as between them and the external environment. Information systems as such are enabling new combinations of people, equipment and places within cities and large metropolitan regions, leading to dramatic changes in the spatial organization of activities.

In the past, researchers focused on the flow of information in colonial American cities through newspapers (Pred 1973), in the urban complex of the Northeastern United States by telephone (Gottmann 1961), and in the information economy by means of office buildings and overnight letter delivery (Sui and Wheeler 1993; Mitchelson and Wheeler 1994).

Geographers' and economists' concern for understanding communications technologies and their impact on the location of human activities is reflected in the early work of many researchers (Gillespie 1983; Gottmann 1983; Goddard and Thwaites 1983; Camagni 1984; Gillespie and Robins 1989; Nijkamp 1986; Camagni and Rabellotti 1986; Giaoutzi and Nijkamp 1988; Davelaar 1991).

All these works systematically describe the way in which telecommunications affect both the centralization and decentralization of activities, increasing the understanding of the new elements that communications technology is introducing in the decision process for the location of activities. Gottmann (1983) claimed that telecommunications 'first has freed the office from the previous necessity of locating next to the operations it directed ... and second has helped to gather offices in large concentrations in special areas'.

A short review of recent findings, in the text below, outlines the evolving context of our discussion.

In his early work, Abler (1970) concludes that 'advances in information transmission may soon permit us to disperse information-gathering and decision-making activities away from metropolitan centers, and electronic communications media will make all kinds of information equally abundant everywhere in the nation if not everywhere in the world'.

More recently Gilder (1995) extended this argument claiming that due to the continued growth of personal computing and distributed organizations advances 'we are headed for the 'death of cities''. Cities are no longer needed to access a wide range of cultural activities and information sources since telecommunications bring the library, concert hall, or business meeting into any home or office (Gilder 1995).

Similarly, Negroponte (1995) states that 'the post information age will remove the limitations of geography. Digital living will include less and less dependence upon being in a specific place at a specific time and the transmission of place itself will start to become possible.'

Nigel Thrift (1996) in his innovative work stresses the importance of face-to-face information in an era of high-speed communications where networks have been generating a demand for instant information in the financial services sector. In this context, Thrift argued that 'since the international financial system generates such a massive load of information power goes to those who are able to offer the most convincing interpretations of the moment. Interpretation of information depends on both face-to-face information and advanced technologies – an activity that is necessarily and increasingly centralized in the leading world financial centers.'

Gaspar and Glaeser (1996) during the same period present strong evidence, in their work on 'information technology and the future of cities', that telecommunications and travel are synergistic, suggesting the intrinsic value of face-to-face interaction beyond what can be communicated at a distance.

Gordon and Richardson (1997) claim that 'rapid advances in telecommunications are now accelerating the decentralization trends … proximity is becoming redundant … entertainment already is … and instruction is more likely to be transmitted over broadband radio frequencies rather than seen in traditional theaters or lecture halls, today's cities continue to become less compact; the city of the future will be anything but compact.'

Peter Hall (1997) in his work on 'modeling the post-industrial city' states that 'the urban world of the 1990s … is a world in which cities deconcentrate and spread to become complex systems of cities linked together by flows of people and information. Thus, while telecommunications enable the evolution of an increasingly complex urban system with multiple linkages and hierarchies, place and centrality remain key issues in the information economy.' Along the same lines, Hall (1997) claims that cities' competitiveness in the global economy 'depends on their capacity to generate process and exchange information'.

Narrowing the Gap

This part of the chapter puts into perspective a reference framework describing the functioning of the urban-cyber context, on the one hand, and the operation patterns of the Integrated Spatial Technology (IST) services, on the other.

The Urban-Cyber Context

Traditional urban services are based on the construction and supply of 'public infrastructure' to citizens, as utilities essential for the operation of the community, such as waste management transportation, and so on. During the past few decades, the notion of 'public', in this context, has evolved to also include services provided to the community by private organizations.

The evolution of ISTs, on the other hand, has created a new context for the operation of urban services. From telephony to multi-party videoconference and from early web services to virtual communities, information and communication technologies are converging to create a unified virtual space where digital services of any type are offered. The provision of such services directly or indirectly enables the evolution of an increasingly complex urban system with multiple linkages and hierarchies incorporating the traditional urban services.

The dynamic interaction between the rapidly evolving technology and the urban structures is highly conditioned by the adoption rates of the users. This makes it necessary to understanding, on the one hand, the urban cyber-context, while, on the other, the behavioural potential of the user.

Since the introduction of digital technology, its developments have followed an entrenched path of miniaturization and decentralization with increasing focus on individual and niche applications. Computer hardware has moved from remote centres to desktop and handheld devices, while being embedded in various material infrastructures. Software has followed the same course. The entire process has converged on a path where various analogue devices have become digital and are increasingly being embedded in machines on a smaller scale. In parallel, there has been a convergence of computers with communications ensuring that the delivery and interaction mechanisms of computer software are now focused on networks of individuals, not simply through the desktop but in mobile contexts. With such a massive convergence and miniaturization, new software and new applications define the cutting edge.

In order to facilitate the understanding of the likely impacts of IST services and their applications on spatial structures, a reference framework could be very useful, as is the case with the ISO 'OSI 7-layer reference model for digital communication networks'. In the next paragraph, such a layered reference framework is presented, which is useful for the study of the relations between physical and virtual space, and has evolved from earlier work in this field (Shiode 2000). In his early work, Shiode proposed a 4-layer framework for the presentation of cyberspace (left-hand side of Figure 5.1) where the '*real world*' *layer* represents the physical space. The '*Internet*' *layer* refers to the tangible infrastructure of the Internet, such as copper and fibre optic cables, where the Internet Protocol (IP) network is implemented. Based on the Internet, the '*web space*' *layer* is an expression for the concept of space where web multimedia content and hyperlinks are located. The most abstract virtual space structures, such as '*cyber-cities*' and '*cyber-places*' are located on the top layer. According to Shiode, cyber-cities appear similar to the real world; however, they operate on the basis of their own rules and provide a quite flexible virtual space for the operation of virtual communities of any type.

One step further for understanding the IST challenge for urban policies is our revised version of Shiode's framework which is discussed in the text below. Two main reasons have motivated such a revision: first, the already realized *integration* of the wired and wireless telecom networks of any physical media, as well as the Internet physical infrastructure; and second, the fact that 'web space' has been expanded to provide more applications than in the late 1990s. Today the term 'web' refers both to a virtual space where content is offered and to many content-independent digital services, including telecommunications and application services. With this in mind, it seems appealing to divide the generic concept of 'web space' into two new layers.

At the bottom of the reference framework (right-hand side of Figure 5.1) remains the *first layer* of the '*physical space*', which is more conveniently now referred to as 'urban space'.

The *second layer* changes from 'Internet' into the '*telecom networks*' layer. In this context 'Internet' is no longer just a virtual digital network based on analogue or early digital telecom infrastructure, used primarily for voice telephony. Today, all modern telecom services are based on advanced digital technologies, some of which have been developed for the Internet, some for voice, and some for data communication.

The same telecommunications infrastructures are used for data, voice (traditional telephony), and media delivery services (for example, video on demand, cable tv). The maturing of technologies such as Voice over Internet Protocol (VoIP) and Quality of Service (QoS), along with the improvement of data line bandwidths, are enabling this development. A typical example of the relation of Internet to telecom systems is that in the early days, voice telephony circuits were used for Internet dial-up access, and Internet protocols were run *on* voice telephony infrastructure; today digital circuits are used for permanent internet connectivity even for the individual home user (for example, cable, digital subscriber line (xDSL)), and voice telephony

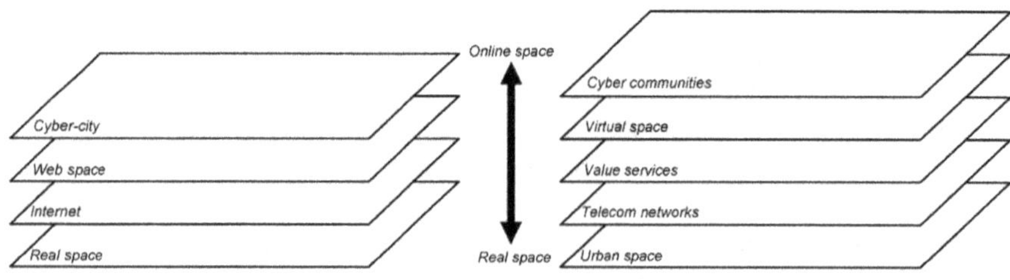

Figure 5.1 Shiode's 4-layer framework (left) and a 5-layer revision (right)

services can be based on technologies introduced by the Internet (for example, VoIP), which is exactly the opposite.

The *third layer* of the revised framework is '*value-added services*' which are offered using services from the 'telecom networks' layer. End-user Internet connectivity and services such as the web, application services, data services, content services, as well as communication services including voice telephony and videoconference, are all typical examples of services on this layer. The thematic range and the types of services offered on this layer are very broad. Indeed, these services may serve as infrastructure for the creation of virtual spaces, as well as for the provision of additional telephony-like services (multi-part teleconference), or even for other autonomous services.

Next comes the *fourth layer* of the revised framework, which is called the '*virtual space*' layer. The services provided on a 'virtual space' make use of value-added services from layer 3 to create 'non-real-space entities'. A typical example of this case is the 'virtual classroom' in e-learning applications. As the Internet has evolved, many ad hoc virtual spaces have been introduced, such as the discussion boards where people who share common interests exchange opinions regardless of their geographical location. In this context, 'virtual space' services are by no means synonymous with 'cyber communities' which, apart from the communication contexts they operate in, are governed by their own rules, depending on the purpose they serve.

Virtual space services are the basis for the creation of '*cyber communities*' in the *fifth layer*, as is the case when virtual classrooms and other services are combined to create integrated space-independent educational structures. A virtual community can be viewed as a virtual space *plus* a set of rules and services that govern the fulfilment of whatever purpose the virtual community serves. E-learning frameworks, in which rules for running educational activities are applied to virtual classrooms, are typical examples of cyber communities. 'Cyber-cities' are a special type of cyber communities that may have a clear relationship with socio-economic patterns in real space as well.

All in all, IST has quietly merged into the existing urban structure and has generated a new urban infrastructure that in itself requires planning and management. This virtual environment, commonly known as 'cyberspace', is a new functional space increasingly used for social and economic activities. In this context, both public and private policy making may directly contribute to the growth in computing power which has contributed to the present proliferation of cyberspace, but there are two other perhaps more important elements that may propel the development of cyberspace: the first is the infrastructure and services that support network communication; and the second is the software and protocols that have made possible the operation of interactive multimedia spaces and the provision of applications based on them.

Operation Patterns

This part of the chapter discusses the evolving patterns of the operation of services in the 'network society' which has developed and, according to Castells (1989, 2000), is changing and restructuring the material basis of society; such an information service has come to dominate wealth creation in a way that information is both a raw material of production and an outcome of production as a tradable commodity. The attitude of decision centres, citizens and service providers towards IST applications, specifically those relating to urban services, is useful both in understanding the impacts and in approaching IST-related urban policy making.

On the basis of the layer framework presented in Figure 5.1, two patterns of operation of IST services that enable the provision of services in urban space are discussed. Focusing on urban policy and management, real urban space is used as the reference layer for all activities. Urban problems are studied in *real space* (Figure 5.1). Any service supplied to meet existing or potential demand should have a real-space reference as well. However, services provided for the solution of certain problems in real space can be acquired on any other of the reference layers.

Two types of interaction patterns for the provision of a type of service in urban space are recognized, as shown in Figure 5.2. The first pattern invokes services from any single layer above real space (Figure 5.2a), while the second invokes services offered in more than one layer (Figure 5.2b). For example, a one-way information flow, regarding urban road traffic load, can be achieved by using a service from the 'value-added services' layer, which ultimately affects citizens' behaviour in real space. The same may occur for any other application where higher-level services need to be invoked in order to satisfy a request or provide services in real space.

An important element of understanding the perspectives of technology and its applications to market positioning is a thorough examination of the patterns of operation in this 5-layer reference framework. Both the nature and the capacity of the provision of such services is a key element in policy making as well as in discovering entrepreneurial opportunities for both private and public sector stakeholders.

In order to implement services which incorporate the multi-layer type of operation, a key requirement is *integration*, which affects both technology and policy levels. On the technology level, integration implies putting together different technologies in order to achieve a specific result, which is not always feasible; therefore, a large amount of added value is required for the integration of diverse and sometimes competing technologies. On the policy level, on the other hand, integration implies understanding the patterns of operation in order to develop the required legal and normalizing frameworks for the operation of all types of such services. The required high degree of service coordination, especially in the case of a pattern of multi-

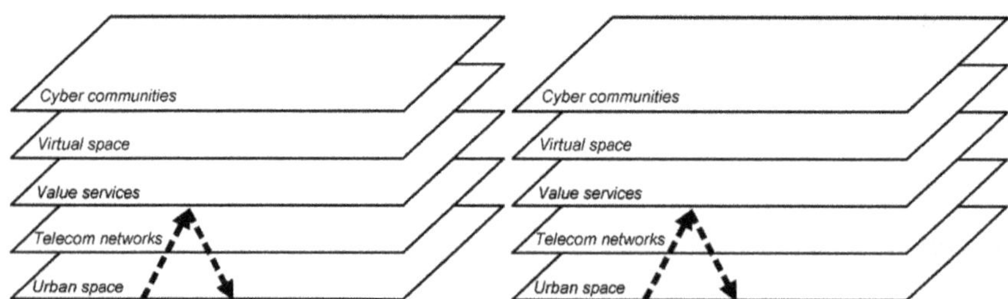

Figure 5.2 Services from one (a) or more (b) layers can be required to satisfy a need in real space

layer service operation, is a motivation for both efficient technical architectures and operational urban frameworks.

Types of IST Applications

In this part of the chapter, a typology of IST applications will be discussed, as these relate to the 5-layer reference model described above. Five main families of applications will be presented: e-work, e-learning, e-commerce, e-government, and Location-Based Services (LBS).

E-work involves all types of teleworking such as home offices, satellite offices, and telecottages. Teleworking describes organizational work performed outside the normal organizational limits of space and time, supported by computer and communication technologies (Olson 1987). Teleworking is also described by Olson as 'work enabled by network technologies, which actually incorporates a shift in performance focus from physical presence to results, empowerment as well as location and time independence'. E-work appears as an option opening new possibilities for business development. E-work has potentially a high impact both directly and indirectly on *traffic congestion*, especially in urban centres, *environmental quality* as a result of the reduction of unnecessary commuting, *energy savings* and maintenance costs for cars by reducing commuting trips, *employment,* substantial new job creation coupled with a very strong job growth in the service sector, *ways of delivering social services* and so on, which in turn have a considerable impact on urban structure (Vescoukis et al. 2004).

E-learning includes all types of computer-supported education, ranging from access to educational content, virtual classrooms, integrated learning frameworks, as well as to virtual universities. The shift from education to 'e-education' may reduce the amount of urban resources assigned to education, especially for those educational activities more effectively carried out over the Internet. Although this is still an open issue, typical cases of such activities could be adult continuing education, as well as types of corporate training. Several considerations have been raised in this respect: How can e-learning serve pedagogical purposes? How early in the educational process should e-learning be applied? How can social consciousness be achieved through virtual spaces? How is (urban) space perceived through e-learning? Could the reduction of floor space requirements for educational activities be a factor affecting the adoption of e-learning and so on?

E-commerce The development of e-commerce – a platform on which to conduct business electronically – has drastically influenced the corporate and consumer market, despite its relatively recent emergence.

It consists of a large set of financial transactions over the Internet, including not only retail sales – Business-to-Customer (B2C e-commerce) – but also Business-to-Business (B2B e-commerce) and Business-to-State (B2S e-commerce). The largest part of e-commerce is B2C, which is a global activity and, as such, although it may have strong impact on the economy and entrepreneurship, its impacts on physical space and urban resources are not easily traceable.

The social and economic developments of cyberspace have raised a number of questions in this context such as: How do conventional retailers compete with electronic markets? Should e-commerce remain tax-exempt? To what extent can and should we protect an individual's privacy? How do we prevent minors from accessing inappropriate contents? These questions relate to policy issues which may touch upon broader socio-economic and ethical issues.

E-government includes all types of information services and of transactions between the state and the citizens, such as electronic tax statements and collection of taxes. A distinction can be made, based on whether e-government services are provided by local authorities or by the State. Although several issues regarding e-government have been raised, such as privacy and identity, e-government is generally considered as a catalyst in saving the time, space and resources in general that we required for the transactions of citizens with the State.

Last, *Location-Based Services* (LBS) is a large and rapidly emerging category of electronic services provided with a high level of integration of technologies and services on any of the layers shown in Figure 5.1. Such services provide real-time, location-dependent information and/or content to clients, according to new business and billing models, such as mobile phone SMSs (short message services). LBS is a category of services with a potential for high impact on urban resources, services and even governance.

The assignment of the services discussed above to the 5-layer reference model (Figure 5.1) is shown in Figure 5.3, where the threshold layer for each service is noted. By 'threshold' is meant the lowest possible layer on which services or applications of a specific type can be offered. *E-work* services may be provided in a context as complex as a virtual community, or as simple as a telecom service that simply connects remote workstations to a central computer. Several intermediate or combined settings are also possible; for example, a third party may provide e-work value-added services to employers wishing to outsource their e-work support, in which case e-work applications reside on the 'value-added services' layer. Provided that the lowest possible layer for an e-work application is the simple case of connectivity, the threshold layer for e-work is the 'telecom networks' layer.

E-commerce is a huge category of applications, none of which can be provided on the 'telecom networks' layer because much more than just connectivity is required.

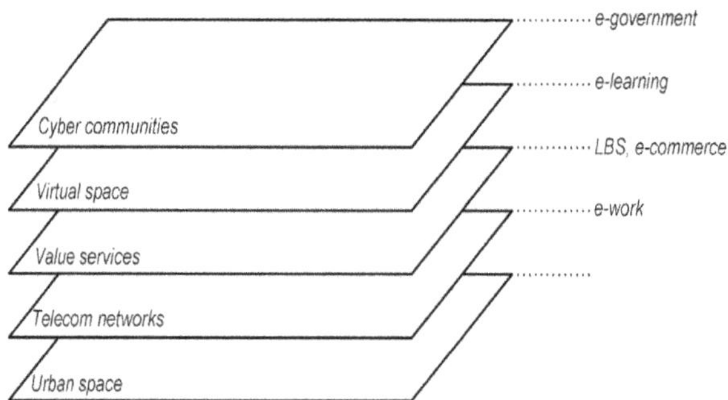

Figure 5.3 The threshold level for the provision of electronic services

Therefore, the 'value-added services' layer is considered the entry (threshold) layer for such applications, where many B2C simple e-commerce services may be provided. Typical examples of e-commerce applications on this layer are third-party maintained sites that offer secure credit-card transaction clearance services to small merchants who cannot afford to maintain their own e-commerce site. Higher-level e-commerce services may be virtual malls, as well as integrated payment systems such as Secure Electronic Transactions protocol (SET).

E-learning applications are introduced on the 'virtual space' layer. In this context, the case of one-way educational content provision is not considered as an e-learning activity. The interactivity among trainees and trainers required for e-learning, as well as the definition of cyber teams such as virtual classrooms, imply that virtual space is the real context of e-learning applications.

Last, *e-government* applications are assigned to the top layer of this reference framework, provided that additional services relating to security and to digital authority are required. Such services are digital certificates and hierarchies of independent certification authorities, as well as high-security communications for the provision of financial, administrative and informational services to citizens.

Location-Based Services are considered as a distinct type of e-services and, as such, the 'value-added services' layer is regarded as the base layer above which they can be implemented, taking into account that such services cannot be implemented using only communication services. In fact, provided that LBS-related evolutions are still ongoing, one can argue that LBS is not a different type of services, but instead, it is a possible characteristic of all or some of the other types of services. The discussion is still open on this issue. However, for the purposes of our approach, LBS will be considered as a distinct family of IST applications and will be further examined later in next-but-one section.

The Demand-Policy-Service Cycle

All urban services, both traditional and 'e'-services, are driven by demand. Space has become the basis for many diverse interactions among physical or virtual, atomic or complex, private or public entities. Phenomena, such as overpopulation, and shortage of resources and activities, such as transportation and road networks, create a new perception of urban structures, which in turn affects the behaviour of individuals, and forms the demand for new or better services.

In order to meet this demand, new policies have to be developed which may contribute to the creation of new urban infrastructures and services. Private organizations are more flexible in this respect and can be more effective in providing both services and infrastructures that would otherwise reach the citizens much later and quite likely at a higher cost. In this context, private organizations play a role in both policy making and the implementation of infrastructures and services, as shown in Figure 5.4.

Because ISTs are an accelerating factor for the 'demand-policy-service implementation' cycle shown in Figure 5.4, it is worth identifying the role of ISTs in generating new demand and their impact on urban policy making.

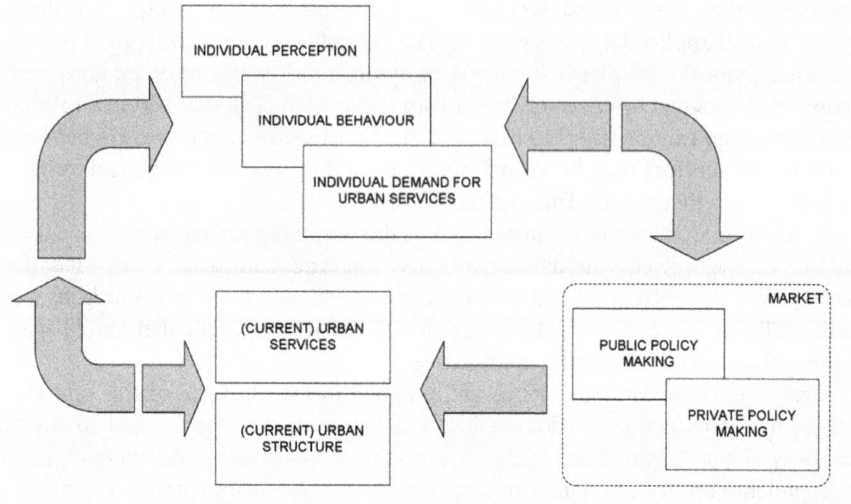

Figure 5.4 A generic 'demand-policy-service' cycle

As technology evolves, urban structures become even more complex. The concept of 'service' is broadened to include, apart from traditional urban services, 'soft' services, as well. 'Soft' services, in this context, do not require traditional infrastructures but instead are based on the flow and process of digital data, produced in huge quantities, as these relate to almost any aspect of urban activity. Value-added services require modern telecom and IT infrastructures that may exist in any of the above-mentioned five reference layers. Another key characteristic of such services is their tendency to relate to the location of the clients, and that is the basis for the evolution of LBS.

The demand for 'soft' services, unlike the demand for traditional, public or private, urban services, follows the supply of the ISTs. There are two main reasons for this: first, the knowledge on ISTs is not generally accessible, even if innovative technologies are mature, which implies that a demand for IST-related urban infrastructures and services, based on such technologies, cannot grow easily. Second, the speed of technological evolutions is maintaining an impressive pace even for those who actively participate in the R&D of ISTs; this implies that citizens are not informed about IST potential unless the IST product reaches them.

As the discussion above suggests, the provision of information on IST possibilities, applications and services, could in itself be a service of value, which has the potential to modify the perception and finally create new demand for IST-based 'soft' services. In such a context, several entrepreneurial opportunities may arise, which are more often timely understood and exploited by the private sector.

IST evolutions may influence urban policy making in two ways: indirect and direct. *Indirect* influence implies that clients, well informed about technology, create demand for IST services before these become available or even part of the urban policy orientation, while *direct* influence implies that policy makers become early adopters and market leaders of IST services, before such services are requested by clients.

As shown in Figure 5.5, the flow of IST-related information to the public at large could alter their perception and eventually create demand for IST-based services. For example, the knowledge that today's technology can enable the integration of satellite positioning technologies with mobile phone equipment could generate demand for location-based services before these are commercially available or even considered by public or private policy makers. This is the indirect way that IST evolutions influence urban policy making.

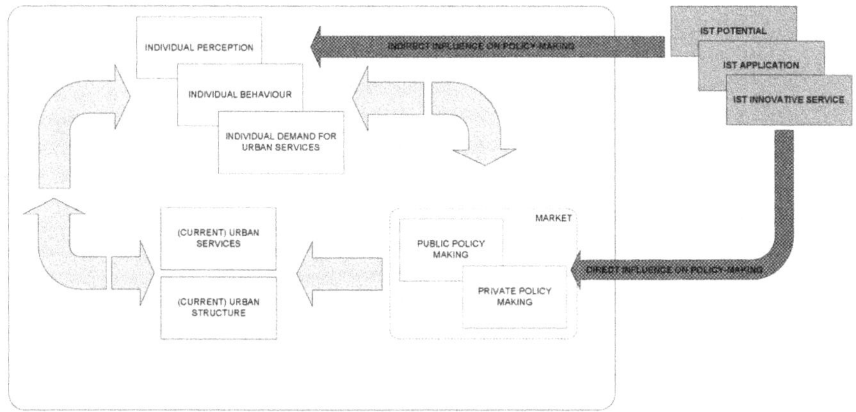

Note: IST = integrated service technologies.

Figure 5.5 Direct and indirect influence of ISTs in demand and policy making

There is also a fair possibility that urban policy making is directly influenced by IST evolutions. Support of innovation by community funds, in order for it to become of value to citizens, is a well-established EU practice. In this context, EU support is available in almost any area of innovative IST applications, including urban applications, with the participation of both private and public sectors. On the other hand, there is always the possibility for purely private involvement in the development and commercialization of any IST-based urban infrastructure and/or service.

A typical example of this is the *deployment of wireless and wired private telecommunications infrastructures and services, where a high degree of integration and convergence of different technologies is noted.* This applies in particular to the case of Location-Based Services which will be discussed in the next section.

Urban Information Location-Based Services

A good example of convergence of technologies is Location-Based Services (LBS). LBS has been around since the 1980s, with the earliest applications, such as Lojack, being in use for auto theft prevention and recovery. More recently in the latter half of the 1990s, global positioning system (GPS)-based systems have been used for early fleet management applications. Since then, three major families of technologies have evolved, creating a world of technological potential and opportunities: the Internet, Mobile personal applications and Spatial Database technologies.

A Short Introduction to LBS

Needless to say, the Internet became the de facto platform for computer communication and applications in the personal, business and entertainment field. Mobile phone technologies became affordable and available practically to anyone, while, at the same time, positioning technologies such as GPS, have allowed accurate personal global positioning using affordable handheld devices. In parallel, in the Computer Science domain, spatial databases which are a special type of databases with regard to *space*, have made possible the representation and storage of geographical information. Where do all three technology families converge?

The answer, as shown in Figure 5.6, is 'Location-Based Services' (LBS), which are information and/or communication services, offered to mobile individuals, with regard to their exact location. The key point in LBS is the location of the client: services of this type are dependent both on the client profile and the client's geographical position. The provision of local information to tourists and vehicle navigation are two typical examples of already commercially available LBSs, while content push (such as promotional commercial content or e-newspapers) is a big new category of LBS that will be seen in the near future.

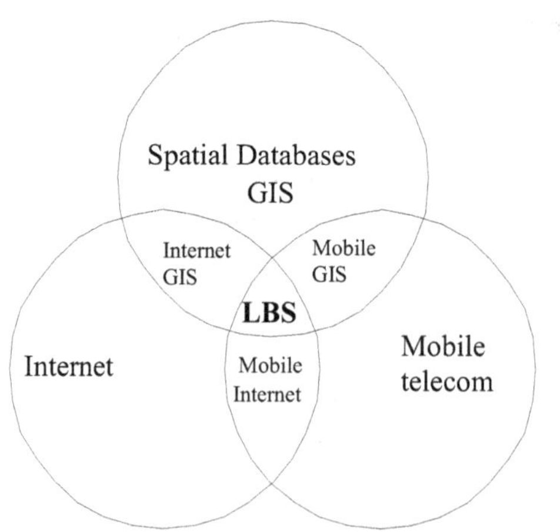

Figure 5.6 LBS and Information Technologies (Shiode 2000)

The next section presents a case study for the provision of an urban Location-Based Service concerning the management of parking spaces.

The Case of Urban Parking Spaces

It is common knowledge that urban parking space is valuable, and its central administration and management could be a challenge for local authorities. Efficient management means that the best use and timely dissemination of information about parking spaces is made, which is the motivation for this case study. Provided that wireless communication network services are widely available as generic Internet 'telecom services' (Figure 5.3), we propose an integration of GSM-based mobile IT services with available geographical information systems (GIS), location technologies and sensing technologies for the management of parking spaces. The objective of this integration is the provision of an LBS, which will provide information on the availability of car parking resources to mobile users requiring access to such resources in real time. In order for this service to be provided, the multiple-layer behavioural pattern (Figure 5.2b) is followed.

The main concept of such services is as follows:

- A mobile user requiring access to parking space, equipped with a GSM phone (or any portable or car-mounted device with wireless connectivity capability) and optionally with a GPS device (either handheld or integrated into a vehicle) sends a special SMS message to a local Urban Resource Information Service Centre (URISC). This SMS message contains a 'request for service' flag, as well as (optionally) exact location information which is acquired from the GPS device. If no such device exists, the user may send a 'location id' code which is made available through the mobile service operator.
- The service centre, which maintains real-time information on parking spaces, retrieves the required information (such as location, cost, availability), and prepares an SMS/MMS (multimedia messaging service) message to be sent to the requesting party.
- If a GPS system with navigational services is available, it can provide guidance to the location of the requested parking space.

The above sequence is shown in Figure 5.7.

Since not all users have access to GPS devices, it is useful to define operation scenarios according to the available user equipment. The following three scenarios can be recognized:

1. User equipped with a mobile phone. This type of user can use URISC services by consulting a map, made available through the mobile operator, where areas of interest are divided into sections, each one of which has a unique code. These codes are made known by the mobile operator using the web and/or the billing system. The user sends an SMS with the section code and the system sends back an SMS containing the requested parking information in that area.

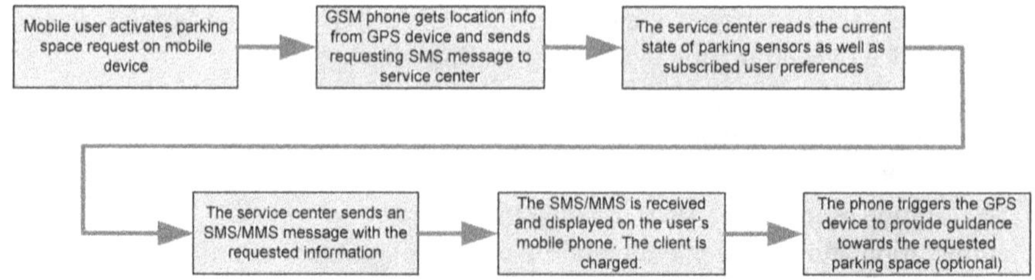

Figure 5.7 Activity sequence for URISC parking space services
Note: URISC = urban resource information service centre

2. User with a mobile phone and a GPS device. This type of user is equipped with a mobile phone and a GPS device. The user simply requests the service using the phone, which interfaces with the GPS and gets location information to form the SMS that will be sent. The local URISC sends back an SMS or MMS with the requested parking information in the user's area.
3. User with a mobile phone and a GPS device with navigation capabilities. Should a more advanced GPS device with a map of the location of interest be available, the message sent back by URISC is used by the mobile phone in order to generate a navigation request to the GPS navigation system which, ultimately, provides routing guidance to the parking place. The situation is similar in the case of an in-vehicle GPS navigation system.

Concluding Remarks

As technology evolves, applications introduced as innovations become commodities, as did electricity and analogue telephony in the early 20th century. Digital technology is evolving rapidly and its developments are not always adopted by the world communities as rapidly as they can be conceived and provided. As IT and telecom technologies converge, a shift towards the integration of mobile telecommunications and advanced information systems is taking place.

It is expected that Location-Based Systems (LBSs) will drive the development and wide adoption of mobile computing devices such as palm or even wearable computers, which will have a profound impact on how people perceive and use their home, workplace, neighbourhood and cities. In an urban context, such wireless developments can gain from the 'professions of location'. So far, technological developments are driven by economic efficiency without taking into account the planning goals at the local or regional level. This turns the urban space into a laboratory experimenting with the behavioural responses of the urban user, and, instead of contributing to the solution of the spatial problems, it aggravates the role of planning. Therefore, IT should be made part of the planning process, while, on the other hand, the planning process must take into account the cyberspaces created by ICTs.

The widespread deployment of advanced telecommunications systems is affecting all urban activities, and illiteracy in information and communications technologies may contribute to an increasing 'digital divide' in the social and political sphere as well; the narrowing of this digital divide should be one of the main goals of urban policy making of the future.

The objective of this chapter was the provision of a reference framework which will enable the dynamic assessment of the impacts of IST-related policy making. The operational aspects of this framework, as well as the direct and indirect influence of ISTs on urban policy making, have also been discussed, focusing on a case of an integrated LBS application for urban parking space management.

Future work needs to further concentrate on the structural and operational aspects of such frameworks, as well as on their integration with decision support systems for urban planning and management. This will enable a better coordination between planning practice and technological developments which may as a result lead to better management of urban and regional resources.

References

Abler, R. (1970), 'What Makes Cities Important', *Bell Telephone Magazine* 49:12, 10–15.

Camagni, R. (ed.) (1984), *Il Robot Italiano* (Milan: Edizioni del Sole 24 ore).

Camagni, R., and Rabellotti, R. (1986), 'Innovation and Territory: The Milan High Tech and Innovation Field', in P. Aydalot (ed.), *Milieux Innovateurs en Europe* (Paris: GREMI).

Castells, M. (1989), *The Informational City: Information Technology Economic Restructuring and the Urban Regional Process* (Cambrige: Blackwell).

Castells, M. (2000), *The Rise of the Network Society*, 2nd edn (Oxford: Blackwell).

Davelaar, E.J. (1991), *Regional Economic Analysis of Innovation and Incubation* (Aldershot: Avebury Press).

Gaspar, J. and Glaeser, E. (1996), 'Information Technology and the Future of Cities', National Bureau of Economic Research (NBER) Working Paper No. W5562.

Giaoutzi, M. and Nijkamp, P. (eds) (1988), *Information Technology and Regional Development* (London: Gower).

Gilder, G. (1995), 'Angst and Awe on the Internet', *Forbes ASAP*, 27 February, 56.

Gillespie, A. (ed.) (1983), *Technological Change and Regional Development* (London: Pion).

Gillespie, A. and Robins, K. (1989), 'Geographical Inequalities: The Spatial Bias of the New Communications Technologies', *Journal of Communications* 39:3, 7–18.

Goddard, J. and Thwaites, A.T. (1983), 'Technological Innovation in a Regional Context, Empirical Evidence and Policy Options'. Paper presented at the OECD Workshop, Research Technology and Regional Policy, Paris.

Gordon, P. and Richardson, H.V. (1997), 'Are Compact Cities a Desirable Planning Goal?', *Journal of the American Planning Association* 63:1, 95–106.

Gottmann, J. (1961), *Megalopolis: The Urbanized Northeastern Seaboard of the United States* (New York: Twentieth Century Fund).

Gottmann, J. (1983), 'Urban Settlements and Telecommunications', in J. Gottman, *The Coming of the Transactional City* (College Park: University of Maryland Institute for Urban Studies), pp. 17–32.

Hall, P. (1997), 'Modeling the Post-industrial City', *Futures* 29: 4/5, 311–22.

Hepworth, M. (1989), *Geography of the Information Economy* (London: Belhaven Press).

Mitchelson, R.L. and Wheeler, J.O. (1994), 'The Flow of Information in a Global Economy: The Role of the American Urban System in 1990', *Annals of the Association of the American Geographers* 84:1, 87–107.

Negroponte, N. (1995), *Being Digital* (New York: Knopf).

Nijkamp, P. (ed.) (1986), *Technological Change, Employment and Spatial Dynamics* (Berlin: Springer-Verlag).

Olson, M.H. (1987), 'Telework: Practical Experience and Future Prospects', in R. Kraut (ed.), *Technology and the Transformation of White-Collar Work* (Hillsdale: Erlbaum), pp. 135–55.

Pred, A. (1973), *Urban Growth and the Circulation of Information* (Cambridge: Harvard University Press).

Shiode, N. (2000), 'Urban Planning, Information Technology and Cyberspace', *Journal of Urban Technology* 7:2, 105–26.

Sui, D.Z. and Wheeler, J. (1993), 'The Location of Office Space in the Metropolitan Office Economy of the United States 1985–1990', *The Professional Geographer* 45, 33–43.

Thrift, N. (1996), 'New Urban Eras and Old Technological Fears: Reconfiguring the Good Will of Electronic Things', *Urban Studies* 33:8, 1463–94.

Vescoukis, V., Stratigea, A. and Giaoutzi, M. (2004), 'Teleworking: From a Technology Potential to a Social Evolution'. Paper presented at Digital Communities 2003: Organizing in a Networked World, Stockholm, June.

Chapter 6

Competition Measurement in Network Industries – The Case of Air Transport

Javier Campos and Manuel Romero

Introduction

During the 1990s, an extensive deregulation and privatization process dramatically changed the market structure and the shape and intensity of regulation in many infrastructure industries around the world. Energy, telecommunications, water, and several transport sectors have all experienced a wide liberalization process mostly aimed at opening them up to competition. As a consequence of this reform the economic literature has recently witnessed a renewed interest in studying the implications for competition policy of these changes. It appears that one of the most relevant contributions is that the network structure of these industries provides new perspectives to be considered in detail.[1]

Within the transport sector, one of the most suitable examples that can be used to study this process is the European air industry, not only because of its natural properties as a network industry – a set of airports and connecting routes – but also because from 1987 to 1997 it was gradually liberalized by means of several deregulation packages enacted by the European Commission. These packages sought to reconcile a more liberal bilateral agreements system, which some members (United Kingdom, The Netherlands and Ireland) had pursued a few years before, with the more traditional and protectionist 'flag-carrier' system heavily defended by other countries (Germany, France, Italy and Spain, amongst others). On 1 April 1997 the formal deregulation of the airline industry within the European Union (EU) was finally completed. Since then, any technically qualified EU airline can operate scheduled flights in any region of the Union, even on wholly domestic routes, without the restraints on fares or capacity that had prevailed during more than 50 years. Following a US-imported pattern, several carriers decided to reorganize their networks from 'point-to-point' (or direct flights) into 'hub-and-spoke' systems and a number of low-cost airlines emerged to serve not only the seasonal and mostly southbound tourist segment that had so far been flying charter but also other high-value clients in search of cheaper seats to selected destinations (Berechman and de

1 References on this issue are growing in the literature. Koski and Kretschmer (2004), and many other contributions in the same issue of the *Journal of Industry, Competition and Trade*, provide an interesting survey. For a more recent analysis, see Martín and Voltes-Dorta (2006).

Wit 1996). Analysts, national governments and officials at the European Commission predicted that competition between airlines would soar, and prices and services would improve in an unprecedented way (see Button et al. 1998; Hakfoort 1999).

According to Fridstrøn et al. (2004), these predictions were overoptimistic. Despite the recent eruption of low-cost carriers on selected routes, average prices across Europe have not decreased very much and, although services have probably improved and new tariffs have emerged, the real impact of the whole process has been relatively smaller than the effect experienced in the US after their air sector liberalization in the 1980s. The overall perception of many travellers is that there is scope for more competition. A few bankrupted companies (*Sabena*, *Swissair*), or those in financial distress (*Air France*, *Olympic Airways*, *Alitalia*), have confirmed that excess capacity was a distinguishing feature of this market before the reform, but also that many governments are still reluctant to abandon the flag-carrier model.

Since 2001 the air transport sector has been hit more hardly by external shocks (the aftermath of September 11, the SARS effect in Asia, and the global economic downturn due to oil prices) than by internal competition forces. As compared with the US, an excessively large number of domestic airlines manage to survive in the newly unified intra-European market. This is so because in most countries little advantage is conceded to foreign companies that come to operate in national airports, which are still dominated by local incumbents which benefit from 'grandfather rights' in the slot assignment process. Since price competition on the major routes is almost ruled out, many companies have refocused their strategies in two directions: on the one hand, the old flag-carriers and trunk-line companies (the 'high-cost' sector) are rearranging their (internal and external) structure in search of consolidating past advantages and of a favourable positioning in the event of future mergers and alliances. On the other hand, new and old companies are increasingly exploiting the low-cost traveller segment on scheduled flights, gaining market share from traditional charter flights.

To examine the competition effects of these strategic movements – mostly related to the evolution of the firms' market shares – and simultaneously provide a tentative evaluation of the effects of liberalization until 2001, we will analyse the evolution of the air market structure in the EU since 1984 using for the first time matrix-defined concentration indices. Despite this new methodology, this approach is not new in the literature: several papers have addressed the same issue before in Europe, but many of them have so far exploited only the geographical implications or the local impact of the results,[2] and few of these works have specifically connected concentration indices with their effects on competition, as it is usually carried out in other types of industrial organization analysis. As a major drawback of this methodology, it should be acknowledged that neglecting the study of the effects of competition on the level and number of airfares, or the overall quality provided by the carriers could be a

2 See for example, O'Kelly (1998), Burghouwt and Hakfoort (2001, 2002) or Burghouwt et al. (2003). A recent exception is provided by Lijesen (2004), who applies the Herfindahl index to the aviation markets, although its methodology differs from ours. A recent paper by Carlsson (2004) also studies prices and departures in European domestic aviation markets using the Herfindahl index as a proxy for market power in oligopolistic city-pair routes.

significant omission. However, it could be also argued that these are often short-term effects, which translate into long-term patterns through changes in market shares.

The major contribution of this chapter is not only methodological. We aim to add new evidence to the discussion of how competition has increased in the air transport sector at the intra-European level, by analysing the changes in concentration and network structure between 1984 and 2001 in the market for scheduled flights. Our analysis attempts to capture both the potential effects of new entries (or exits) on specific routes and the route reorganization of the main companies in response to these challenges.

After this introduction, the structure of this chapter is as follows. In the next section we first review the problem of measuring concentration in the air industry from the point of view of standard industrial organization by using concentration indices. In the section after this wediscuss how to implement these indices in the European air industry. We point out that market definition is the key issue in the implementation process and propose two alternatives: the route or the airline. For each of these two approaches a specific methodology is then fully developed focusing on three key variables for competition: number of flights, passengers carried (demand side) and seats supplied (supply side). Following this, we briefly describe the data selection process and then use the methods presented in the previous section to empirically analyse the evolution of air transport concentration in Europe from different approaches. The final section is a conclusion, summarizing the results and providing a digression on their policy implications for the near future.

Measurement of Concentration in Air Transport

The measurement of market concentration is always an attempt to answer with a single value the question to what extent the economic activity within a particular industry is controlled by a handful of firms. This will be always open to subjective interpretations, since any comparison of different individuals often requires the judgement of whether or not their differences allow a valid comparison. The usual approaches range from providing a diversity of inequality measurements (together with standard concentration indices), to create new (or adapt existing) indices to take into account the specificities of the sector under scrutiny.[3] We will follow these two approaches: first, in this section, we adapt standard concentration indices to the air transport sector. Then, later in the next section, a new set of tools will be proposed to take into account some of the network characteristics of this industry.

Adapting Traditional Concentration Indices to the Characteristics of the Air Industry

A basic feature associated with the traditional study of market concentration is that it often incorporates two relevant aspects of industrial structure: namely,

3 Annex 6.A provides a quick review of the most usual concentration indices used in industrial organization.

size inequalities and the number of firms. For this reason, it is now customary to distinguish between the *relative concentration indices* that focus on how much is the difference (or inequality) of firm sizes within a given distribution of firms, and the *absolute concentration indices* that emphasize the relative importance of each firm within its industry (as a proxy for market power).

In markets where firm sizes greatly differ amongst compared individuals, such as the European aviation market (where 'high-cost' carriers currently co-exist with smaller 'low-cost' airlines), relative versus absolute concentration is often a very relevant discussion. For example, the existence of a large number of small operators that gather an important share of output on a given route yields concentration results totally opposed to the case of the same market share being under the control of one or two large companies. These criticisms suggest that there is a need to redefine standard concentration indices to consider the differences among firms.

In the case of air transport, the network properties that characterize this industry include the number of airports (and, possibly, the distance between them), the flights or seats offered by the companies (even distinguishing between different time intervals during the day or the week), and the actual demand (passengers carried over specific routes). Depending on how these elements are implemented in the empirical work, two major approaches can be followed to measure the level of concentration in the air industry. The first method, more traditional, defines the intercity route (as a single pair of origin-destination airports) as the basic unit of analysis, and then applies (variations of) the standard concentration indices. The second approach (in the next section) uses each airline as the basic unit of analysis and tries to specifically explore the network structure of each airline. Each one of these approaches implicitly conveys a different definition of the market where concentration is measured, as Veldhuis (1997) points out. However, both analyses are relevant to figure out the true level of concentration in the industry.

The Intercity Approach: The Use of Traditional Concentration Indices

Consider a (national or international) city-pair route as described in Figure 6.1, where 0 and 1 (arbitrarily) define the origin and destination airports. This route is served by n different carriers, all of them offering their customers a two-way direct transport service. To avoid further complications we assume no significant differences between the quality of each company (same air and ground services), and fully concentrate on the quantity of services provided, denoted by q_i for firm i ($i = 1, \ldots, n$).[4]

Variable q_i usually refers to the number of flights per company (on a daily, weekly, monthly or annual basis), but it might reflect the number of passenger seats offered (thus capturing differences in each company's types of aircraft and fleet composition). Furthermore, since demand can be below supply, q_i could be also defined as the actual number of passengers carried by each company. In any

4 Note that a city may be served by more than one airport. In this case, to avoid aggregation problems, an inter-airport approach should be followed.

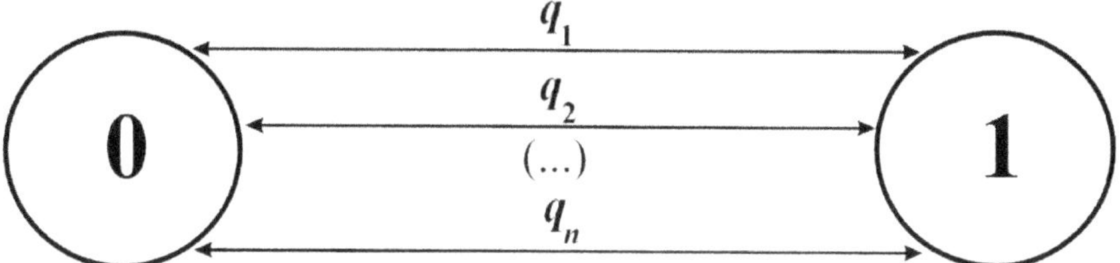

Figure 6.1 The intercity (route) approach to concentration measurement

case, $Q = q_1 + ... + q_n$ would represent the total traffic on any given route, which is implicitly considered as the 'market' where the *n* firms are competing.[5]

Once each market is defined in this way, the use of the concentration indices in Table 6.1 is straightforward. Many studies simply calculate the variance and put it together with a couple of absolute concentration measures. Early works frequently relied on the CR4-index, which is defined as the sum of market shares of the four largest suppliers. This index, along with its family of CR8, CR20 and CR50-indices, has been widely used for market power analysis and regulatory policy. The main problem with this indicator is the arbitrary character of its cut-off point. In some markets, for example, the two or the five largest suppliers may be more relevant.

This feature is especially important in measuring changes in the level of competition. Most recent studies on market competition are based on indices derived from economic theory, like the Herfindahl index (*H*), which is based on oligopolistic competition. The *H*-index links the relative sizes of the firms to industry profits in a standard Cournot model, and its inverse, $1/H$, may be interpreted as the number of equally sized firms that an equally competitive market would have.[6] As a drawback, it tends to give excessive weight to larger firms; therefore several adjustments have been proposed in the literature. The most successful is the entropy index (*E*), which improves the influence of smaller firms. Since it can be shown that *H* and *E* are particular cases of a general class of index (Hannah and Kay 1977), they are both still routinely reported in most studies on market concentration. By applying these indices to each route, it could then be easily analysed how concentration evolves over time. Despite being a necessary preliminary task, this approach has at least three major limitations:

5 By recalling that these are direct flights (see Figure 6.1), note that passengers and passenger-kilometres would yield the same result, since the *n* companies in the route would operate on the same distance. Alternatively, revenue data could be also used in the definition of q_i. However, this raises a number of methodological objections, such as the need to consider the effect of changing prices and pricing strategies or the need to adjust monetary values in cross-country and across-time comparisons. Hence, we stick to *quantity* data.

6 The link between this index and the underlying theory was established by Cowling and Waterson (1976). Its solid theoretical foundations favoured its adoption in 1984 by the US Department of Justice as a concentration measure for merger reviews. Ever since, this practice has been followed by several other regulatory bodies.

1. First of all, it is critically dependent on the route-sampling process. The choice of unrepresentative city-pairs would invalidate any analysis of, for example, the evolution of concentration in the European air industry. However, since an excessively large number of routes would probably disperse the results, a compromise between these two effects should be achieved. In addition, the selection of time intervals for comparison purposes is also critical, although the annual nature of the databases used in empirical work often preclude other possibilities than year-by-year comparisons.
2. A second criticism lies in the aggregate nature of the definition of q_i. In air transport networks, concentration of traffic may be important in three dimensions: production (how flights and passengers are distributed among different carriers), time (how flights are concentrated on specific time intervals within the day or week), and quality (how passengers with different preferences choose between different types of services – first class, business, tourist – offered by the carriers within the same flight). Although all these three dimensions are crucial for competition analysis, the approach described so far only allows us to deal with the first of them. Time concentration is a continuous source of controversy in many congested airports, where the most valuable slots are assigned to companies enjoying 'grandfather rights'. Quality concentration is a useful indicator to detect consumers' patterns, providing a unique insight into the prospects of new business segments, such as 'low-cost carriers'. Unfortunately, most air transport databases (see the next but one section) do not allow empirical studies to be carried out with these two levels of aggregation.
3. The final shortcoming of the intercity approach is that it completely misses the network properties of air transport discussed so far. Although it certainly exists, competition between airlines is seldom limited to a single route (especially in many European domestic markets). The intercity approach does not consider the network definition – in terms of number of airports served and the distance between them – of each carrier's output, which should be a key factor, and therefore should be complemented with a different approach.

These drawbacks, particularly the last one, do not necessarily reject the use of the intercity approach for the measurement of competition in the airline sector. In fact, in Lijesen et al. (2002) it is explicitly advocated that this approach should be used to study the competition in civil aviation in Europe, also recalling that, ultimately, the critical factor that carriers compete for is the traveller. If a traveller wants to get from city 1 to city 2, he or she is not interested in the domestic market as a whole, nor in either of the individual airports in it, since this traveller can only choose from carriers offering a service between cities 1 and 2. If competition is to benefit the consumer, it is bound to do so at a city-pair level.

However, the city-pair may not be the level the regulator is interested in, as he would prefer to know about the region over which he has regulatory power, a bundle of routes where a certain carrier is believed to be dominant, or the region where a certain event has effects. If, for instance, two companies are to merge, a regulator would be interested in all the city-pair markets on which these carriers are active. In this case, competition indicators measured at the city-pair level can be aggregated: for instance, by taking the weighted average of the indicators for all routes on which the future mergers are active. In any case, these concentration indicators should be used with caution, and great attention should be paid to the definition of the market and the importance of imperfect substitutes. The simultaneous use of several indicators, along with economic intuition and common sense reduces the risks involved in this approach.

Concentration Measurement at the Airline Level

As an alternative means to overcome the limitations described above, a completely different approach to market concentration measurement can be suggested by exploiting the characteristics of each carrier's network.

Consider, as depicted in Figure 6.2, an airline (say, i) serving a network composed of N airports, where qhj ≥ 0 denotes the total volume of air traffic between airports h and j (with h, j = 1, …, N and h ≠ j). Note that, since the distance between airports may differ, the definition of q should now specifically include either the flights or the seats (both multiplied by the flown distance), or the usual passenger-kilometre data. By collecting all the qhj pairs together we can easily construct the origin-destination matrix for airline i for any given period of time t (namely, OD_i^t), which simultaneously serves to summarize the airline's network structure and the actual distribution of its traffic between airports:

$$OD_i^t = \begin{bmatrix} 0 & q_{12} & q_{13} & \cdots & q_{1N} \\ q_{21} & 0 & q_{23} & \cdots & q_{2N} \\ q_{31} & q_{32} & 0 & \cdots & q_{3N} \\ \cdots & \cdots & \cdots & 0 & \cdots \\ q_{N1} & q_{N2} & q_{N3} & \cdots & 0 \end{bmatrix}, \tag{6.1}$$

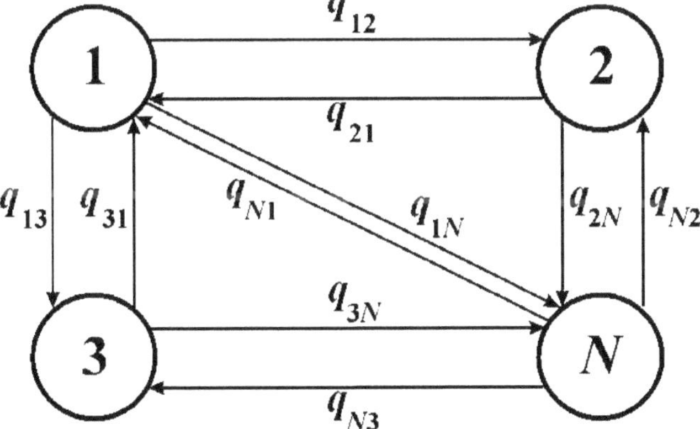

Figure 6.2 The airline approach to concentration measurement

where, by convention, we assume that the rows will represent the origins and the columns the destinations.

Properties of the Origin-Destination (OD) Matrix

It is worth noting several interesting properties of the *OD* matrix. First, it is always an $N \times N$ square matrix, although some of its elements may be zero (*incomplete* network, with empty routes). Particularly, by construction, all the elements in the main diagonal are zero, that is, $q_{hh} = q_{jj} = 0$. When no other element is zero, the network is said to be *complete*. The *OD* matrix is asymmetric, unless incoming and outgoing traffic coincides, $q_{hj} = q_{jh}$. In such a (rare) case, the matrix can be simply defined by the elements above (or below) the main diagonal.

A second property is that the economic characteristics of an airline's network are embedded in the internal structure of its *OD* matrix. For example, as mentioned above, it is usual to distinguish between two major types of network: fully connected ones (FC) and hub-and-spoke (HS) networks, as depicted below.

Figure 6.3 illustrates an FC network where there are $N = 3$ airports denoted 1, 2, and 3, where all passengers fly non-stop from origin to destination through three

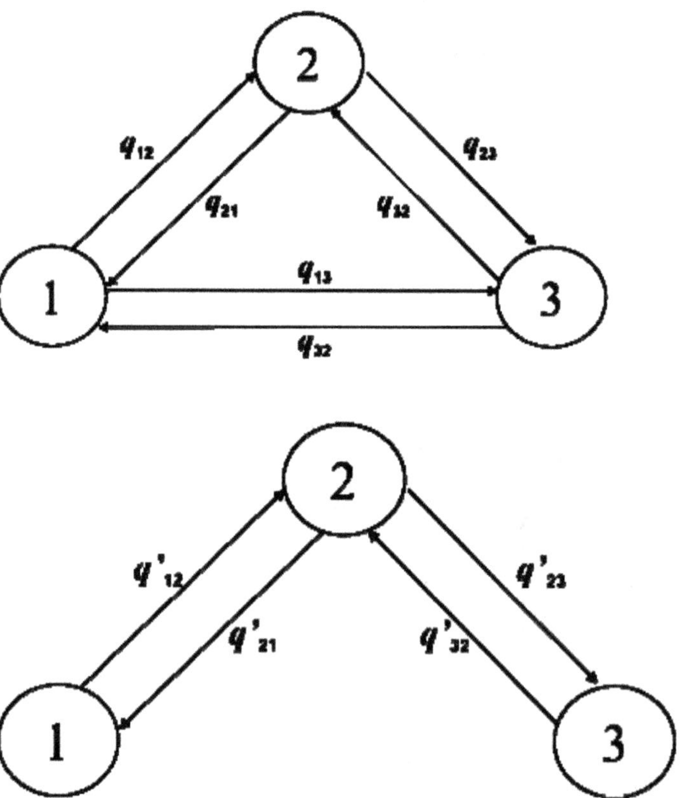

Figure 6.3 Fully connected vs. hub-and-spoke networks

possible (two-way) routes: route 1-2, route 2-3, route 1-3. Figure 6.3 (right) illustrates a typical HS network, where route 1-3 no longer exists. Passengers between 1 and 3 now fly through the hub airport, 2. The OD matrices of these networks clearly differ. Traffic will be assigned to certain rows or columns (relative to the others) thus reflecting the central or peripheral nature of certain airports. In addition, the hub-and-spoke network would be an incomplete one – with zero elements other than the main diagonal – since small airports ('spokes') would mostly have only direct flights with the hub(s). When the network has plenty of direct connections among all its N airports, the OD matrix would be complete.

By adapting Equation (6.1), it is easy to show that the structure of the OD matrices of FC and HS networks would be, respectively:

$$OD_{FC} = \begin{bmatrix} 0 & q_{12} & q_{13} \\ q_{21} & 0 & q_{23} \\ q_{31} & q_{32} & 0 \end{bmatrix}; \; OD_{HS} = \begin{bmatrix} 0 & q'_{12} & 0 \\ q'_{21} & 0 & q'_{23} \\ 0 & q'_{32} & 0 \end{bmatrix}, \tag{6.2}$$

where q'_{12}, q'_{21}, q'_{23} and q'_{32} should now include the travellers who used to fly on route 1-3 (provided that the change from an FC to a HS network does not alter the total number of travellers). Furthermore, note that different companies, with different network structures should have different OD matrices.

Comparing Networks

How can these characteristics be summarized into a single (and comparable across different matrices) value? One obvious answer is to use some kind of matrix operator that captures the information encapsulated in the elements (rows and columns) of a matrix. According to the definitions presented in Annex 6.B, the best candidates are the *permanent* and the *norm*.[7]

The permanent of a matrix, perm(OD), adds all multiplicative permutations of its elements, thus condensing the information they contain. If the number of zeros is large, it will yield a lower value than in the case of a complete network. Thus, the lower the value of the permanent, the more likely that the structure represents a 'hub-and-spoke' style network. To confirm or discard this result we can also calculate either the 1-norm or the ∞-norm (only one of these), which respectively take into account the information by destinations (columns) or origins (rows). When comparing two OD matrices, a larger norm means a more concentrated network around a particular airport. The Frobenius norm of the OD matrix, given by the square root of the sums of the squares of the terms of the matrix, resembles the Herfindahl index defined in Annex 6.A, its advantage lies in considering both rows and columns simultaneously.

---

7 The *trace* operator is obviously discarded, since the elements of the main diagonal are all zero in our OD matrices. The *determinant* operator is less informative than the *permanent*, since it subtracts part of the traffic and could yield (meaningless) negative values.

Finally, note that, since the dimension (N) of any given OD matrix may differ among firms or across time (for the same firm) it is useful to denote it also as N_i^t, and any cross-section or temporal comparison using the matrix operators must take into account this feature. An easy adjustment would be simply to redefine each q_{hj} as a relative value $s_{hj} = q_{hj}/Q_i^t$, where Q_i^t represents the overall sum of the total traffic by columns and rows:

$$Q_i^t = \sum_{h=1}^{N_i^t} \sum_{j=1}^{N_i^t} q_{hj}. \tag{6.3}$$

Competition in European Air Transport: Network Analysis

This section is devoted to the empirical analysis of the methodologies described above. We start by characterizing the properties of the databases used in our investigation, highlighting their main advantages and limitations. We then proceed to describe how the sampling procedure was carried out, and then present a summary of the most relevant results obtained using the concentration measurement procedures discussed above.

Data Characteristics and Sample Selection

As in many other empirical papers, the objectives of our analysis have been downsized and adjusted to data availability and reliability. In the European air transport sector the main source of information is the official statistics collected by the International Civil Aviation Organization (ICAO), since data directly obtained from the companies is seldom comparable across them (and even over time). The European Union also provides some information on an EU-wide basis through *Eurostat*, while the airports sometimes publish some traffic data. In all cases we rely on aggregate route information (but it never distinguishes different types of passengers within each route) collected on annual terms (monthly data should be sometimes preferable to evaluate the competitive strategies of the firms and to take into account the seasonality of air travel and the existence of peak-load periods through the year).

For this study, focused on the intra-European air transport market we have selected a representative sample of routes and companies from *Traffic By Flight Stage*, the detailed traffic database elaborated by ICAO, using data directly provided by the airlines. Our sample includes information, for most relevant European city-pair routes, (disaggregated by company) of volume of scheduled passengers, total capacity and available seats for the period 1985–2000. It also collects the type of plane and the number of flights per plane. We have chosen the 24 busiest European airports (Amsterdam, Athens, Barcelona, Berlin, Brussels, Copenhagen, Dublin, Düsseldorf, Frankfurt, Geneva, Lisbon, London, Madrid, Manchester, Milan, Munich, Newcastle, Nice, Paris, Rome, Stuttgart, Venice, Vienna and Zurich) and the 100 densest routes between them every year. The result has been 16,386 observations corresponding to 95 different airlines which, in terms of flights, passengers and seats – our three target variables – represent on average almost 80 per cent of the (international) intra-European air transport market. Thus, for the airline-centred approach, we just selected 13 companies representing 70 per cent of scheduled flights in the intra-European market between 1985 and 1990. They included the largest (current or former) flag-carriers (Air France, British Airways, Iberia, Lufthansa, SAS, Alitalia, KLM, Finnair, Sabena, and Swissair), two non-flag-carriers (Air Europa, Spanair) and even a low-cost operator (easyJet), whose data were available for at least three consecutive years.

Results and Discussion

In this section we will follow the same methodological order described above: first, the intercity approach (where the unit of analysis is the city-pair route), and then the airline approach (where the study is focused around the OD matrix of each airline) (see Tables 6.C1 to 6.C3 in Annex 6.C).

The intercity approach The intercity approach examines the evolution of concentration on the 100 densest city-pair European routes from 1984 to 2000.[8] For each route and year, we have the number of operators, their number of flights, the passengers they transported, and the seats offered by each of them. With this information we can carry out a standard concentration measurement analysis for each route and year using the tools described in Annex 6.A.

Our results are summarized in Table 6.C1, where we first report the *average* number of operators (over the 100 densest routes) and then – for the number of flights, carried passengers and seats – the *average* of one absolute concentration measure (standard deviation) and the *averages* of two relative ones (CR-4 and Herfindahl indices). Starting from 1984, the evolution of the average number of competitors on the densest routes exhibits a progressive increase that is consistent with the progressive introduction of liberalization policies (what are called the 'packages', released by the European Commission in 1987, 1990 and 1992, and the liberal bilateral agreements that several countries had agreed upon before). Note, however, that between 1984 and 1987, the average number of competitors on the 100 densest routes is very low (scarcely above 2.00), thus indicating that many of these routes were only operated by two companies, possibly one flag-carrier from the origin, and another flag-carrier from the destination city. The effect of the 1987 and 1990 packages seems very low: the average number of carriers on a particular route increases at the same pace as before. After 1992, however we observe a slight acceleration culminating around 1997–98. This is the period with a bigger increase in competitors on European routes. Afterwards, it seems that the market becomes more stable – maybe because of airports and skies saturation – and the possibility of continued growth is exhausted.

When we look at the columns reporting the standard deviations (in terms of flights, passengers and seats), the picture is less clear. These values reflect that, on average, the dispersion around the mean has increased over time in all cases. This

8 From year to year, the routes included in the list change only very slightly (± 2). In the 17-year period of our study we used 114 different routes.

growth was particularly acute until 1997 (especially in flights and passengers), but the final years in our sample show a slower trend. The conclusion could be the same: competition after 1997 does not seem to be growing fast, although the imperfect nature of absolute concentration measures such as the standard deviation requires further analysis.

The use of concentration indices allows a greater precision, since larger values of these can be directly interpreted as more concentration (a value of 1 is tantamount to monopoly power). In 1984, both the CR-4 and Herfindahl indices depart from very high values, suggesting that the four largest operators on each route controlled over 99 per cent of traffic. On many routes, as mentioned above, there were just one or two operators, and only minor and marginal companies 'competed' with them. Again, the 1987 and 1990 packages seemed ineffective, although concentration was to some extent reduced after 1992. After 1997, particularly in the case of passengers, the level of concentration grows again – perhaps the leading companies introduced larger aircraft – and this trend continues until the end of our sample.

To confirm our results, we also performed the calculation of the Entropy index and the Relative Entropy (see Annex 6.A), but its interpretation did not qualitatively change the results from the Herfindahl index. In addition, we compared our results with those from Lijesen et al. (2002) and Lijesen (2004), where concentration indices are also used. Even though their sample and methodology differs from ours and our sample period is larger, they also detect a U-shape turn in the evolution of concentration in air transport markets in Europe measured at the city-pair route level. During the 1980s and until 1992, there was a slow decrease in concentration which accelerated after 1992. Since 1998, despite the formal culmination of this process after 1 April 1997, on average competition has not increased on the 100 densest European routes in terms of market shares. Just the opposite is true: in fact, the figures show a growing trend towards concentration, since few routes have seen the emergence of a relevant number of new competitors.

The airline approach: matrix analysis The airline approach shifts the focus of the analysis from the route to the airline, thus allowing the characteristics and structure of each carrier's physical network to be specifically considered. As described earlier, we first elaborated the OD matrix for each company and year. We took into account their network size (or total number of airports served, N), which changed across companies and over the years (but very slightly). Then we filled each OD cell with the number of flights (multiplied by the distance in kilometres between airports) and the passenger-kilometres flown. We then calculated the permanent and the norm of the resulting 220 matrices, whose values are summarized in Tables 6.C2 and 6.C3. For simplicity, we only report the value for the Frobenius norm.

Consider first the permanent in Table 6.C2, where the values have been normalized between 0 and 1. By recalling that the lower the value of the permanent, the more likely that the network represents a 'hub-and-spoke', our results show – not surprisingly – that most of the flag-carrier companies (particularly Air France, Iberia, KLM and Lufthansa) concentrate most of their carried passengers on a single or a pair of airports. The values calculated for the number of flights are, on average, larger (see final column). This result indicates that companies fly between several airports, but only a handful of them are relevant from the point of view of carried passengers. A closer look at the type of aircraft used on several routes (for example, for Iberia or Air France, although this is not reflected in the table) confirms that smaller, feeder routes, are served by smaller planes, thus reinforcing the hub-and-spoke model. It is also quite interesting to note that non-flag-carrier companies (Air Europa and Spanair) rely less on this sort of network model and, as can be expected, the low-cost carrier (only easyJet in our sample) has – in general – higher values for its permanent. Finally, note also that the evolution of permanents over time is mostly decreasing. Our results suggest that most companies readjusted their network configuration during the 1980s and 1990s, thus positioning for a more competitive market. In particular, again the flag-carriers seemed to intensify their hub-and-spoke network strategy.

Looking at the results in Table 6.C3 we can somehow compare *OD* matrices across companies through their Frobenius norms. A larger norm means a more concentrated network around a particular city. In terms of flights, this is the case for British Airways (London), Air France (Paris), Iberia (Madrid), and KLM (Amsterdam). Some of the companies (Iberia, Air France) tend to increase the value of their norms over time, whereas others reduce it. Interestingly, note that the non-flag-carriers (Air Europa and Spanair), and the only low-cost carrier (easyJet) have smaller values. The remaining companies do not exhibit significant trends. The values in terms of carried passengers provide a similar explanation.

Conclusions

Market power measurement has always been a traditional issue in many empirical industrial organization studies. The extensive deregulation and privatization process that many infrastructure industries have undergone around the world during the 1990s has brought it back to the forefront of the tasks of many competition agencies. However, the specific economic characteristics of many of these industries demand an explicit treatment of their network properties when dealing with them. This has been the main aim of this chapter, which has been particularly focused on the measurement of competition in the European air transport market.

Our work has provided both new theoretical tools and empirical findings about the measurement of competition in a market where its network characteristics often preclude the open confrontation of rivals on an equal basis. Our analysis of the 1984–2000 period suggests that despite the fact that the EU air travel markets have been gradually deregulated and competition has therefore increased, a number of market imperfections still persist, possibly preventing an economically optimal resource utilization from a pan-European viewpoint.

As in many other sectors with former public operators, flag-carriers depart from an advantageous situation. Since entry is limited because of a high level of congestion in main airports and a slot assignment system based on 'grandfather rights', not surprisingly high concentration has emerged on many routes even after the liberalization. Furthermore, in reply to the pro-competition measures taken by the Commission, many companies have opted for strategies that favour alliances or mergers (as the recent case of *KLM* and *Air France*, or *Lufthansa* and *SWISS*), in order to exploit returns to scale and gain a better positioning in the global markets.

In a route-level analysis, we have shown that the number of carriers on each route is still lower than the maximum allowed by the current regulation; the concentration indices show a decline in the early 1990s, but then an increase afterwards. On an airline level analysis, the hub-and-spoke network structure dominates the industry, giving advantages to some carriers and preventing a greater degree of competition. Until 2001 the low-cost carriers made small attempts to exploit the niches left by the larger companies. However, these larger companies responded by cutting costs (and services) and turning their interest also to the low-cost segment.

In summary, the liberalization packages of the 1990s did not end the story in the transport sector. There are reasons to fear that, without a more vigorous competition policy both at the national and EU levels, the welfare losses stemming from insufficient competitive pressure on airlines are likely to increase rather than diminish over time. We observe signals suggesting that the European aviation industry may be facing a period of major consolidation. The merger of some relevant airlines has already been forecast by many analysts as merely a matter of time. This may in some circumstances have positive effects on economic efficiency. But in combination with the anti-competitive effects of the hub-and-spoke operations, frequent flyer programmes, and other restrictions on competition, the horizontal and vertical concentration in the aviation industry represents a formidable challenge to competition authorities at the national and European level. A forceful competition policy will be required in order to enhance, or even preserve, the present degree of competition in the air travel markets.

Acknowledgements

The authors acknowledge comments and suggestions by the participants. Javier Campos also gratefully acknowledges financial support from the Spanish Ministry of Science and Education through grant SEJ2004-00143 and from the BBVA Foundation.

References

Berechman, J. and de Wit, J. (1996), 'An Analysis of the Effects of European Aviation Deregulation on an Airline's Network Structure and Choice of a Primary West European Hub Airport', *Journal of Transport Economics and Policy* 30:3, 251–70.

Burghouwt, G. and Hakfoort, J. (2001), 'The Evolution of the European Aviation Network, 1990-1998', *Journal of Air Transport Management* 7:5, 311–8.

— (2002), 'The Geography of Deregulation in the European Aviation Market', *Tijdschrift voor Economische en Sociale Geografie* 93:1, 100–106.

Burghouwt, G., Hakfoort, J. and van Eck, J.R. (2003), 'The Spatial Configuration of Airline Networks in Europe', *Journal of Air Transport Management* 9:5, 309–23.

Button, K., Haynes, K. and Stough, R. (1998), *Flying into the Future. Air Transport Policy in the European Union* (Cheltenham: Edward Elgar).

Carlsson, F. (2004), 'Prices and Departures in European Domestic Aviation Markets', *Review of Industrial Organization* 24:1, 37–49.

Cowling, K. and Waterson, M. (1976), 'Price Cost Margins and Market Structure', *Economica* 43, 267–74.

Cullen, C. (1990), *Matrices and Linear Transformations*, 2nd Edition. (New York: Dover).

Curry, B. and George, K. (1983), 'Industrial Concentration: A Survey', *Journal of Industrial Economics* 31:3, 203–55.

Fridstrøn, L., Hjelde, F., Lange, H., Murray, E., Norkela, A., Pedersen, T.T., Rytter, N., Talén, C.S., Skoven, M. and Solhaug, L. (2004), 'Towards a More Vigorous Competition Policy in Relation to the Aviation Market', *Journal of Air Transport Management* 10:1, 71–9.

Hakfoort, J.R. (1999), 'The Deregulation of European Air Transport: A Dream Come True?', *Tijdschrift voor Economische en Sociale Geografie* 90:2, 226–33.

Hannah, L. and Kay, J.A. (1977), *Concentration in Modern Industry* (Cambridge: Cambridge University Press).

Koski, H. and Kretschmer, T. (2004), 'Survey on Competing in Network Industries: Firm Strategies, Market Outcomes and Policy Implications', *Journal of Industry, Competition and Trade* 4:1, 5–31.

Lijesen, M.G. (2004), 'Adjusting the Herfindahl Index for Close Substitutes: An Application to Pricing in Civil Aviation', *Transportation Research Part E: Logistics and Transportation Review* 40:2, 123–34.

Lijesen, M.G., Nijkamp, P. and Rietveld, P. (2002), 'Measuring Competition in Civil Aviation', *Journal of Air Transport Management*, 8:3, 189–97.

Martín, J.C. and Voltes-Dorta, A. (2006), 'A New Methodology to Measure Hubbing Practices in Airline Networks: Avoiding Common Pitfalls'. Mimeo, University of Las Palmas.

O'Kelly, M.E. (1998), 'A Geographer's Analysis of Hub-and-Spoke Networks', *Journal of Transport Geography* 6:3, 171–86.

Veldhuis, J. (1997), 'The Competitive Position of Airline Networks', *Journal of Air Transport Management* 3:4, 181–8.

Annex 6.A A Quick Review of Standard Concentration Indices

Concentration indices are traditionally used to provide a synthetic summary of market structure, as well as to evaluate the existing degree of competition in particular industries. They are simply defined over any output variable (for example, physical production, sales, employment, assets and so on), which is then related to the industry as a whole. Thus, if q_i denotes the output of firm i within an industry of n firms, then total industry output is given by:

$$Q = q_1 + \ldots + q_n = \sum_{i=1}^{n} q_i, \qquad (6.A1)$$

where firms are usually ranked by size. Correspondingly, the share of total output of the ith firm is defined by $s_i = q_i/Q$, and the mean output by $\overline{q} = Q / n$ (see Table 6.A1).

Table 6.A1 Traditional concentration measures in industrial organization

	Relative concentration measures		
	Definition	*Complete equality*	*Complete inequality*
Variance	$\sigma^2 = \sum_{i=1}^{n} (q_i - \overline{q})^2 / n$	0	∞
Standard deviation	$\sigma = \sigma^{1/2}$	0	∞
Coefficient of variation	$CV = \sigma / q$	0	$(n-1)^{1/2}$
Gini coefficient	$G = (1/n)\sum_{i=1}^{n} (n - 2i + 1)s_i$	0	$(n-1)/n$
	Absolute concentration measures		
	Definition	*Minimum (equal shares)*	*Maximum (one firm)*
Concentration ratio (of k-order)	$CR_k = \sum_{i=1}^{k} s_i$	k/n	1
Herfindahl index	$H = \sum_{i=1}^{n} s_i^2$	$1/n$	1
Entropy index	$E = \sum_{i=1}^{n} [s_i \ln(1/s_i)]$	$\ln n$	0
Relative Entropy index	$RE = E / \ln n$	1	0
Hannah-Kay index	$HK(a) = \left[\sum_{i=1}^{n} s_i^{1+a}\right]^{1/a}$		1

Source: Adapted from Curry and George (1983).

Annex 6.B Matrix Operators for Network Comparisons

Let $A = \{a_{ij}\}$ define a square $N \times N$ matrix. The following operators' definitions are standard in matrix algebra (see, for instance, Cullen 1990):

- *Trace*: The trace of an $N \times N$ matrix is the sum of the diagonal elements.
- *Permanent*: The permanent of an $N \times N$ matrix a_{ij} is the sum of certain products of the entries. Specifically,

$$\mathrm{perm}(a_{ij}) = \sum_{\sigma} a_{1\sigma(1)} a_{2\sigma(2)} \ldots a_{N\sigma(N)}, \tag{6.B1}$$

  where σ ranges over all the permutations of $\{1, 2, \ldots, N\}$.

- *Determinant*: The determinant of an $N \times N$ matrix a_{ij} is the sum and difference of certain products of the entries. Specifically,
-

$$|A| = \det(a_{ij}) = \sum_{\sigma} (-1)^{\mathrm{sgn}(\sigma)} a_{1\sigma(1)} a_{2\sigma(2)} \ldots a_{N\sigma(N)}, \tag{6.B2}$$

  where σ ranges over all the permutations of $\{1, 2, \ldots, N\}$ and $(-1)^{\mathrm{sgn}(\sigma)} = \pm 1$, depending on whether σ is an even or odd permutation.

- *Norm*: For a vector, its 2-norm, or *Euclidean norm*, is defined by the Euclidean length:

$$\left\| \begin{matrix} a \\ b \end{matrix} \right\| = \sqrt{a^2 + b^2}. \tag{6.B3}$$

  In the case of matrices, the *2-norm*, or *Euclidean norm*, of a matrix A is its largest singular value,[9] defined by the number

$$\|A\| = \max_{x \neq 0} \frac{\|Ax\|}{\|x\|}. \tag{6.B4}$$

  There are three particular cases of relevance:
  – The *1-norm* of a matrix is the maximum amongst the sums of the absolute values of the terms in a column:

9 Any $m \times n$ real matrix A can be factored into a product $A = UDV$, with U and V real orthogonal $m \times m$ and $n \times n$ matrices, respectively, and D a diagonal matrix with positive numbers in the first rank-A entries on the main diagonal and zeros everywhere else. The entries on the main diagonal of D are called the singular values of A. This factorization $A = UDV$ is called a singular value decomposition of A.

$$\|A\|_1 = \max_{1\le j\le N}\left(\sum_{i=1}^{N}|a_{ij}|\right). \tag{6.B5}$$

For example, for a 2 × 2 matrix: $\left\|\begin{matrix} a & b \\ c & d \end{matrix}\right\|_1 = \max\left(|a|+|c|, |b|+|d|\right).$

- The ∞-*norm* of a matrix is the maximum amongst the sums of the absolute values of the terms in a row:

$$\|A\|_\infty = \max_{1\le i\le N}\left(\sum_{j=1}^{N}|a_{ij}|\right). \tag{6.B6}$$

For example, for a 2 × 2 matrix: $\left\|\begin{matrix} a & b \\ c & d \end{matrix}\right\|_\infty = \max\left(|a|+|b|, |c|+|d|\right).$

- The *Hilbert-Schmidt norm* (or Frobenius norm)[10] of a matrix A is the square root of the sums of the squares of the terms of the matrix A:

$$\|A\|_F = \sqrt{\sum_{\substack{1\le j\le N \\ 1\le i\le N}} |a_{ij}|^2}. \tag{6.B7}$$

Any of the matrix operators defined above summarizes into a single value (part of) the information embedded in a matrix. For concentration measurement purposes we will use as input the origin-destination matrix for each major European carrier at different time periods (OD_i^t), whose general term is $\{a_{hj}\}$ (where h arbitrarily represents the origin, and j the destination). Refer to the section entitled 'Concentration Measurement at the Airline Level' for a discussion of the advantages and disadvantages of each operator for the measurement of the internal structure of each company's network.

10 This is also the norm of a *Frobenius form* (also called a rational canonical form), which is a block diagonal matrix with each block the companion matrix of its own minimum and characteristic polynomials. Each of the minimum polynomials of these blocks is a factor of the characteristic polynomial of the original matrix. The polynomials that determine the blocks of the rational canonical form sequentially divide one another.

Annex 6.C Concentration in European Air Transport: Some Results

Table 6.C1 Intercity approach: evolution of concentration by city-pair routes*

	Number of operators per route	Number of flights			Passengers			Seats		
		Standard deviation	CR-4 index	Herfindahl index	Standard deviation	CR-4 index	Herfindahl index	Standard deviation	CR-4 index	Herfindahl index
2000	4.22	1958	0.9866	0.5419	212378	0.9836	0.5809	287973	0.9802	0.5589
1999	4.24	1975	0.9787	0.5306	214892	0.9881	0.5718	287037	0.9863	0.5451
1998	4.32	1993	0.9720	0.5002	213420	0.9829	0.5262	240895	0.9828	0.5071
1997	4.34	1951	0.9768	0.4890	192880	0.9740	0.5231	216788	0.9815	0.4987
1996	4.48	1528	0.9632	0.4735	172232	0.9687	0.5084	212889	0.9862	0.4948
1995	4.31	1510	0.9570	0.4207	166813	0.9588	0.4518	208266	0.9641	0.4292
1994	4.84	1471	0.9523	0.4308	154557	0.9670	0.4445	205436	0.9797	0.4214
1993	4.81	1495	0.9682	0.5472	148773	0.9716	0.4635	210380	0.9652	0.4318
1992	3.44	1364	0.9787	0.6229	139995	0.9871	0.4731	183071	0.9789	0.4138
1991	3.05	1322	0.9763	0.6211	136749	0.9812	0.4806	162955	0.9708	0.5139
1990	2.99	1283	0.9796	0.7186	125353	0.9842	0.4981	163625	0.9826	0.5947
1989	2.97	1177	0.9716	0.7158	136481	0.9907	0.5281	166376	0.9858	0.5841
1988	2.80	1120	0.9842	0.7339	132426	0.9932	0.5393	156294	0.9800	0.6065
1987	2.35	1006	0.9999	0.7454	110011	0.9979	0.5507	131109	0.9903	0.6148
1986	2.34	919	0.9980	0.4217	104536	0.9781	0.4321	128089	0.9654	0.3925
1985	2.07	930	0.9917	0.4181	104095	0.9680	0.4199	124858	0.9522	0.3826
1984	1.98	906	0.9960	0.4658	99745	0.9758	0.4640	125551	0.9646	0.4207

* The analysis was carried out on the 100 densest city-pair routes each year. Reported values refer to the averages over these routes.

Table 6.C2 Airline approach: permanent of the OD matrices by company and year

	2000	1999	1998	1997	1996	1995	1994	1993	1992	1991	1990	1989	1988	1987	1986	1985	1984	Mean
Flights																		
Air Europa	0.32	0.31	0.34	0.28	0.22	0.27	0.18	–	–	–	–	–	–	–	–	–	–	0.27
Air France	0.12	0.11	0.12	0.10	0.11	0.10	0.12	0.13	0.10	0.11	0.09	0.10	0.12	0.18	0.11	0.10	0.12	0.11
Alitalia	0.15	0.14	0.18	0.22	0.16	0.16	0.15	0.17	0.12	0.12	0.18	0.15	0.13	0.14	0.18	0.21	0.22	0.16
British Airways	0.12	0.11	0.12	0.22	0.21	0.20	0.20	0.17	0.22	0.20	0.24	0.24	0.17	0.16	0.24	0.23	0.16	0.19
easyJet	0.64	0.74	0.82	–	–	–	–	–	–	–	–	–	–	–	–	–	–	0.73
Finnair	0.21	0.20	0.16	0.19	0.22	0.26	0.22	0.21	0.14	0.19	0.15	0.18	0.18	0.16	0.13	0.12	0.12	0.18
Iberia	0.20	0.21	0.22	0.20	0.20	0.23	0.21	0.22	0.21	0.21	0.23	0.21	0.22	0.22	0.22	0.23	0.22	0.22
KLM	0.32	0.30	0.30	0.31	0.31	0.36	0.32	0.32	0.22	0.31	0.33	0.34	0.31	0.31	0.34	0.32	0.33	0.31
Lufthansa	0.22	0.24	0.21	0.20	0.24	0.24	0.23	0.22	0.21	0.22	0.33	0.24	0.22	0.21	0.21	0.24	0.22	0.23
Sabena	–	–	0.13	0.09	0.10	0.12	0.12	0.11	–	–	–	–	–	–	–	–	–	0.11
SAS	0.24	0.23	0.25	0.25	0.24	0.26	0.27	–	–	–	–	–	–	–	–	–	–	0.25
Spanair	0.29	0.32	0.28	0.32	0.34	0.33	0.38	–	–	–	–	–	–	–	–	–	–	0.32
Swissair	0.13	0.18	0.18	0.17	0.18	0.16	0.17	–	–	–	–	–	–	–	–	–	–	0.17
Passengers																		
Air Europa	0.13	0.11	0.12	0.08	0.06	0.05	0.03	–	–	–	–	–	–	–	–	–	–	0.08
Air France	0.01	0.00	0.01	0.00	0.00	0.00	0.01	0.00	0.00	0.01	0.01	0.00	0.02	0.00	0.01	0.00	0.02	0.01
Alitalia	0.02	0.04	0.02	0.02	0.06	0.06	0.04	0.07	0.02	0.02	0.03	0.03	0.03	0.02	0.02	0.02	0.02	0.03
British Airways	0.01	0.01	0.02	0.02	0.01	0.02	0.02	0.03	0.02	0.03	0.04	0.04	0.05	0.06	0.04	0.03	0.03	0.03
easyJet	0.45	0.34	0.22	–	–	–	–	–	–	–	–	–	–	–	–	–	–	0.34
Finnair	0.11	0.10	0.12	0.09	0.07	0.06	0.11	0.14	0.17	0.09	0.08	0.08	0.08	0.06	0.03	0.02	0.02	0.08
Iberia	0.00	0.00	0.00	0.00	0.00	0.00	0.01	0.02	0.01	0.01	0.01	0.01	0.02	0.02	0.02	0.01	0.02	0.01
KLM	0.02	0.00	0.00	0.01	0.01	0.00	0.02	0.02	0.02	0.01	0.03	0.02	0.01	0.01	0.02	0.02	0.02	0.01
Lufthansa	0.02	0.00	0.01	0.00	0.00	0.00	0.03	0.02	0.01	0.02	0.03	0.04	0.02	0.01	0.01	0.02	0.02	0.02
Sabena	–	–	0.00	0.00	0.00	0.00	0.02	0.01	–	–	–	–	–	–	–	–	–	0.01
SAS	0.04	0.03	0.04	0.05	0.03	0.06	0.08	–	–	–	–	–	–	–	–	–	–	0.05
Spanair	0.19	0.12	0.08	0.08	0.08	0.16	0.18	–	–	–	–	–	–	–	–	–	–	0.13
Swissair	0.09	0.08	0.08	0.07	0.07	0.06	0.18	–	–	–	–	–	–	–	–	–	–	0.09

Note: Companies are listed in alphabetical order. Values have been normalized between 0 and 1 (a lower value reflects a hub-and-spoke network).

Table 6.C3 Airline approach: Frobenius norm of the OD matrices by company and year (in millions)

	2000	1999	1998	1997	1996	1995	1994	1993	1992	1991	1990	1989	1988	1987	1986	1985	1984	Mean
Flights																		
Air Europa	57.26	63.93	84.33	23.37	18.86	13.04	11.96	–	–	–	–	–	–	–	–	–	–	38.96
Air France	314.88	282.77	255.28	249.80	259.82	253.73	237.58	231.01	216.59	208.95	195.67	163.76	163.28	130.29	131.33	126.50	125.30	208.62
Alitalia	112.45	114.37	100.60	86.32	85.01	80.07	79.65	79.32	70.45	12.93	57.67	59.93	60.19	47.46	57.45	55.30	50.07	71.13
British Airways	260.96	242.66	185.64	178.28	176.95	182.90	148.79	129.42	137.71	137.43	146.38	139.74	135.36	127.38	122.95	122.23	115.57	158.26
easyJet	77.53	61.63	50.93	39.57	–	–	–	–	–	–	–	–	–	–	–	–	–	57.42
Finnair	146.45	123.89	132.10	123.92	118.49	111.42	96.66	101.21	102.62	83.78	79.26	83.40	52.39	41.64	47.05	53.07	87.19	93.21
Iberia	304.76	286.66	290.74	281.88	286.02	281.40	272.73	285.93	218.73	205.23	232.43	195.12	207.40	156.41	150.92	153.54	148.66	232.86
KLM	235.45	215.35	220.75	140.36	165.86	162.88	142.39	135.21	90.77	101.76	117.47	120.91	136.22	135.02	118.79	105.16	101.21	143.86
Lufthansa	199.01	203.04	146.42	133.87	122.20	142.88	129.52	121.92	116.64	113.52	110.21	97.08	82.36	79.53	72.71	72.06	87.70	119.45
Sabena	–	–	160.37	147.48	147.48	147.84	139.69	147.76	–	–	–	–	–	–	–	–	–	148.44
SAS	102.32	98.99	100.20	114.36	98.32	88.84	90.03	–	–	–	–	–	–	–	–	–	–	99.01
Spanair	28.21	12.27	31.80	32.46	20.48	18.00	21.40	–	–	–	–	–	–	–	–	–	–	23.52
Swissair	160.27	152.77	175.01	123.50	106.66	95.41	101.97	–	–	–	–	–	–	–	–	–	–	130.80
Passengers																		
Air Europa	2413.4	1281.8	509.7	1347.8	1024.7	5108.4	5371.6	–	–	–	–	–	–	–	–	–	–	2436.8
Air France	28658.0	24880.0	21280.0	20904.0	18917.0	17756.0	20697.0	19197.0	19386.0	19576.0	20850.0	21420.0	20312.0	14178.0	17193.0	17570.0	16869.0	19979.0
Alitalia	16398.0	13920.0	13421.0	11030.0	9297.7	9370.8	8855.0	8784.9	8872.2	7251.8	8879.9	8751.3	6428.7	5873.2	5318.5	6002.1	7474.0	9172.3
British Airways	35087.0	32887.0	32816.0	31151.0	30555.0	31292.0	33450.0	32897.0	27105.0	23959.0	26397.0	23182.0	22942.0	19792.0	17544.0	18507.0	17172.0	26866.8
easyJet	9296.1	7240.0	5164.3	916.8	–	–	–	–	–	–	–	–	–	–	–	–	–	5654.3
Finnair	9022.7	8810.0	8517.7	7622.6	7042.0	6733.1	6340.6	5870.0	5351.0	833.5	4703.4	4638.6	4472.3	2472.6	3209.4	4123.6	2871.8	5449.1
Iberia	27195.0	24920.0	19804.0	18269.0	16799.0	17920.0	14536.0	12452.0	12803.0	12248.0	12839.0	11680.0	11111.0	10515.0	9801.2	9522.8	9801.2	14836.3
KLM	2284.4	1937.8	1567.5	1297.5	1507.1	1683.2	1478.0	1207.9	1761.1	1610.1	1656.3	2377.5	1931.4	1912.9	2454.6	3873.7	3881.5	2024.9
Lufthansa	11572.6	11536.8	11347.8	11748.6	762.8	14223.0	13209.0	13175.0	11571.0	10204.0	10669.0	9299.0	8044.6	8038.6	7260.2	7209.6	9125.5	9941.0
Sabena	–	–	16571.0	15928.0	14579.0	13632.0	13396.0	13071.0	–	–	–	–	–	–	–	–	–	14529.5
SAS	3192.8	2234.0	3278.2	3216.7	3458.3	3545.5	2238.2	–	–	–	–	–	–	–	–	–	–	3023.4
Spanair	1009.1	1455.0	1551.2	1748.4	1962.2	2126.6	2156.2	–	–	–	–	–	–	–	–	–	–	1715.5
Swissair	13856.0	13362.0	9346.4	9096.3	9166.1	9868.9	10257.0	–	–	–	–	–	–	–	–	–	–	10707.5

Note: Companies are listed in alphabetical order.

Chapter 7

Enabling Technologies and Spatial Job Markets

Vassilios Vescoukis

Introduction

Educational systems are facing the challenge of new tools offered by enabling Network Technologies. The speed of the technological evolution in communication networks is impressive, but traditional vocational training systems are still not taking advantage of this evolution either as a training tool or as a subject itself. Putting this together with the slow pace of evolution of vocational education, we witness a gap between the classroom and the job practice, which is even wider in non-urban regions that are less privileged and sometimes of low priority for central administrations. To deal with this, Network Technologies can be regarded as *enabling technologies* that support new knowledge delivery approaches using independent-of-time-and-space learning paradigms. Their independence from space enables education actions to take place in any geographical location, and, thus, the inherent handicaps of non-urban areas in this field may be treated.

On-the-job training, which has been recognized as a very important educational activity, is usually carried out in 'the old way'. In this chapter, an alternative to traditional on-the-job training activities in technical vocational training is presented, based on distributed network technologies, which favour remote areas. This approach is based on business simulation activities, conducted asynchronously and independently of physical location, over the Internet. Students involved in such activities 'practise' simulated tasks virtually in order to master skills and thus gain understanding of the real aspects of the job for which they are being trained. Real work environments are simulated using network services that help students perform several tasks of a job just as they would do in the real world. A distinguishing characteristic of this approach is the independence from time and space, which enables students of non-urban and remote regions to 'practise' jobs that may not even exist in their own geographical location.

This chapter is organized as follows: In the next section, the motivation and rationale behind the job simulation concept is presented, followed by an examination of the educational context. Having defined the scope of our approach, a job simulation schema is then presented, together with all the relevant concepts and procedures. There is then a discussion on the case of applying this schema to a geographically distributed network context. Finally, the key points are summarized and a few concluding remarks are made.

Motivation

Educational systems all around the world are being restructured. Although enabling telecommunication technologies have motivated this restructuring, it is apparent that the evolution of technology alone is not the main reason for reconsidering education and knowledge delivery. Instead, technology is a tool for developing the way knowledge is delivered, and, as such, it is indeed powerful. New technologies enable us *to set higher educational goals, to accelerate learning and experience fewer shortcomings than in the past*. The urge to implement these new possibilities is creating a 'knowledge gap' which has to be bridged if we want to materialize the goals set. This is being done by working people of different ages and backgrounds, professionally based at different levels of organizations and professions. Hence, *one of the most dominant challenges that current educational systems need to face is the relation of the knowledge delivery domain to the knowledge application domain*, that is, the job market.

A gap between the classroom and job practice is evident. To fill this gap, the European Union is supporting the restructuring of the vocational education systems. Using network and distributed systems technologies, synchronous and asynchronous content delivery is possible; collaborative learning paradigms independent of time and space have been introduced; and, last but not least, assessment has been made much more sophisticated. In some cases, new technologies enable the complete restructuring of the learning process, especially at the post-secondary education level.

However, no global uniform approach exists that can be applied to diverse educational contexts using technology-assisted learning. This work focuses on the technical professional education of graduate students at secondary level (high school) educational institutions. On-the-job training is part of the curriculum of such institutions. It is common knowledge that, in the real world, on-the-job training is seldom practised as it should be in order to produce the required educational outcome. In more than just a few cases, students practising on-the-job training in companies affiliated with their school end up being just temporary assistants doing trivial things. From the company's point of view, under certain conditions this may be understandable, but, still, the outcome of the educational process is much less than expected and the effectiveness of the whole system suffers as well. Table 7.1 summarizes some interesting facts about network learning (Sage 2002).

In the following section we present a computer-assisted educational structure for Asynchronous Job Simulation Activities (JSAs). JSAs immerse students in a simulated business environment where they can practise tasks to master skills and gain understanding of the job concerned; thus JSAs become an alternative to traditional on-the-job training activities. Operating over a computer network, they do not require a real company, and therefore students located in border areas can participate in such activities whether or not there is a real company in their geographical location. In job simulation, students can perform activities such as starting a company, creating a production strategy, and defending an investment plan to a board of directors. Unexpected events or changes in the business climate can arise, which the student is challenged to resolve (Linser and Naidu 1999). Thus, students learn by doing in a

Table 7.1 Facts on network learning

• 40 per cent of every dollar spent on training is spent on travel costs. • Motorola projects that for every $1 spent on training, there will be $30 in productivity gains in three years. • 50 per cent of employee skills become outdated in 3–5 years (*Source*: *Merrill Lynch's The Book of Knowledge)*
• Adult learning effectiveness drops off after 30 minutes. • Without use and practice, people forget 25 per cent of what they know in six hours and 33 per cent within 24 hours. (*Source*: *Training and Development*, 2001)
• Education Development Center (EDC) study says that 70 per cent of what people know about their jobs comes informally and through the people they work with. • Senge – formal training accounts for only a fraction of the learning that occurs. • The greatest learning opportunities for employees are each other. (*Source*: *Training*, 2000)
• Average Fortune 500 Company loses $64 million per year because of ineffective knowledge sharing. • 38 per cent of a professional's time is spent looking for information. • Less than 20 per cent of the knowledge available to a company is used . (*Source*: *Training and Development*, 2001)

simulated (virtual) work environment where they perform tasks similar to those of the real-world job, learning from both successes and failures (Schank 1997).

Network technologies are enabling technologies for making JSAs possible and close to those of the real world, mainly because they support open, flexible and learner-centred patterns of study and facilitate collaborative work (ALTP 1997). However, the use of such technologies is not straightforward in all educational contexts where simulation activities take place. The definition of a complete educational structure is, therefore, necessary.

The Educational Context

Secondary level technical schools in Greece provide vocational training to students aged 15 to 18. Usually, on-the-job training activities are part of the curriculum of such institutions; their necessity, especially in the technical job domains, is evident. These activities take place in real workplaces of companies affiliated to schools. For a short period of time (one to a few weeks), students leave the classroom and practise in a real-world environment the technical job they learn in school. Of course, the location of the company's site is within a reasonable distance from the location of students. When integrating on-the-job training with traditional learning as in this case, several difficulties arise, ranging from practical to legal issues. Still, a well-performed, on-the-job training activity is indeed considered of great educational value. However, there are a number of obstacles in applying on-the-job training activities. These obstacles can be classified in four categories: practical, legal, geographical and human-factor related.

Practical problems are related to the coordination and management of on-the-job training activities along with traditional learning. Time schedules and duration should be suitable and appropriate for students, schools and participating companies. Extra-curricular activities undertaken by students, as well as the dominant constraint to finish the school period by the end of June, make the use of alternative solutions such as practice in out-of-school hours or during the summer practically impossible.

Legal problems relate to the legal status of trainees when working on a company's premises: Do they have social security? Who is responsible for what, and to what extent? And are the tasks relevant? The legal responsibility and the insurance issue can lead to the rejection of any idea for on-the-job training activities, especially in jobs with a high risk of industrial accident.

Geographical limitations are amongst those obstacles that are practically impossible to overcome. On-the-job training requires that there is a real company that actually offers the job to learn. This is not the case in peripheral border areas, where students have considerably fewer opportunities to learn by actually practising the job of their choice.

However, even if the aforementioned issues could be resolved, one quite essential problem would still remain: that of the *human factor*. As mentioned above, students practising on-the-job training might end up doing trivial things irrelevant to the purpose of their training.

One can easily come to the conclusion that, no matter how important on-the-job training is, putting it into practice in secondary education technical training presents large obstacles for the majority of professional specialties. In order to overcome such obstacles, several alternatives have been considered, the *job simulation activities* concept, being one of them. This idea is currently being applied in corporate training, in graduate studies, and, generally, in educational environments where older and more mature audiences than those of our focus, are trained. JSAs face several obstacles, too. Unless they are properly organized and supported, they fail to meet the challenge of matching the real-world situation as closely as possible. Setting up JSAs is therefore by no means easy and straightforward, involving both a fair knowledge and monitoring of each profession. This, in turn, leads to the need to define a suitable educational structure, rules, tutor and learner roles, as well as communication and assessment procedures.

Computer network support is essential in order for JSAs to be undertaken by distributed teams rather than individually, especially in locations where physical constraints do not allow full local deployment of such activities. In the next section we present such a structure, and proceed towards the development of a generic JSA model.

A Job Simulation Environment

Definitions and Structure

A *virtual enterprise* is a job simulation activity where real-world professional activities of specific technical specialties are reproduced in a network-based learning environment. Each virtual enterprise has its own structure, objectives and operating scenarios that correspond to those of a real-world company. Operating scenarios and educational material are contained in the *virtual enterprise handbook*. One such handbook corresponds to a specific class of virtual enterprise.

A *virtual marketplace* is the network-based simultaneous operation of virtual enterprise simulation activities. Virtual enterprises participating in a virtual marketplace can be of the same or of different types. A virtual market has its own rules, common for all the types of virtual enterprises in each specific context. These rules, as well as coordination, communication and evaluation procedures are contained in the *virtual market handbook*.

In order to set up and run virtual enterprise simulation activities in a specific educational environment, the following steps need to be taken.

Step 1. *Define a vision for the virtual marketplace*. This should be viewed as a high-level system structuring and organization that monitors and acts as a generator of rules, constraints and assumptions, which comprise the simulated environment. A realistic approach to the definition of such an environment is to choose which characteristics of the real world are considered most important and relevant to each specific educational context. Rules and constraints may be valid everywhere in the virtual marketplace or may depend on each enterprise type. An operational structure corresponding to the virtual marketplace should be created. Roles, responsibilities and communication procedures need to be defined for an effective implementation of JSAs. The outcome of this step is the *virtual market handbook*.

Step 2. *Create virtual enterprise handbooks*. Each virtual enterprise type has its own documentation handbook which contains educational material, implementation details, rules, procedures and scenarios for the simulation activities. Part of the virtual enterprise handbook should be read only by supervising trainers.

Step 3. *Define auditing, assessment and feedback procedures*. Needless to say, continuous tracking and evaluation of the simulation activities is imperative both for making the necessary adjustments at the correct time and for the quality assurance of the simulation.

Step 4. *Create a network infrastructure to support the simulation activities*. Operations handbooks, learning material, operating scenarios, communication paths, auditing and assessment should all be implemented in a distributed network learning environment, so that accessibility of the simulation activities from distributed sites can be achieved.

In order to define an operational structure for the implementation of simulation activities, we have taken into account that in a real workplace, in order to take any action, a group of co-workers is stimulated by complex real-world events and conditions. One of the most important benefits of real on-the-job training is the interaction between trainees and workers. A JSA should be organized in a manner to develop the skills for such interactivity, even though it does not occur in a real workplace. In order to achieve that, we make a clear distinction between the virtual workplace and the traditional source of knowledge.

An organizational structure for performing business simulation activities (BSAs) is shown in Figure 7.1. The *control environment*, shown in Figure 7.1a, is where the simulation is conceived, managed and supported. It is the generator and the tutor of the simulated environment. The *simulation environment*, shown in Figure 7.1b, is where the simulation activities actually take place. Trainee communities are organized in cooperating, competing or simply coexisting virtual enterprises, which perform several simulated tasks by accepting stimuli from the communications centre.

The control environment is organized into *units*; units are panels of trainers and consulting practitioners. In geographically distributed environments, units can operate by network-based communication, such as teleconferencing, asynchronous messaging or any groupware platform. Four units operate in the *control environment*: the *directing board*, the *command unit*, the *auditing and assessment unit* and the *documentation unit*.

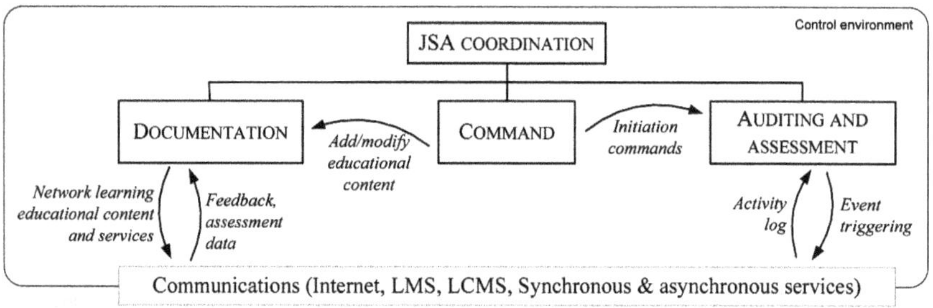

(a) The control environment

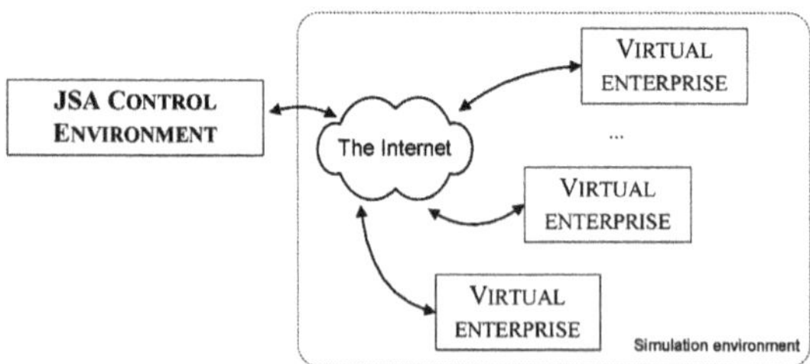

(b) The simulation environment

Figure 7.1 An operational structure for simulation activities in technical professional education

The *directing board* is the policy-making centre of the JSAs. Trainers and industry experts participate in the directing board, setting the rules of the activities, as well as the constraints and assumptions for the virtual marketplace, the simulation of time, and so on. The *command unit* is the management centre of the simulation activities. The creation of virtual enterprise instances is triggered here, according to the simulation requirements and the available resources. Students are assigned to positions in each virtual enterprise by decision of the command unit. The execution of all simulation activities is initiated at the command unit.

The *auditing and assessment unit* supports the implementation of the simulation activities by processing and communicating the orders of the command unit to the simulation environment through the communications centre. Auditing, reporting, logging, feedback, as well as the assessment of activities are undertaken by the auditing and assessment unit. During simulation activities, trainees can address the control environment only through the auditing and assessment unit.

The *documentation unit*, is the 'centre of knowledge' of the job simulation environment. It is composed of a panel of domain experts, as well as of network learning infrastructure that supports distributed learning and asynchronous communication. Reference material, textbooks, rules, simulation scenarios and documentation and case studies are all offered to the simulation environment by this unit. An important responsibility of the documentation unit is to keep the knowledge content up-to-date as developments in the various professions occur.

The *simulation environment* is where the simulation activities take place. There are several virtual enterprise instances, operating independently, cooperatively, or even competitively, according to the virtual market rules. Each instance operates according to the virtual enterprise handbook of its specific virtual enterprise class. Communication services are offered by groupware-messaging services which, depending on the available resources, can support group coordination, videoconferencing, voice communication, synchronous and asynchronous messaging.

Simulation Activities

The building element of JSAs is the *simulated job task*, or simply *task*. It is a single activity or a sequence of activities that usually occur in the corresponding real-world workplace. The granularity of the task definition depends on the specialty domain of each virtual enterprise class. A rule of thumb is that a *simulated job task* corresponds to the same action(s) that are repeated together in the real world, each time with different instances of initiation conditions, which are an atomic, building block for more complex sequences of activities. The following are good examples of tasks: placement of an order, issuing of an invoice, installation of an appliance, maintenance of a car battery, and so on. A task has discrete initiation requirements (pre-conditions), as well as discrete results (deliverables and post-conditions), and is carried out by assigning resources to action(s). A typical task record is shown in Figure 7.2.

A sequence of tasks is called a *simulation scenario*. A task can be part of many scenarios, not always in the same order. A scenario is a sequence of simulation

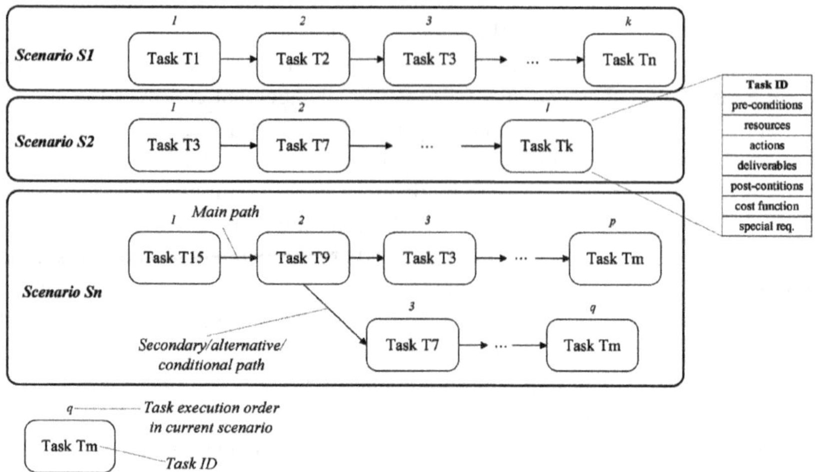

Figure 7.2 Tasks and scenarios

activities that produce a result visible to the virtual client. The following are examples of scenarios: execution of a commercial order, servicing an automobile, development of a custom software application, installation of an alarm system, and provision of a customer-requested service. Scenarios may have a single or several alternative execution paths, as shown in Figure 7.2.

Two scenarios are defined for all the virtual enterprise classes: the first is the 'foundation of the enterprise', and the second is the 'estimation of the profit/loss result'. The foundation scenario is always executed first by all virtual enterprise instances in order to bring up the legal, technical and procedural issues that relate to the creation of 'their own' new company. The profit/loss scenario is usually executed last, in order to estimate the (virtual) financial result of the simulation activity. Neither of these activities usually takes place in real on-the-job training environments, although it is indeed important for students to be guided towards addressing such issues in order to be able to start developing their own entrepreneurship aspirations. This is considered a major advantage of this approach.

Building on the concepts of *task* and *scenario*, four phases for preparing and executing simulation activities are defined: set-up, preparation, run and assessment.

> *Set-up*. First, the control environment needs to be set up. The completion of this phase is marked by the creation of the virtual market textbook and of one virtual enterprise textbook for each virtual enterprise class, as well as the installation of the computer and communication resources.
>
> *Preparation*. Before initiating the business simulation activities, participants are notified about the context, the procedures, the rules and the available tools and resources. Decisions taken by the *directing board* are carried out by the *command centre* and virtual enterprise instances are put together. Students are

informed about their assignments, roles and communication procedures, more or less in the same way that newly recruited personnel are introduced to a real company's environment. If the location of participants has a geographical distribution, all the necessary communications are properly established and verified.

Run. Under the supervision of trainers and using support from the computer-based learning infrastructure, the virtual enterprises' staff run scenarios by executing *tasks*. Students assign (virtual but limited) resources, decide on priorities and run tasks. Paper-based tasks are always executed from scratch, while technical-oriented practical tasks are usually run in a real or virtual lab and may, or may not be, executed in their entirety because of practical constraints. The outcome of each task/scenario is a 'product' or 'service' offered to the virtual market and interim auditing documents, as well as a profit/loss financial result.

Assessment. Assessment takes place upon completion of the job simulation activities, based on the answers to electronic questionnaires completed by all stakeholders involved. Questionnaires contain a common section and an addressee-dependent section. Every stakeholder evaluates his/her experience in the simulation environment, the learning material and the simulation fidelity; students are evaluated on a personal and a team basis. Personal evaluation addresses their performance in executing tasks and in producing quality products, whereas team evaluation takes into consideration the whole image of the performance as well as the virtual financial result that has been achieved.

Geographical Distribution and Network Support

The business simulation environment presented in this chapter has been conceived with its implementation in a network-based environment in mind. Design decisions concerning such an environment should be taken only after careful consideration of several issues, such as the scale of implementation, the geographical distribution, the computer-assisted learning framework and the communication requirements. If the scale of the implementation is small, as is the case in a single school, the computer and network infrastructure requirements are minimal. One application server loaded with communication and computer-assisted learning software, as well as a few client workstations for trainees, seem fairly adequate. Larger implementation scales involve larger computational and communication loads, and thus have higher requirements.

The geographical distribution of JSAs is a key factor in defining the network and application requirements. If all the activities take place in the same building, we have a simple case where a local area network (LAN) is fairly sufficient. This is the case in most small-scale implementations occurring within school premises. The geographical distribution of the simulation activities in cooperating lab centres, companies or other geographically dispersed schools, generates several new technical requirements which will not be further discussed here.

Another key factor is that of the computer-based learning framework. The available choices are numerous, take quite different pedagogical approaches, and have varying network and software requirements. A long discussion on this topic can be found in the literature. The JSAs described here impose no special constraints or requirements from the computer-based learning environment, apart from the electronic provision of the learning content. Integrating the communication services of the learning framework with those of the job simulation activities is good practice; it is not, however, an imperative requirement.

With respect to communications, several technical choices also exist, ranging from feature-loaded groupware applications to plain old e-mail. The definition and management of the concept of time should be taken into consideration if time is to be simulated too. Virtual time management is not an issue if the simulation activities occur in real time, which is common for implementations in a homogeneous, centralized school environment. Instead, time management becomes an issue where the geographical distribution of the game is high, that is, when the game occurs in different sites, cities or even countries.

In Figure 7.3, a deployment diagram for a full-scale distributed implementation of simulation activities is shown. Each virtual enterprise has its own application server, which has custom software application services relating to the virtual enterprise type. The network learning environment, as well as the communication services, are provided by different nodes. A separate node is dedicated to the management of the rules and constraints of the virtual market.

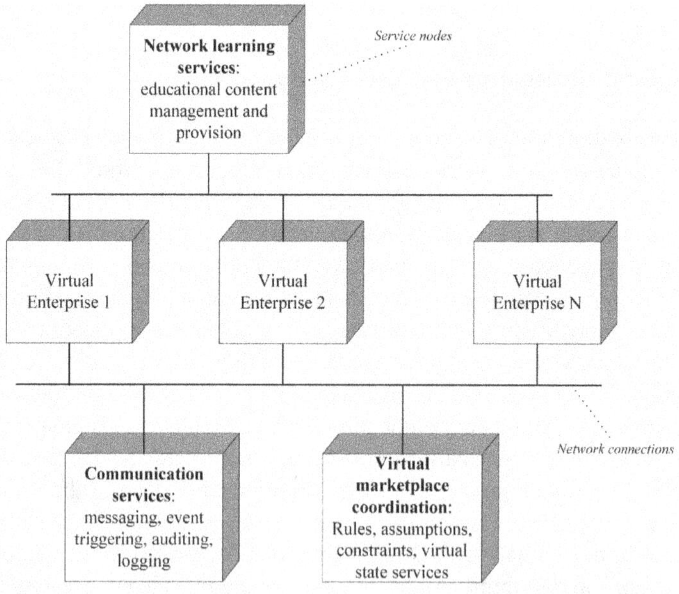

Figure 7.3 A deployment diagram for supporting distributed JSAs

Non-urban and especially border regions are less privileged in terms of educational infrastructure, services and opportunities. Several constraints ranging from cost to lack of student critical mass and to limited computer network infrastructure create significant educational disparities. In more than a few cases young persons living in such areas do not have the opportunity to study any technical job that the educational system has to offer; instead, they can only choose from whatever is available in their region, regardless of whether they like it or not.

Furthermore, even if some technical specialty of their interest is offered in their area, the availability of a company site where they can do their on-the-job training is not assured. This is a typical case of educational inequality within the scope of this chapter. As discussed elsewhere (Vescoukis and Retalis 1998, 1999), in such cases the situation can be improved using network-based educational tools. In most typical computer-assisted learning environments, this improvement is focused on the provision of access to educational resources such as libraries, virtual labs and communication with distant colleagues.

Using such a deployment, most of the aforementioned shortcomings of on-the-job training opportunities in border and non-urban areas are eliminated. Remote, non-privileged students have the opportunity to participate in JSAs of any technical specialism they wish, no matter whether the required infrastructure exists in their area or not. From the educational viewpoint, this is an improvement on the issue of equal opportunities in geographical areas that do not offer the opportunity of practice in all technical specialism.

Discussion

The approach described in this chapter has been followed in implementing a series of simulation activities within a suitable educational context using network learning support, in a secondary level technical professional education school in Greece.

Following the steps described in the section called 'A Job Simulation Environment, an operational structure was set up as in Figure 7.1. One virtual market handbook and virtual enterprise handbooks for six virtual enterprise types were prepared. Within a time period of 18 months, 11 instances of six virtual enterprise types have participated in simulation activities that have been run according to the process mentioned earlier.

Following the run of the simulation activities, assessment questionnaires were filled out by participants, both trainers and tutors. Several interesting points were noted, which have led to three useful conclusions.

First, JSAs offer trainers the opportunity to transfer to trainees their personal entrepreneurial experience by playing the roles of business consultants within the simulation environment. In such an environment, trainers get to know better how their students think by monitoring their actions in two capacities: one as the consultant who is a member of the same team operating within the simulation environment, and the other as the scenario and problem designer as a member of the control environment. Such a setting balances the impersonal context of computer-based learning, the traditional classroom and traditional on-the-job training; hence, it improves the communication and human relationships between trainers and trainees and enhances their motivation for mutual understanding.

Second, when participating in JSAs, trainees have the opportunity and the responsibility to take business decisions, applying them in practice, and learn from the outcomes of their actions, even if their choices are not optimal. As a result, they become more self-confident and become better prepared to deal with employers in an increasingly competitive job market. Computer and network support enable the accumulation of written knowledge and experience derived from the problems that each team has to cope with during the JSAs, into future learning material to be used in similar activities.

Third, the operation of JSAs eliminates location, space and time limitations. This is a very important characteristic of the concept which gives the opportunity to students located in non-urban areas to (virtually) practise the job they *want* to learn instead of a job they *can* learn, as a result of restrictions related to infrastructure or even to local cultures. When applied on a reasonable scale, this can lead to the normalization of educational inequalities and lack of opportunities in non-urban and border areas.

As mentioned at the beginning of this chapter, one of the most dominant challenges that current educational systems fail to face is the relation of the school to the job market. New technologies alone cannot bridge such a gap: they need to be well utilized in proper educational contexts. However, new technologies, especially network technologies, when properly deployed, can narrow the gap of educational inequalities. Towards this goal, a network-based job simulation activities schema has been presented. This schema can be regarded as an alternative to traditional on-the-job training, focusing on secondary level technical education. Although JSAs have not been implemented in a geographically distributed environment, they appear as a promising and viable solution to educational isolation and limitation.

References

ALTP (1997), 'Networked Open Learning: Course Material'. University of Lancaster, Advanced Learning Technology Programme, Lancaster.

Linser, R. and Naidu, S. (1999), 'Web-Based Simulations as Teaching and Learning Media in Political Science', in: *AusWeb99. Conference Proceedings* <http://ausweb.scu.edu.au/aw99/papers/naidu/>.

Sage (2002), 'Facts and Figures from the Worlds of E-Learning, Training, Work' <http://www.sagelearning.com/papers/Facts%20and%20Figures%20from%20the%20Worlds%20of%20e.doc>.

Schank, R.C. (1997), *Virtual Learning: A Revolutionary Approach to Building a Highly Skilled Workforce* (New York: McGraw-Hill).

Vescoukis, C.V. and Retalis, S. (1998), 'Thesis and Antithesis on the Use of Network Technologies in Education', in: Proceedings of Networked Lifelong Learning (NLL)'98 Conference, Sheffield, April.

Vescoukis, C.V. and Retalis, S. (1999), 'Networked Learning Using User-Enriched Learning Material', *Journal of Computer Assisted Learning* 15:3, 211–20.

PART 2
European Integration Policy and Network Development

Chapter 8

The Enlargement of the European Union and the Emerging New TEN Transport Patterns

Christos Dionelis and Maria Giaoutzi

Introduction

Transport is one of the most important factors in European economic, social and political development, with a critical strategic role in the integration process. In addition, the enlargement of the European Union (EU) is expected to have serious consequences in the structure of the existing European transport infrastructure.

Generally in Europe, but particularly in the non-Member States, the transport sector is characterized by a certain complexity, regarding:

- the economic context in which it operates;
- the policy context shaping its development; and
- the political context likely to constrain the direction and pace of future change.

Particularly in the past, and even when a general consensus on the main objectives existed, there was uncertainty and confusion regarding the nature and extent of the proposed solutions for certain transport-related problems. The problem is becoming sharper, when we move towards major challenges, such as enlargement.

The new evolution underlines the need for immediate critical strategic decisions in several fields, such as a new framework of major transport networks, modal shift and funds allocation. Currently, transport questions in the Union are handled by a coherent set of rules and provisions (the *transport acquis communautaires*), which form the Common Transport Policy (CTP). With its origins dating back to the Treaty of Rome, CTP, after the 1980s, has emerged as one of the most dynamic policy areas in the Union. In more detail, the EC Treaty refers to Transport Policy in Articles 70–80. The *transport acquis* also includes the secondary legislation, that is, several hundreds of Regulations, Directives and Decisions. This legislation covers a wide area of social, technical, fiscal, safety and environmental requirements.

As transport problems directly reflect the segmentation of Europe's transport systems and the markets themselves, the primary role of the CTP was to remove the existing technical and institutional barriers between the Member States.

Sustainable development can be defined as development that meets the needs of the present without compromising the ability of future generations to meet their own needs (Baker et al. 1997).

The goal of the EU's transport policy is to ensure sustainable mobility for people and goods and create a coherent global transport system, which produces the best possible returns, not only in terms of investment, but also in securing safety and other environmental and social priorities.

Because of its potentially detrimental impact on the environment and public health the transport sector presents one of the greatest policy challenges for sustainable development within the EU (Banister et al. 2000).

The environmental impacts of transport activity include:

- emissions of greenhouse gases that are widely perceived as the main cause of global warming;
- emissions of compounds that make the ozone layer thinner and thereby cause damaging infiltration of ultraviolet radiation;
- production of persistent organic pollutants that damage the biological systems;
- in addition, transport usually contributes more than half of all local and regional air pollution and noise, particularly from road transport, and is the major source of external acoustic nuisance in urban areas.

All transport networks, either regional or international, are systems incorporating compatible physical infrastructure and operational standards that ensure sustainability.

However, having to face the problem of a non-sustainable increase of transport, the EU is trying to take certain necessary measures that can reduce the environmental damage caused by the prevailing trends in the transport sector. This is one of the main goals that have been set by CTP.

Thus, the EU policies try to decrease the use of road and air transport, by promoting alternative, more energy efficient and environmentally friendly modes of transport such as rail and maritime transport. The White Paper European Transport Policy for 2010: Time to Decide stresses the importance of a balanced growth amongst the different modes of transport (European Commission 2001a).

In the same context, the existence of an integrated transport network, continuously monitored in its progress, comprising infrastructure for the different modes of transport, operation standards, and traffic management systems is a basic step in serving the European policies, and thus can strongly facilitate sustainable mobility.

Reflecting the importance of the existing problems, a critical agenda point in the discussions between the Union and the acceding countries was the acceptance of *sustainable mobility* as the fundamental objective, and the adoption of the *acquis communautaires* of the CTP.[1]

---

1 Agenda 2000: Agenda 2000 was an action programme whose main objectives were to strengthen Community policies and to give the European Union a new financial framework for the period 2000–06 with a view to enlargement.

Written just before the final act of the historic enlargement[2] of the Union, the present chapter focuses on the main characteristics of the transport scene in Europe, and the main expressions of the CTP in terms of infrastructure development, that is, the transport networks and the corridors. Some of the implications that the enlargement is expected to cause on the EU's transport infrastructure are also tackled in this chapter.

The Transport Infrastructure in the Union – The Trans-European Network

The fundamental objective for an efficient transport policy is either to provide an adequate global transport system between the various parts of a unified territory or to secure the existence of efficient links between weakly communicating regions. The European Union includes many 'difficult-to-reach' regions, such as the island nations (Britain and Ireland); a country without natural borders with the rest of the Union (Greece); and long mountain barriers, especially the 'Alpine wall', physically separating north and south Europe (the Pyrenees do much the same for the Iberian states). The challenge of unifying these far-flung regions into a single market, much less into a political and economic union, has induced greater EU activism in transport issues, if only to prevent further decoupling of the periphery from the centre.

The main tasks of the CTP have been set to promote a well-balanced link between the development of transport and economic activities in the regions of the Member States (European Commission 1992).

In particular, for every 'phase' in which the European transport policy is expressed, there has been the objective of strengthening economic and social cohesion by the contribution which the development of transport infrastructure can make to reduce disparities between regions and linking island, landlocked and peripheral regions with the central regions of the Community.

Policy has evolved along three main fronts: harmonization initiatives to standardize the Members States' technical, social and fiscal policies; liberalization, using the discipline of market forces to strengthen transport industries; and most recently, structural policies, via commitments to developing a cross-border European transport infrastructure. The financing of the various steps of the CTP involves the use of significant structural funds in well-structured global programmes[3] to finance infrastructure, operations, innovations and research, to attain the Union's main objective of sustainable mobility.

Efficient transport is a key determinant for growth for any economy, be it the economy of the European Single Market or of any other politico-economic entity. The main goal of the EU's CTP is to ensure sustainable mobility for people and goods and create a coherent transport system, capable of giving the best possible

2 The official enlargement of the Union materialized in May 2004. This was the first time that an enlargement of the Union incorporated so many new countries, with different economic and political background.

3 The Community Support Frame and the Cohesion Fund are good examples of mobilizing funds for transport infrastructure development.

returns, not only in investment but in securing safety and other environmental and social priorities as well.

In seeking an appropriate transport system, which could serve its economic, social and political needs, the European Union has worked for the establishment of a transport network on its territory, which could contribute to the attainment of major Community objectives, such as the smooth functioning of the internal market and the strengthening of economic and social cohesion.

In order to realize the great expectations of its citizens and markets, the European Union has engaged in a major infrastructure programme for a trans-European network in the field of transport (TEN-T), which actually comprises infrastructure and traffic management and navigation systems.

In July 1996, the European Parliament and Council adopted Decision N° 1692/96/EC on Community guidelines for the development of the trans-European transport network (TEN-T). These guidelines covered roads, railways, inland waterways, airports, seaports, inland ports and traffic management systems, which can serve the entire continent, carry the bulk of the long-distance traffic, and bring the geographical and economic areas of the Union closer together.

The legal basis for the TEN-T was provided by the Treaty on the European Union itself. Under the terms of Chapter XV of the Treaty (Articles 154, 155 and 156), the European Union must aim to promote the development of various trans-European networks (transport, energy, communications) as a key element for the creation of the Internal Market and the reinforcement of Economic and Social Cohesion. This development includes the interconnection and interoperability of national networks, as well as access to such networks.

In accordance with these broad objectives, the Community developed the above-mentioned guidelines, that is, a general reference framework for the implementation of the network and identification of projects of common interest. The time horizon to complete the network was set at 2010. The maps in the Annexes of Decision 1692/96/EC illustrated the outline of the trans-European transport network as planned for 2010.

Decision No 1692/96/EC of 23 July 1996 on *Community guidelines for the development of the trans-European transport network (TEN-T)* sets out broad lines for the measures necessary to set up the network. The Decision established characteristics of the networks for each mode of transport, eligible projects of common interest, and priority projects. Emphasis was placed on environmentally friendly modes of transport, in particular rail projects. In total, 14 projects were designated as priority projects. The Essen European Council (1994) had already attributed particular importance to this list of priorities.[4]

Decision 1692/96/EC was primarily addressed to the Member States, which were primarily responsible for realizing the TEN-T. A number of financial instruments were set up at Community level, each with their own legal basis, in order to conduct

4 In more detail, an initial list of proposed priority projects had been evaluated and defined in further detail by the *Christophersen Group* and had been shaped by the principles established at the Essen Council in 1994.

the development of the TEN-T and to support Member States financially in specific cases.

In May 2001, the European Parliament and the Council adopted Decision N° 1346/2001/EC which amended the TEN-T guidelines as regards seaports, inland ports and intermodal terminals. It also specified in more detail the criteria of projects of common interest in relation to these infrastructures. With this amendment the multimodal dimension of the network was emphasized, as seaports and inland ports become fully part of the network.

In order to reduce the bottlenecks on major routes and in a small number of major projects, in October 2001 the Commission initiated *a major revision of TEN-T Guidelines* in the lines of the *White Paper on a European Transport Policy for 2010* (European Commission 2001a), in order to tackle the new challenges facing transport and to help meet the objectives of the new transport policy as described in the White Paper. It aimed at reducing the bottlenecks in the planned or existing network without adding new infrastructure routes, by concentrating investments on a few horizontal priorities and a limited number of new specific projects.

Thus, the expressed will of the Union, together with the impending enlargement of the EU in 2004, also coupled with serious delays and financing problems in the realization of the TEN-T – particularly its cross-border sections – led to the realization of a thorough revision of the TEN guidelines. On the basis of proposals made by a specially appointed working group headed by former European Commissioner *Karel van Miert*, this revision was adopted in the form of Decision 884/2004/EC of 29 April 2004. The revision comprised the following main elements:

- The number of priority projects was increased to approximately 30, with a total cost estimated at €250 billion. Many of these projects involve the Member States that joined in 2004;
- EU environmental rules were brought into sharper focus, particularly through a strategic environmental impact assessment as a complement to the conventional environmental impact assessment.

The guidelines for the development of the TEN, embracing a major infrastructure programme, were always adopted by the Parliament and the Council.[5] The decisions

5 A complete list of the European *acquis* on the TEN-T is as follows:
- *Council Regulation (EC) No 2236/95* of 18 September 1995, laying down general rules for the granting of Community financial aid in the field of trans-European networks, amended by *Regulation (EC) No 807/2004* of the European Parliament and of the Council of 21 April 2004;
- *Decision 1692/1996/EC* of the European Parliament and the Council of 23 July 1996, on Community Guidelines for the development of the TEN-T, amended by:
  - *Decision 1346/2001* of the European Parliament and the Council of 22 May 2001, as regards seaports, inland ports and intermodal transport,
  - *Decision 884/2004* of the European Parliament and the Council of 29 April 2004, on Community Guidelines for the development of the trans-European Transport Network.

are addressed primarily to the Member States and constitute a general reference framework for the implementation of the network and identified projects of common interest, which are now to be completed by 2020. The real significance of these guidelines are in fact that they are the first attempt to describe a vision of how an integrated network in the Union's territory should develop in a certain horizon (now set in 2020).

The TEN-T provides the physical infrastructure for the internal market, and it has been planned and defined to become a fundamental tool for growth and competitiveness. A number of financial instruments were set up at Community level and by the European Investment Bank in order to conduct this policy, in particular a specific budget for the funding of the TEN-T. If the necessary infrastructure were not completed and interlinked to enable trade to be conducted, the internal market and the territorial cohesion of the European Union would remain ideas that have failed to come to full fruition.

The TEN-T accounts for almost half of total goods and passenger traffic in the EU. This network serves as a reference point for other Community legislation and is highly visible to EU citizens. It has been a successful land-use planning instrument and has enabled links to countries bordering the EU to be improved.

The Future Enlargement

The European transport debate has reached a critical stage, in part because of the rapidly changing geopolitical and spatial dimension of the European Union. Once confined to an economic grouping of only six neighbouring, industrialized states in west-central continental Europe, the European Economic Community/European Community/European Union has grown in all four directions. In the 1970s it expanded to the north and northwest, adding its first non-contiguous territories; in the 1980s it spread southward across the north shore of the Mediterranean basin from the Atlantic to the Balkans. The 1995 expansion added vast new territory to the heart of Central Europe – with Austria – and additionally, towards north and east, extending to the Arctic Circle in Sweden and Finland, and even bordering Russia.

To these concerns were recently added the vast needs of the countries of Central and Eastern Europe (CEEC), which EU leaders have pledged to incorporate, eventually, into the EU fold. Membership negotiations were undertaken for about a decade; the negotiations gradually included 10 to 13 candidate countries for accession.[6,7]

Central Europe is now a dynamic region where travel is both a major component of lifestyle and a crucial element for economic growth. The transport sector accounts for between 6 and 9 per cent of GDP, constituting a market for services and investment in Western and Central Europe worth €400–600 billion annually, of which Central Europe's share would be in the order of €20 billion (TINA 1999).

6 These countries are: Estonia, Latvia, Lithuania, Poland, Czech Republic, Slovakia, Hungary, Slovenia, Romania, Bulgaria, Cyprus, Malta and (lately) Turkey.

7 Out of the 13 candidate countries for accession, ten became full members in 2004, and two (Bulgaria and Romania) in 2007. Negotiations on Turkey's accession are in progress.

After the 2004 and 2007 enlargement, the population of the European Union increased by about one third, from 375 to 480 million. However, the GDP of the enlarged Union has only increased, in year 2000 levels, by around 4 per cent, or €244 billion at 2000 prices. Although the average per capita GDP of the new Member States, €2,300 in the year 2000, should, according to conservative estimates, double over the next 15–20 years, it will still remain well below the 2000 EU-15 average of €17,200 (TINA 1999).

Enlargement and Transport Needs

The transport infrastructure networks in the territory of the 12 new Member States are vital to competitiveness, economic growth and employment throughout Europe, not least in the Union itself. It is important to underline the need for the new Member States to adopt the principles and logic of the TEN-T. This means that not only new financing schemes will be used to develop the existing transport infrastructure, but also this infrastructure must follow the rationale of the TEN-T, paying specific attention to basic provisions of the TEN-T, such as the promotion of accessibility, interoperability, effectiveness and environmental protection.

If the new Member States continue to keep the structure of national transport networks, then this means that European traffic passes through connected national subsystems rather than through a proper European system operating as a function of the single market.

It is precisely in the transport sector, where the shortcomings of the communications system are most evident, that the concept of *trans-European network* is being developed as an instrument for market integration and economic and social cohesion, since a more integrated transport system can help ensure that the peripheral, remote or less developed regions can be fully integrated into European economic life. The aim of this policy is not to redesign the European transport system in such a way as to organize it into a uniform whole, but instead to achieve interconnecting national subsystems which, for obvious reasons related to economics and the internal operability of the system, remain the 'essential carriers' of traffic.

If things do not change, the order of growth in road traffic can lead to capacity problems on main transport routes, as well as to a substantial increase in accidents, and can negatively affect the environment. It will therefore be important that in the new territory of the Union, infrastructure development should aim to achieve *sustainable mobility*, balancing economic efficiency, safety and minimal environmental damage. In the long run, this requires an efficient and balanced multimodal network for the whole of Europe, and specifically for the enlarged Union. This transport network should develop in line with present and future traffic demand,[8] and should enable European transport services to utilize each mode according to its comparative advantage.

---

8 This was one of the fundamental principles during the TINA process; it has the meaning that future needs in transport infrastructure cannot be reported solely by each country in an arbitrary way, but should be consistent with traffic demand.

Increasing trade and economic development will generate continued growth in traffic throughout Europe. This will particularly affect Central Europe's transport network.

Traffic forecasts (PHARE 1999) show that freight transport will grow considerably between 1996 and 2015, with domestic transport growing by 40–70 per cent, exports by 90–150 per cent, and imports by 80–140 per cent. Passenger transport will also grow, with international traffic doubling or even quadrupling.

As regards modal split, in the next 15–20 years the share of road freight transport in the new Member States will grow:

- from the year 2000 level of 85 per cent to 89–93 per cent for domestic movements;
- from the year 2000 level of 29 per cent to 36–43 per cent for exports; and
- from the year 2000 level of 18 per cent to 29–37 per cent for imports.

For road passenger transport, the increase in car travel between 2000 and 2015 will for some countries be in the order of 150 per cent (TINA 1999).

In this framework, the 2004 Study TEN-STAC[9] shows that completing the networks will considerably reduce journey time for passengers and goods, through a 14 per cent reduction in road congestion and improved rail performance. For interregional traffic alone, the benefits are estimated to be almost €8 billion per year.

In addition, freight transport in the EU is expected to increase by more than two-thirds between 2000 and 2020, and to double in the new Member States. Without TEN-T, this increase in transport would be impossible to handle, and our rate of economic growth significantly slowed.

Completing the networks will also bring important dividends for the environment. According to the study, on the basis of current trends, CO_2 emissions from transport would be 38 per cent greater in 2020 than in 2003. However, completing the 30 priority axes should slow down this increase by about 4 per cent, representing a reduction in CO_2 emissions of 6.3 million tonnes per year.

With EU enlargement, the countries of Central Europe will increasingly adopt a dual role, both as the constituent parts of the wider European Union and as an interconnection with the New Independent States in Eastern Europe and the littoral countries of the Mediterranean and the Black Sea.

The efficiency of transport systems will play an absolutely critical role in determining whether the future expansions succeed in truly incorporating these new states, or whether they will become burdensome territorial appendages.

9 'TEN-STAC: Scenarios, Traffic forecasts and Analysis of Corridors on the trans-European Network', a European Commission Project, NEA Transport research and training BV, January 2003 – May 2004.

The TINA Process

In 2000, the extension of the existing TEN in the candidate countries for accession was based on the results of the Transport Infrastructure Needs Assessment (TINA) process.

The purpose of TINA was to identify and assess the transport infrastructure needs of the future Member States. In this respect, TINA should define a multimodal transport network in the acceding countries, which should become the extension of the 1996 TEN-T, in order to identify the necessary measures to construct/upgrade this network and to calculate its cost.

In preparation for the enlargement of the European Union to the east, the European Commission set up TINA to oversee and coordinate the development of an integrated transport network in the 11 countries that had then applied for EU membership: Bulgaria, the Czech Republic, Estonia, Hungary, Latvia, Lithuania, Poland, Romania, Slovakia, Slovenia and Cyprus, and to ensure coherence with the trans-European Transport Network (TEN-T) within the EU. As in the EU, the idea is to upgrade existing or build new infrastructure in order to create a coherent network out of the current patchwork of transport links and achieve safe and speedy connections between countries that are necessary to increase the efficiency of the Single Market and maximize the potential of European trade.

The TINA exercise was undertaken during 1996–99 and gave concrete results in terms of transport infrastructure needs in the candidate countries. It was a significant stage in the work to develop a coherent transport network in the countries that had applied for European Union membership, with the identification of an outline network for these countries, linking them to the TEN-T within the EU itself. On this basis, the TINA Network formed the basis for the extension of the TEN-T in the enlarged Union.[10]

In the TINA process, the Transport Ministries, the Commission and the TINA Secretariat[11] have worked to define the precise transport links and nodes that should comprise the extension of the TEN-T into the territory of the new Member States. The result of this exercise was the TINA network. This multimodal network, accepted by all the acceding countries, in its final form comprised 18,683 kilometres of roads, 20,924 kilometres of railway line, 4,052 km of inland waterways, 40 airports, 20 seaports, and 58 river ports. The TINA network also included 86 terminals, out of which 20 were situated in seaports and river ports, and 66 were nodes of the railway or road network.

The cost of completing this network was estimated at nearly €90,000 million (COWI Consult 1999; TINA 1999), to be spent between 1999 and 2015.[12] Broken down on a country-by-country basis this translates to a level of around 1.5 per cent

10 A major revision of the TEN-T guidelines took place in 2004.

11 A permanent body, established in Vienna between 1997–2000, to provide technical support in the TINA process.

12 It should be mentioned that the reported costs per mode cannot be compared as unit costs, as they represent a wide range of projects and interventions such as upgrading, new construction of various types of infrastructure (such as expressways, motorways, double railway lines, and so on), electrification, signalling, transport control systems, and so on.

of forecast gross domestic product up to 2015, compared with spending on transport infrastructure in the EU of some 2 per cent of GDP per year.

In general, the TINA network was built on the following assumptions:

- the network should be in line with the criteria laid down in the EU Guidelines for the Development of the TEN-T (Council Decision 1692/96/EC), according to the objectives described in Article 154 of the Treaty; (these criteria refer to accessibility, interconnectivity, efficiency, interoperability, elimination of bottlenecks, environmental consideration, and so on);
- the technical standards of the future infrastructure should ensure consistency between the capacity of network components and their expected traffic. To achieve this, it was accepted that these standards should be in line with the recommendations of the UN/ECE Working Party on Transport Trends and Economics (WP.5) on the definition of transport infrastructure capacities (Trans/WP.5/R.60);
- the time horizon for achievement of the network should be 2015;
- the cost of the network should be consistent with realistic forecasts of financial resources, so that average costs should not exceed 1.5 per cent of each country's annual GDP over the period up to 2015.

The TINA multi-modal transport network was the main field for major transport infrastructure investments in the candidate countries of Central and Eastern Europe. The basic fund that finances transport schemes on the TINA Network between 2000–06 is Instrument for Structural Policies for Pre-Accession (ISPA), administered by the European Commission-DG REGIO.[13]

The Pan-European Transport Corridors and Areas

Another part of the Pan-European transport infrastructure scene is the 'Corridors concept', which has developed over recent years along with the Pan-European Transport Conferences. More specifically, during the Conferences in Crete in 1994 and Helsinki in 1997, the UN-ECE,[14] the ECMT,[15] the European Commission, and all European States supported the development of ten PECs, (pan-European Transport Corridors)[16] and four pan-European Transport Areas (PETrAs). The idea is that when transport links are needed, a first step before the creation of transport networks is the establishment of transport corridors. The linear shape of these multimodal transport routes can attract investments and traffic, in a coordinated effort to quickly develop the necessary, suitable[17] links to promote trade and social and political relations between European regions that historically have remained rather isolated.

13 Directorate General for Regional Development.

14 United Nations – Economic Commission for Europe.

15 European Conference of Ministers of Transport.

16 Also known as the 'Crete' or 'Helsinki Corridors'.

17 This could be translated into modern, safe, environmentally friendly and interoperable infrastructure and services.

The 'character' of a transport corridor is different from the transport network, as the former mostly focuses on the linkage and the transport itself, while the latter focuses on the regional development of the area served by the network. The corridors are simply the first step, and future coordinated actions for network development will certainly follow.

The ten Helsinki Corridors are:

i. *Corridor I* is a multi-modal transport link, running in North-South direction, from Helsinki (Finland) to Warsaw and Gdansk (Poland).
ii. *Corridor II* connects Berlin (Germany) with Minsk, Moscow and Nizhny Novgorod (Russia).
iii. *Corridor III* links Berlin/Dresden (Germany) to Kiev via Lviv (Ukraine).
iv. *Corridor IV* is a multi-modal East-West transport link running from Dresden/Nürnberg (Germany), via Praha (Czech Republic), to Vienna (Austria)/Bratislava (Slovakia), Budapest (Hungary) and to Romania, and further to Bulgaria, Greece and Turkey.
v. *Corridor V* runs from the South-West in Slovenia towards the North-East in the Ukraine. It has several other branches to Italy, Hungary, Croatia and Bosnia Herzegovina.
vi. *Corridor VI* runs from the North to the South, connecting the Polish Baltic Sea ports of Gydnia and Gdansk with Slovakia and the Czech Republic.
vii. *Corridor VII* is the Danube river, which connects the North Sea with the Black Sea crossing Germany, Austria, Slovakia, Hungary, Croatia, FR Yugoslavia, Romania, Bulgaria, Moldova and the Ukraine.
viii. *Corridor VIII* links the Adriatic-Ionian Sea with the Black Sea, crossing Albania, Former Yugoslavian Republic of Macedonia (FYROM), Bulgaria, with a branch to Greece.
ix. *Corridor IX* links Helsinki (Finland) with Alexandroupolis (Greece), and with several other branches.
x. *Corridor X* is a link running from the North-West to the South-East. It connects Salzburg (Austria) to Thessaloniki (Greece), with several other branches.

These transport Corridors comprise 25,500 km of railway lines and 23,030 km of roads; that is, about 48,530 km of long-distance links. The multi-modal nature of the corridors is complemented with 33 airports, 9 sea ports and 49 river ports as nodes (European Commission 2001b).

The four PetrAs are:

1. The Arctic PETrA
2. The Mediterranean Basin PETrA
3. The Adriatic-Ionian Sea PETrA
4. The Black Sea PETrA

The four PETrAs mostly aim to develop the short sea shipping services in the relevant regions, and the land networks that efficiently connect the ports with the hinterland of the region.

As their good functioning is a serious prerequisite for the better interconnection of the European markets, the various international financial institutions paid specific attention to the development[18] of these transport Corridors and Areas. Although the needs for financing are still huge, the large investment schemes that were undertaken for the Corridors were a reason for the significant progress to date.

These multimodal transport Corridors and Areas have provided an important focus for investment by the international financial institutions, and significant progress has been achieved in their development. The establishment of the Corridors and Areas is part of the actions, which form the Pan-European Transport Network Partnership endorsed at the Third Pan-European Transport Conference at Helsinki (ECMT 1997). This Partnership aims to accomplish an infrastructure set-up all over the European continent, which allows transport services to follow the principle of sustainable mobility as set out in the Common Transport Policy (CTP) of the European Union. Therefore, the Corridors' development should be seen as a means of achieving the goals of the CTP. These transport Corridors of transnational character, play a very important role in European transport and economic integration. There is not only the infrastructure linkage between regions, but also the interoperable operational-institutional framework along these arteries that help to bring together the various economies and societies.

In conclusion, the Pan-European Corridors concept, even with its weaknesses, has been very helpful for the promotion of the CTP. This statement can be supported by the following facts:

- The cooperation along the PECs has been organized through non-binding Memoranda of Understanding (MoU), signed by Ministers of Transport and the European Commission's Transport Commissioner.
- The Commission continues to monitor, synchronize and participate in the works of the Corridors.
- For the period 1996–2004, the projects on the TINA network that were part of the Corridors had priority over other projects in the acceding countries.

The Way Ahead

Until rather recently, European transport networks have tended to develop based on separate national perspectives, with the emphasis being placed upon individual modal networks rather than on integrated transport systems.

The Commission's transport policy seeks measures to establish an interoperable transport network, based on common rules and provisions, capable of promoting the principle of sustainable mobility. In this respect, the main task remains a more intensive use of environmentally friendly modes of transport with spare capacity, such as rail and inland waterways. By improving the potential of these modes, common transport policy can create and offer effective alternatives to single mode road journeys. The forthcoming enlargement is a great challenge for the Union's Common Transport Policy. As a major first task, the trans-European transport

18 Mostly meaning radical rehabilitation and upgrading schemes.

network (TEN-T) must be revised, in order to include all the national networks of the new Member States, in a coherent manner.[19] This was the task undertaken during the TINA exercise. Further action is needed on the new needs, linking the new Union to its neighbouring countries towards the western Balkans, the newly independent states (NIS) and the Caucasian and Asian states.[20]

In this context, the new funding needs are expected to be a major field for technical and political debate during the coming years.

This is the main message of the Commission's latest White Paper. In September 2001 the European Commission expressed the main directions of its Transport Strategy, issuing the White Paper: *European Transport Policy for 2010: Time to Decide*. In the White Paper, more than 60 measures are proposed to promote the establishment of transport systems capable of providing efficient transport for goods and people throughout the Union, with special emphasis on safety, comfort and environment. The concept of 'sustainable mobility' remains at the heart of the approach.

In the field of infrastructure planning, the Trans-European Network (TEN-T) and, subsequently, the TINA network and the Pan-European corridors, delineate an ambitious plan of structural integration. In this respect, the elimination of bottlenecks in the network is as important as the promotion of regional development and cohesion.

In conclusion, and in order to make a clear distinction between the various concepts that were mentioned in this chapter, Table 8.1 can be useful.

Table 8.1 Basic information regarding various infrastructure concepts

Concept	States affected	Decision making	Remarks
TEN-T–1996 Decision 1692/96/EC	EU-15	EU	It defined 14 priority projects. Horizon: 2015
TEN-T–2004 (revised) Decision 884/2004/EC	EU-27	EU	It defined 30 priority projects. Horizon: 2020
TINA Network	11 acceding countries	EU	The TINA network was defined through the TINA exercise between 1996–2000, in order to define what would be the extended TEN-T, after the EU enlargement. Horizon: 2015
Pan-European Corridors They were decided during the Pan-European Conferences of Crete (1994) and Helsinki (1997)	All European countries	Steering Committees of the Corridors	For the acceding countries, the Corridors were the backbone network for the TINA Network in their territories. For the third countries, they are the main linkages to the Union.

19 This was achieved with Decision 884/2004 of 29 April 2004, which amended the *Decision 1692/1996/EC* of the European Parliament and the Council of 23 July 1996, on Community Guidelines for the development of the trans-European Transport Network.

20 In this context, the Helsinki Corridors' alignment maybe has to be re-assessed.

References

Baker, S., Kousis, M., Richardson, D. and Young, S. (eds) (1997), *The Politics of Sustainable Development: Theory, Policy and Practice within the European Union* (London New York: Routledge).

Banister, D., Stead, D., Steen, P., Åkerman, J., Dreborg, K., Nijkamp, P. and Schleicher-Tappeser, R. (eds) (2000), *European Transport Policy and Sustainable Mobility* (London New York: Spon Press).

COWI Consult (1999), 'Unit Costs in Transport Infrastructure in Acceding Countries', PHARE Project, 1998.

ECMT (1997), 'Towards a European Wide Transport Policy, a Set of Common Principles'. Third Pan-European Transport Conference, Helsinki, 23–25 June: Helsinki Declaration: ECMT, 25 June.

European Commission (1992), *The Future Development of the Common Transport Policy* (COM (92) 494).

—— (2001a), *European Transport Policy for 2010: Time to Decide* (COM (2001) 370).

—— (2001b), *Pan-European Transport Corridors and Areas: Annual Report 2001*. DGTREN- B2, TINA Vienna.

PHARE (1999), 'Traffic Forecast on the Ten Pan-European Transport Corridors of Helsinki'. NEA-INRETS-IWW, July.

TINA (1999), 'Transport Infrastructure Needs Assessment'. Final Report, TINA Secretariat, November 1999.

Chapter 9

Spatial Development and Cohesion Impacts of European Transport Investment and Pricing Policies

Lars Lundqvist

Introduction

The ESPON Programme (2002–06) was launched after the preparation of the European Spatial Development Perspective (ESDP), adopted in May 1999 in Potsdam (Germany) by the Ministers responsible for the Spatial Planning of the EU and calling for a better balance and polycentric development of the European territory (ESDP 1999). 'The ESPON 2006 Programme – Research on the Spatial Development of an Enlarging European Union' is being implemented in the framework of the Community Initiative INTERREG III. The programme consists of thematic projects, policy impact projects, coordinating cross-thematic projects, data and supporting projects, and networking activities.

As one of the policy impact studies, ESPON project 2.1.1 (*Territorial impact of EU transport and trans-European Network (TEN) policies*) studied whether EU transport and TEN policies (TEN-T) are likely to affect territorial development along the lines suggested in the ESDP. Methods, concepts and indicators were developed to assist in this mission. The project should indicate the influence of transport and TEN policies on spatial development and recommend policy developments in support of territorial cohesion and a polycentric and better balanced EU territory. This chapter draws on results reported in the Final Report of ESPON 2.1.1 (Bröcker et al. 2005).

Some hypotheses concerning relations between infrastructure and regional development, based on earlier research, are presented in the next section. The ESPON 2.1.1 analysis was based on three models: two dealing with the impacts of transport policies and one dealing with the impacts of ICT policies. This chapter focuses on transport policies using two models for an impact analysis of the European spatial system: SASI and CGEurope. Then the models are briefly introduced and the 13 policy scenarios of ESPON 2.1.1 are defined. The major part of this chapter concentrates on five forward-looking scenarios and one backward-looking scenario. The main impact of these scenarios on EU efficiency in terms of total welfare or total production and EU cohesion is subsequently reported. A methodology for analysing the potential for polycentric development is then outlined, and the results of policy scenarios are also reported. Finally, the potential conflict between the efficiency and the cohesion aspects of policy impacts is discussed. Some conclusions are provided at the very end.

Hypotheses on Infrastructure and Regional Development

Earlier research on the relationships between infrastructure and regional development and earlier experiences from applications of SASI and CGEurope could be summarized in a number of hypotheses on the outcome of the analysis. Most of these were listed in the third interim report of ESPON 2.1.1 (Bröcker et al. 2003):

- Social and technical macro-trends (population, productivity, and so on) tend to be more important for regional socio-economic development than transport infrastructure scenarios.
- Relatively large improvements in accessibility will translate into small increases in regional economic activity.
- A slight cohesion effect of transport investments in terms of accessibility and GDP cannot reverse the general trend towards economic polarization. The cohesion effect is likely to occur only if cohesion is measured in relative terms.
- Transport policy can lead to considerable effects for certain regions or for certain aspects of development, mainly as a result of impacts on relative positions in regional competition within a context of generally increased accessibility and economic performance in absolute terms.
- The impact of transport investments will depend on the competitiveness of the regional economies: a peripheral area may benefit from better market access, but its production may, on the other hand, be subject to a higher degree of competition from imports.
- The effects on polycentric development are likely to depend on the specific combinations of pricing and investment policies. Pricing policies raising the general level of private transport cost can be expected to harm peripheral regions more than the centre.
- The impact of transport investments on economic development can be expected to be greater in regions with less developed networks than in regions with dense and well-developed networks.
- The impact of transport investments on industrial and social restructuring is a slow process. Synergies with other types of social infrastructure can be crucial.

Many of these hypotheses, based on earlier research, may be tested or further elucidated by projecting the regional development impacts of selected transport policy scenarios.

Projecting Transport Policy Impacts: Models and Scenarios

The SASI model is a recursive simulation model of socio-economic development of regions in Europe subject to exogenous assumptions about the economic and demographic development of the European Union as a whole and transport infrastructure investments and transport system improvements (Wegener and Bökemann 1998). For each region the model forecasts the development of accessibility, GDP per capita, and unemployment. Although SASI is basically distributive, it can also be applied in a generative mode (by applying the exogenous constraints only to the reference scenario and applying the adjustment factors computed for the reference scenario to the policy scenarios).

The CGEurope model evaluates the impact of changes in transport and travel cost and travel times on regional welfare in a spatial computable general equilibrium model that is constructed on a consistent theoretical basis of microeconomic reasoning (Bröcker 2002). According to the model, transport cost changes affect inter-firm interaction through changing costs for goods transport, as well as changing costs for passenger business travel that is assumed to be closely tied to trade flows between firms. The model permits substitution in production and consumption. The production technology represented by production functions is not affected by transportation policies (for example, through knowledge externalities).

For the analysis of transport policy impacts, 13 *policy scenarios* have been defined: three retrospective scenarios of infrastructure investments (A1–A3) and ten prospective scenarios of infrastructure investments (B1–B5), pricing (C1–C3) and combinations of investments and pricing (D1–D2). The policy scenarios are defined in Table 9.1. In this chapter, the focus is on six of these:

- *Impacts of transport investments 1991–2001*: difference between the effects of scenario A3 and those of the retrospective reference scenario A0;
- *Impacts of priority transport investments 2001–21*: difference between the effects of scenario B1 and those of the prospective reference scenario 00;
- *Impacts of TEN-T/TINA/national transport investments 2001–21*: difference between the effects of scenario B2 and those of the prospective reference scenario 00;
- *Impacts of marginal cost pricing of all modes*: difference between the effects of scenario C3 and those of the prospective reference scenario 00;
- *Combined impacts of priority transport investments and marginal cost pricing of all modes*: difference between the effects of scenario D1 and those of the prospective reference scenario 00;
- *Combined impacts of TEN-T/TINA/national transport investments and marginal cost pricing of all modes*: difference between the effects of scenario D2 and those of the prospective reference scenario 00.

Marginal cost pricing is simulated in a simplified way by increasing the costs of all modes by 10 per cent.

It should be pointed out that, in this implementation, neither SASI nor CGEurope accounts for the financing of transport investments or the use of income from

Table 9.1 Transport policy scenarios (the analysis focuses on policies in italics)

Time horizon	Policy type	Scenario characteristics	
Retrospective 1991–2001	Reference	A0	Do-nothing
	Investments	A1	Only rail projects
		A2	Only road projects
		A3	*Rail and road projects*
Prospective 2001–21	Reference	00	Do-nothing
	Investments	*B1*	*Priority projects (new list)*
		B2	*TEN-T/TINA projects (incl. some large national projects)*
		B3	TEN-T/TINA projects except cross-border corridors
		B4	TEN-T/TINA cross-border corridor projects only
		B5	TEN-T/TINA projects only in Objective 1 regions
	Pricing	C1	Reduction of the price of rail transport
		C2	Increase of the price of road transport
		C3	*Social marginal cost pricing (SMCP) of all modes*
	Combination	*D1*	*Priority projects plus SMCP (B1+C3)*
		D2	*TEN-T/TINA projects plus SMCP (B2+C3)*

increased transport costs (through fees or taxes). Therefore, the models tend to exaggerate the benefits of investments and the negative effects on the economy of cost increases.

Efficiency and Cohesion Impacts of Transport Policy Scenarios

The total impact of the six transport policy scenarios outlined in the previous section on the total economic development in EU27+2 (15 former EU nations, 12 accession countries, Norway and Switzerland) is reported in Table 9.2. It can be seen that the economic growth impacts have the expected signs and that the size of impacts in scenarios A3-C3 is 4–12 times larger in SASI as compared with CGEurope (the difference is particularly large in the investment scenarios). This can be explained by the recursive structure of SASI and the larger flexibility in terms of population and employment changes (volume and location). CGEurope is calibrated on a base year and the transport policy scenarios are the only factors of change determining the new spatial equilibrium. The results of the combined policies (D1–D2) show, of course, intermediate impacts between the positive effects of investments and the negative effects of marginal cost pricing. The lack of closure of the models (in terms of investment funding and use of income from cost increases) should be stressed once more.

Table 9.2 Impacts of the six policy scenarios on total economic development*

Model	A3	B1	B2	C3	D1	D2
SASI	+0.63	+1.60	+2.62	–1.42	+0.14	+1.16
CGEurope	+0.16	+0.14	+0.26	–0.36	–0.21	–0.09

* Percentage change of GDP/capita (SASI) and Equivalent variation (CGEurope) as compared with the relevant reference scenario.

Table 9.3 Impacts on relative (G/A) and absolute (AC) measures of economic cohesion (GDP/capita) of six policy scenarios

Model	A3		B1		B2		C3		D1		D2	
	G/A	AC	G/A	AC	G/A	AC	G/A	AC	G/A	AC	G/A	AC
SASI	•	—	•	—	•	—	•	++	•	•	+	—
CGEurope	+	–	+	–	+	–	–	++	+	+	+	+

+/++ Weak/strong cohesion effect
G/A Geometric/arithmetic average
–/— Weak/strong divergence effect
AC Correlation between absolute change and level
• Very weak effect

The qualitative impact of the six transport policy scenarios on economic cohesion (that is, on the distribution of GDP/capita) is reported in Table 9.3 above. For each scenario one indicator of relative cohesion (G/A) and one indicator of absolute cohesion (AC) are used to illustrate the cohesion results. Other indicators are described in Bröcker et al. (2005). It can be seen that CGEurope tends to generate outcomes that are somewhat more cohesion-oriented in general than those of SASI. The most pronounced difference between the models appears in scenario D2. We also observe that the cohesion impact depends to a large extent on how cohesion is measured. Scenario B2 weakly improves the relative cohesion, while the absolute cohesion is reduced. An equal percentage income increase in all regions does not affect the relative cohesion but reduces the absolute cohesion (increases absolute inequality). Scenario C3 works in the opposite direction: absolute cohesion is strongly improved while relative cohesion is mildly reduced. Hence, it does matter how cohesion is perceived and exactly measured in terms of performance indicators.

Development Potential: Methodology

Polycentric development has emerged as a key concept during the European Spatial Development Perspective (ESDP) process. In ESDP (1999, p. 20), this is motivated in the following way: 'The concept of polycentric development has to be pursued, to ensure regionally balanced development, because the EU is becoming fully integrated in the global economy. Pursuit of this concept will help to avoid further excessive economic and demographic concentration in the core area of the EU.'

The concept '*development potential*' has been used in analyses of the role of infrastructure in regional development and for characterizing urban regions during recent decades. Biehl (1991) used indicators for infrastructure supply, location, agglomeration and sectoral structure to explain the income of European regions. In the 'Study on the construction of a polycentric and balanced development model for the European territory' (Azevedo and Cichowlaz 2003), key conditions for polycentric development were measured by a set of indicators representing mass, competitiveness, connectivity and development trend. Subsets of these indicators were used to produce typologies of urban systems in peripheral European regions. Similar types of indicators have been used by ESPON 1.1.1 for synthesizing the typologies of Functional Urban Areas (FUAs) and Metropolitan European Growth Areas (MEGAs).

Drawing on these ideas, a methodology for analysing the impacts of EU transport policies on polycentric development has been suggested. Building on the tradition of analysing key conditions for development, it combines indicators for the dimensions *mass*, *competitiveness*, *connectivity* and *development trend* into a composite indicator of *development potential*. This composite indicator is used to compare the impacts of transport policy scenarios on development potential at the NUTS-3 territorial level. By analysing the spatial pattern of these impacts, the effects on the potential for polycentric development may be indicated. This methodology for illustrating polycentricity effects is strongly related to the CPMR and ESPON 1.1.1 approaches for producing urban typologies but is adapted to the output available from the models used in ESPON 2.1.1.

The following four indicators have been selected on the basis of the relevance and the availability of endogenous model results (for retrospective or prospective scenarios, respectively):

- *Mass*: population density in the horizon year (2001 or 2021);
- *Competitiveness*: gross domestic product (GDP) per capita in the horizon year (2001 or 2021);
- *Connectivity*: multimodal accessibility in the horizon year (2001 or 2021);
- *Development trend*: difference in GDP/capita between the horizon year and the base year (2001 and 1991 or 2021 and 2001).

Each indicator is measured at the NUTS-3 spatial level and computed from the results of the SASI model. The values of each indicator in each scenario are normalized by their respective maximum values in the reference scenario (A0 for the retrospective analysis, and 00 for the prospective analysis) and multiplied by 100. Hence, for the relevant reference scenario, each indicator obtains positive values for each NUTS-3 area with a maximum value of 100. For the other scenarios the maximum values can deviate from 100.

For each scenario a composite indicator of development potential is computed:

- *Development potential*: geometric average of the values of the four indicators (mass, competitiveness, connectivity, and development trend).

The development potential is calculated for each of the 1321 NUTS-3 areas. Absolute and relative differences in development potential between any policy scenario and the relevant reference scenario are used to illustrate the effects of specific policies.

Development Potential: Impacts of Policy Scenarios

In this section, the impacts of investment and pricing policies on the development potential will be reported. The spatial patterns of these impacts are displayed in map form and shares of the total development potential located in the European core region and in categories of an urban/rural typology are calculated. Some comparisons with model results in terms of relative welfare or GDP/capita are outlined.

Impacts on Total Development Potential and Total GDP

The total impacts of the six policy scenarios on GDP and development potential based on the results of the SASI model are shown in Table 9.4.

Table 9.4 Impacts of policy scenarios on total GDP and total development potential*

Indicator	Scenario					
	A3	B1	B2	C3	D1	D2
Total GDP	+0.63	+1.60	+2.62	–1.42	+0.14	+1.16
Total development potential	+1.68	+2.88	+5.13	–2.59	+0.20	+2.45

* SASI model: percentage change as compared with the relevant reference scenario.

The efficiency results for the SASI model in Table 9.2 are repeated for comparison. The total impacts on GDP are small for all scenarios (between –1.5 per cent and +2.6 per cent). All policies lead to larger impacts on total development potential (sum of development potential over all NUTS-3 regions) than on total GDP. The impacts range between –2.6 per cent and +5.1 per cent. The first combined scenario (D1) leaves the total development potential and the total GDP almost unaffected as compared with the reference scenario. The second combined scenario (D2), however, results in positive impacts on both total development potential (+2.5 per cent) and total GDP (+1.2 per cent).

Spatial Distribution of Absolute Impacts on Development Potential

The resulting spatial patterns of the absolute impacts of the prospective policy scenarios on development potential are shown in Figures 9.1 to 9.5.

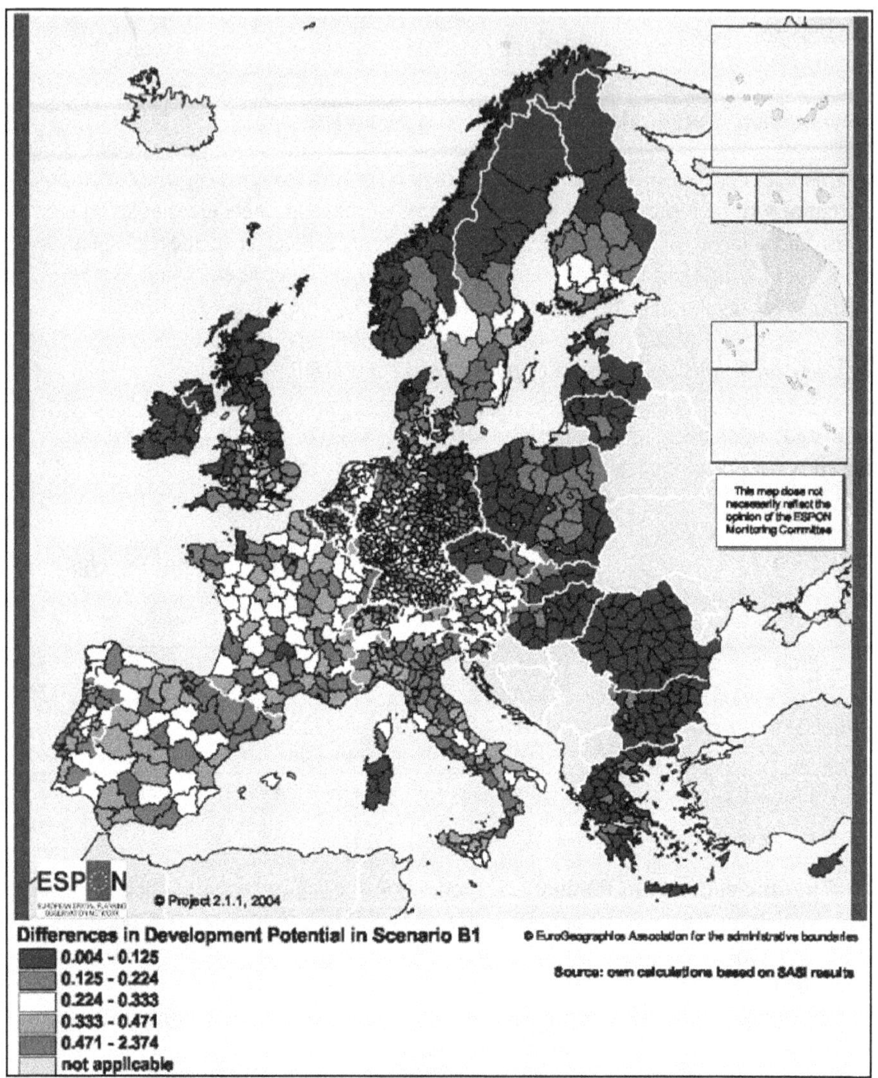

Figure 9.1 Differences in development potential (computed as geometric average) between scenario B1 and the reference scenario 00

Source: Bröcker et al. (2005).

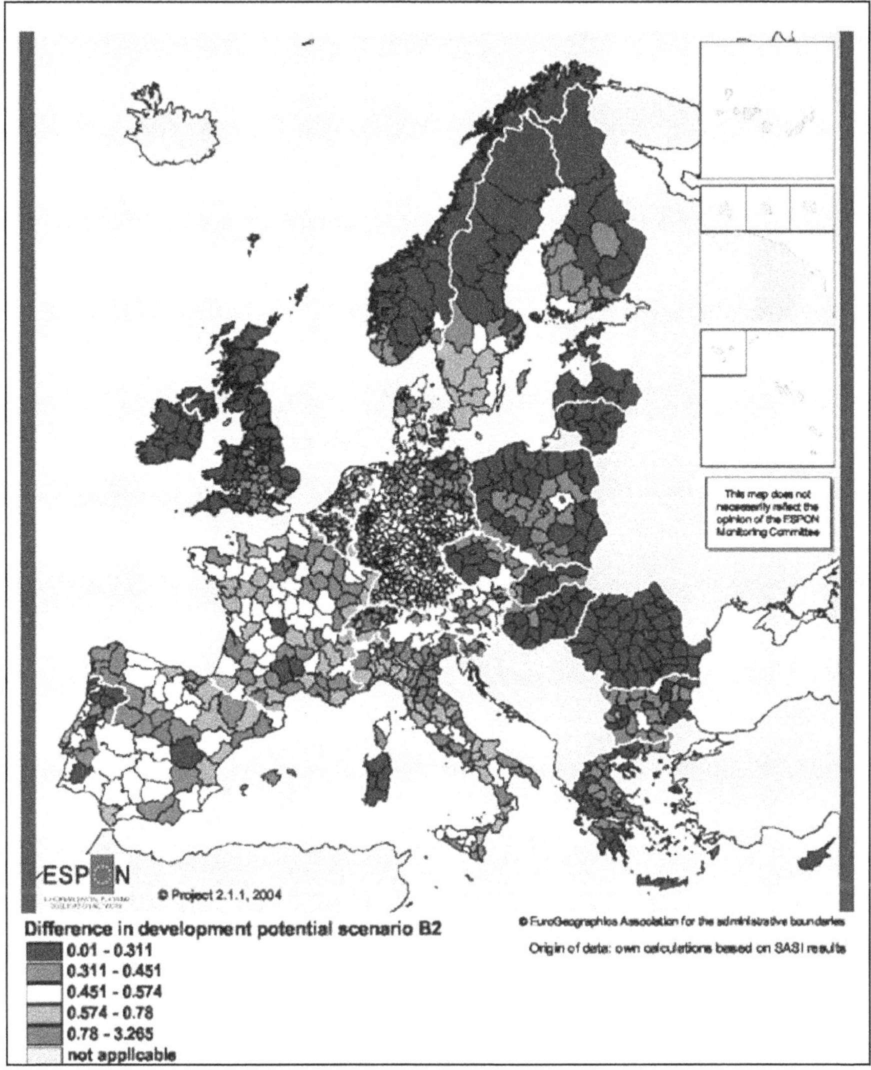

Figure 9.2 Differences in development potential (computed as geometric average) between scenario B2 and the reference scenario 00

Source: Bröcker et al. (2005).

According to Figures 9.1 and 9.2 transport investments have considerable positive effects on the development potential of many regions outside the 'pentagon' (the polygon defined by the metropolises of London, Paris, Milan, Munich and Hamburg). Large positive impacts are observed in the north-eastern part of Spain and along the coastal region to Italy, in many Italian regions (particularly on the east coast), and in southern Scandinavia.

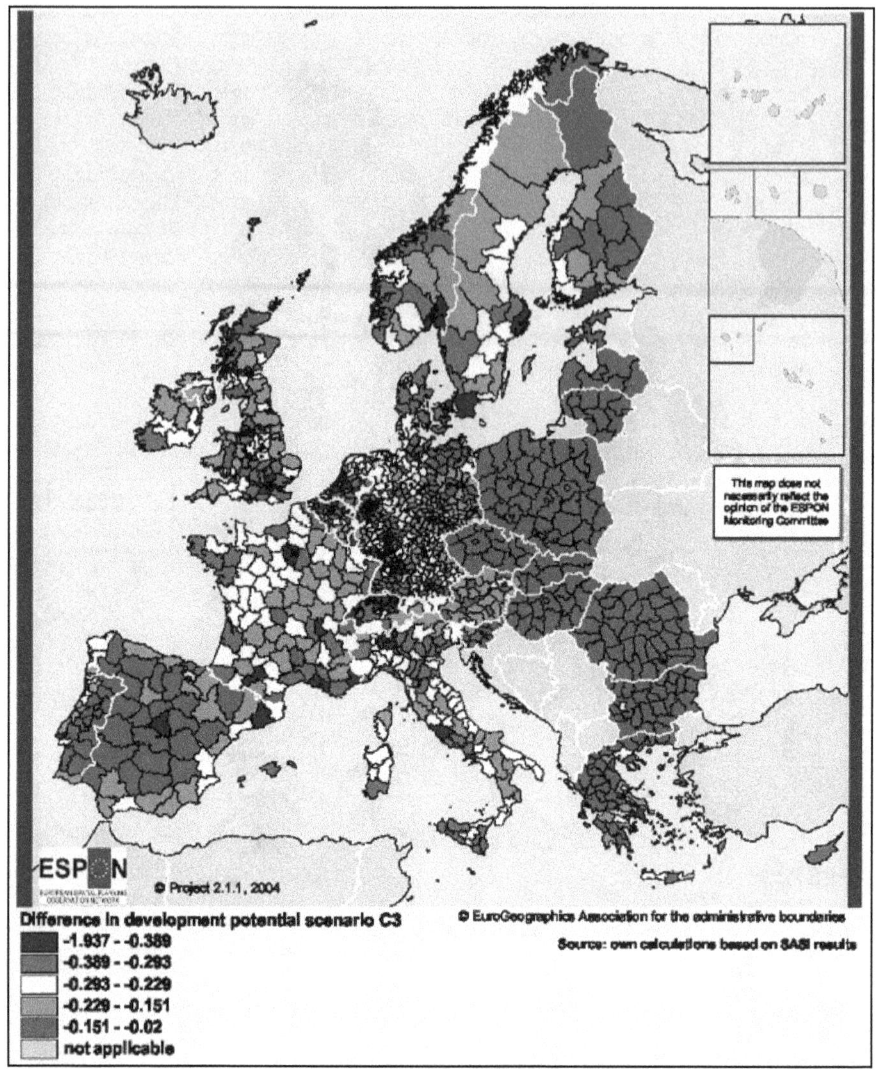

Figure 9.3 Differences in development potential (computed as geometric average) between scenario C3 and the reference scenario 00

Source: Bröcker et al. (2005).

The marginal cost scenario leads to the least negative absolute impacts on the development potential in the eastern part of EU27, in the Iberian Peninsula and in parts of Wales and Scotland (see Figure 9.3). The most negative impacts in absolute terms are found in a dispersed set of highly urbanized and mainly metropolitan regions, many of which are located in the European core region.

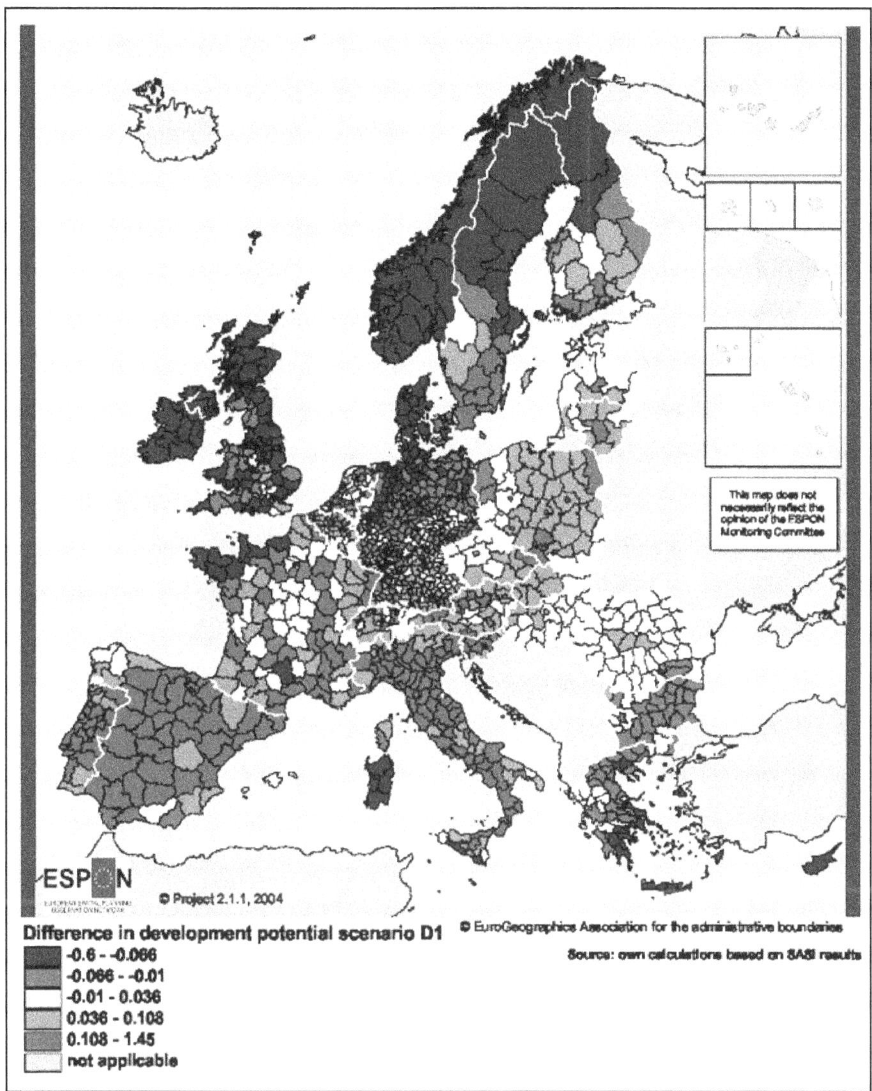

Figure 9.4 Differences in development potential (computed as geometric average) between scenario D1 and the reference scenario 00

Source: Bröcker et al. (2005).

Finally, the effects of the combined investment and marginal cost pricing scenarios (Figures 9.4 and 9.5) are similar to the impacts of investments but with relatively improved positions of regions in East Central Europe.

It can be concluded that marginal cost pricing improves the relative position of some peripheral regions and most accession countries in terms of the absolute level of their development potential. The transport investments improve the relative position of semi-central regions, mainly outside the 'pentagon'. The combined

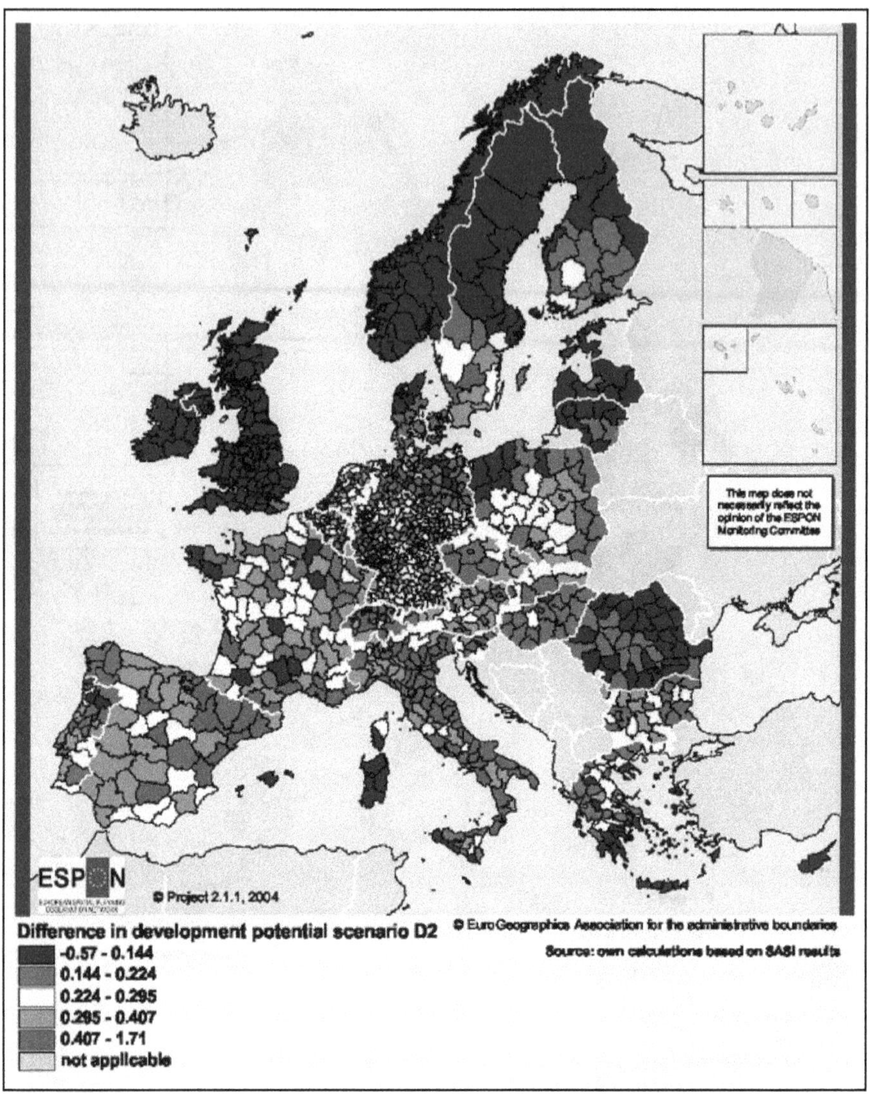

Figure 9.5 Differences in development potential (computed as geometric average) between scenario D2 and the reference scenario 00

Source: Bröcker et al. (2005).

investment and marginal cost pricing scenarios mainly extend the regions with improved relative position from transport investments to areas in the eastern part of the 'pentagon' and the East Central European regions outside the 'pentagon'. The results indicate that transport policy (investments and/or pricing) can potentially be used to encourage various forms of polycentric development.

Table 9.5 summarizes for eight different scenarios the share of the total development potential located within the 'pentagon' and the share located in regions

Table 9.5 Share of total development potential located in certain region types in reference and policy scenarios*

Region	Scenarios							
	A0	A3	00	B1	B2	C3	D1	D2
Pentagon	57.41	57.37	57.14	56.98	56.83	57.27	57.10	56.94
Low rurality	64.15	64.11	63.87	63.83	63.69	63.91	63.86	63.72
Medium rurality	4.70	4.71	4.77	4.78	4.80	4.76	4.77	4.80
High rurality	31.15	31.18	31.36	31.39	31.50	31.33	31.36	31.48

* SASI model: percentage of total development potential.

with various degrees of rurality. The typology of relative rurality has been defined by ESPON 1.1.2. The table mainly shows a high stability in these aggregate shares. There is a weak decentralization trend (reduced share for the 'pentagon' and for urban regions), which is reinforced by transportation investments and slightly counteracted by marginal cost pricing.

Spatial Distribution of Relative Impacts on Development Potential

When analysing the impacts on development potential in relative terms (percentage change of zonal values in relation to the relevant reference scenario), a quite different picture emerges. The impacts of the investment scenarios are more decentralized, occurring in particular in Southern and Eastern Europe.

The marginal cost pricing leads to the least negative impacts in East Europe and in the eastern part of Central Europe. The combined scenarios lead to the worst outcome for many of the 'pentagon' regions and for the UK, Ireland and northern Scandinavia, while most regions in the southern and eastern part of Europe experience highly positive relative impacts.

Comparisons of the Spatial Distribution of the Relative Impacts on Welfare and Income

The impacts on regional welfare (CGEurope) and GDP/capita (SASI) resulting from the models are displayed in relative terms in the final report of ESPON project 2.1.1. The results are structurally similar to the spatial distributions of the relative impacts on development potential. This is especially true for the SASI results, on which the analysis of relative impacts on the development potential is based. The observation is not surprising since two of the indicators included in the development potential are based on GDP/capita. Population density is very stable and the main feature of SASI is a linkage between accessibility and GDP/capita.

Analysis of Potential Goal Conflicts between Efficiency and Cohesion

Efficiency and Cohesion

Two of the underlying objectives of the European Spatial Development Perspective (ESDP) are:

- a more balanced competitiveness of the European territory; and
- economic and social cohesion.

The aim for a polycentric and balanced spatial development is related to both efficiency and equity criteria. It can be seen as an ambition to reconcile potential conflicts between European competitiveness and territorial cohesion.

Method

Each policy scenario results in forecasts of socio-economic development measured in, for example, GDP, equivalent variation, or development potential. These results may be aggregated to European totals (or averages) reflecting efficiency (see Tables 9.2 and 9.4). The equity or disparity can be measured in terms of any type of cohesion or inequity indicator. The analysis focuses on the following results based on the SASI model:

Efficiency indicators:

1. Total GDP in the horizon year (2001 or 2021);
2. Total development potential in the horizon year (2001 or 2021).

Inequity indicators:

1. Ratio between the arithmetic and geometric means (A/G) for GDP/capita in the horizon year (2001 or 2021) (measures relative differences);
2. Difference between maximum and minimum zonal GDP/capita in the horizon year (2001 or 2021) (measures absolute differences);
3. Ratio between the arithmetic and geometric means (A/G) for development potential in the horizon year (2001 or 2021) (measures relative differences);
4. Difference between maximum and minimum zonal development potential in the horizon year (2001 or 2021) (measures absolute differences).

The analysis of trade-offs can be performed for any regional subdivision of EU27+2 as a starting point for developing typologies based on the existence and character of goal conflicts.

While the analysis of the impacts of transport and TEN policies on polycentric development in the previous section focuses on the balancing of spatial structures and/or urban systems, the purpose of this section is to outline explicit analyses of potential trade-offs between the aims for efficiency (competitiveness) and equity (cohesion). It should be mentioned, however, that the indicators of absolute inequity are very simple and may give an extreme evaluation of disparities.

Results

The results for scenarios A3, B1, B2, C3, D1 and D2 are reported in Table 9.6 in terms of the percentage difference from the corresponding indicator value for the relevant reference scenario (that is, A0 for the retrospective scenario A3, and 00 for the five prospective scenarios). The results reported in Table 9.4 are repeated for convenience.

Table 9.6 indicates that there are trade-offs between efficiency and the indicator of *absolute inequity* (max-min). Transport investments increase both efficiency and inequity in comparison with the reference scenarios. Marginal cost pricing decreases both efficiency and inequity. The combined policies behave like the transport investment scenarios but the conflicts are milder. The percentage impact of policy scenarios on efficiency is, in most cases, about the same size as the percentage impact on absolute inequity.

If, instead, the policies are evaluated by the indicator of *relative inequity* (ratio between arithmetic and geometric means) the results become different. With only a few exceptions all policy scenarios lead to higher degrees of cohesion (diminishing inequity) than the corresponding reference scenarios. The main exception is the impact of marginal cost pricing (C3) on the distribution of development potential. Only when considering the impact of marginal cost pricing on GDP and GDP/capita is there a non-marginal goal conflict between efficiency and cohesion if inequity is measured in a relative sense.

Two aspects mentioned earlier should be kept in mind when judging these results. First, the SASI model is basically redistributive (although it is applied here in a generative mode) which may lead to underestimation of the impacts on efficiency. Secondly, the funding of transport investments and the utilization of income from pricing policies are not taken into account in the models. This tends to overestimate the impacts on efficiency.

Table 9.6 Indicators of efficiency and equity in the horizon year for six scenarios in comparison with the relevant reference scenario*

Indicator	Scenario					
	A3	B1	B2	C3	D1	D2
Total GDP (I)	+0.63	+1.60	+2.62	–1.42	+0.14	+1.16
Total development potential (II)	+1.68	+2.88	+5.13	–2.59	+0.20	+2.45
Inequity (A/G) GDP/capita (1.)	+0.03	+0.02	–0.49	–0.17	–0.18	–0.70
Inequity (Max-Min) GDP/capita (2.)	+0.77	+1.98	+3.89	–1.21	+0.74	+2.60
Inequity (A/G) Development potential (3.)	–0.30	–0.66	–2.27	+0.47	–0.32	–1.98
Inequity (Max-Min) Development potential (4.)	+1.74	+3.45	+4.71	–2.89	+0.47	+1.78

* SASI model: percentage change from the relevant reference scenario.

Conclusions

Overall, the results support the hypotheses outlined earlier. The magnitude of impacts of transport investments on total GDP found in this project (0.2–2.6 per cent) is somewhat high compared with the experience of earlier studies. However, the investment scenarios are comprehensive and funding of these investments is not represented in the models.

We have focused on the spatial distribution of absolute changes in development potential between prospective policy scenarios and the reference scenario. The spatial patterns become different if the maps are displaying relative differences. A small absolute change in development potential may in some regions with low development potential correspond to a large relative change. A proper choice of mode of display has to be made with regard to the policy context concerned. Impacts of transport policies in terms of relative differences in GDP/capita or welfare are reported for the SASI and CGEurope models in Bröcker et al. (2005).

Transport policy can assist in the spatial redistribution of economic development along different lines depending on the specific mix of policy instruments (for example, investment and pricing). In this sense, transport policy can influence polycentric development.

The aim for cohesion has to be qualified. Transport policies may very well increase cohesion in a relative sense while simultaneously reducing cohesion in an absolute sense. There is a clear goal conflict between the ambition of increasing total income and the objective of reducing regional income disparities in an absolute sense (at least if measured by the range of income variation). The same holds true for development potential. If cohesion is measured in a relative sense, there are hardly any clear conflicts between the goals of efficiency and equity. In our six scenarios and with our two performance measures we only found one exception of some substance.

In any policy context the actual meaning of equity or cohesion needs to be clarified so that relevant indicators can be selected and a proper analysis of goal conflicts and trade-offs can be conducted. Absolute and relative cohesion indicators can lead to very different policy assessments. Many popular equity indicators do measure cohesion in a relative sense.

References

Azevedo, R. and Cichowlaz, P. (2003), 'Study on the Construction of a Polycentric and Balanced Development Model for the European Territory'. Paper presented at the Conference of Peripheral Maritime Regions of Europe (CPMR), Peripheries Forward Studies Unit.

Biehl, D. (1991), 'The Role of Infrastructure in Regional Development', in R.W. Vickerman (ed.), *Infrastructure and Regional Development* (London: Pion), pp. 9–35.

Bröcker, J. (2002), 'Spatial Effects of European Transport Policy: A CGE Approach', in J.G. Hewings, M. Sonis and D.E. Boyce (eds), *Trade, Networks and Hierarchies:*

Integrated Approaches to Modelling Regional and Interregional Economics (Berlin: Springer), pp. 11–28.

Bröcker, J., Capello, R., Lundqvist, L., Pütz, T., Rouwendal, J., Schneekloth, N., Spairani, A., Spangenberg, M., Spiekermann, K., Vickerman, R. and Wegener, M. (2003), 'Territorial Impact of EU Transport and TEN Policies'. Third Interim Report of Action 2.1.1 of the European Spatial Planning Observation Network – ESPON 2006, Institute of Regional Research, Kiel.

Bröcker, J., Capello, R., Lundqvist, L., Meyer, R., Rouwendal, J., Schneekloth, N., Spairani, A., Spangenberg, M., Spiekermann, K., van Vuuren, D., Vickerman, R. and Wegener, M. (2005), 'Territorial Impact of EU Transport and TEN Policies'. Final Report of Action 2.1.1 of the European Spatial Planning Observation Network – ESPON 2006, Institute of Regional Research, Kiel.

ESDP (1999), 'European Spatial Development Perspective'. Document adopted by the Informal Council of EU Ministers responsible for Spatial Planning, Potsdam, 10–11 May.

Wegener, M. and Bökemann, D. (1998), 'The SASI Model: Model Structure'. SASI Deliverable D8, Berichte aus dem Institut für Raumplanung 40, Institut für Raumplanung, Universität Dortmund, Dortmund.

Chapter 10

The Network versus the Corridor Concept in Transportation Planning

Christos Dionelis, Maria Giaoutzi and John Mourmouris

Introduction

The geopolitical and spatial demarcation of the European Union went through a new phase with the expansion of May 2004. This has added vast new territories to the heart of Central Europe (CEECs), as well as important Mediterranean centres such as Malta and Cyprus.

In this context the fundamental objective for transport has been to provide efficient links between the various regions, unifying these far-flung regions into a political and economic union.

Due to the existing segmentation of the European transport systems and markets and the problems these may create, transportation planning is bound to attract, in the near future, significant institutional and political attention.

Recent developments in transport research and policy making call for sound tools and operational concepts in order to cope with the complexity of the existing problems in infrastructure patterns. The competition and cooperation between the various modes of transport, the conflicting interests and the allocation of limited funds for transport infrastructure developments are issues requiring rational decision making.

The present chapter focuses on the evolving nature of such problems, dealing with the shift in emphasis, from the *network* to the *corridor* concept, in the Common Transport Policy (CTP), which will enable a broader range of more specialized attributes to be incorporated into the analysis and evaluation phase of the planning process.

The chapter is organized as follows. The first section presents the framework of the CTP, its objectives and some of its main achievements. The emphasis is placed on the Trans-European Transport Network (TEN-T), the Pan-European Transport Corridors and the TINA Network in order to follow the developments of the context that has created the need for the evolution of the study concepts.

The next section then presents a set of indicators used in practice for the study and evaluation of a transport corridor or network. This is followed by an examination of the appraisal framework for the evaluation of transport corridors and networks. Then, the next section elaborates on the process of the recent TEN-T revisions and comments on the factors that have led to such an approach. Finally, some conclusions are drawn, based on the previous analysis.

The Common Transport Policy

Transport, both as an industry and as a policy area, has played an important role in the post-war European integration process. One of the Union's key commitments was to develop an integrated transport policy, expressed by appropriate institutional and legislative provisions that would contribute towards the achievement of what has been expressed as the goal of 'sustainable mobility'. European goods and people are more mobile than ever before and the Common Transport Policy (CTP) continues to be a major vehicle for an effective investment policy.

In accordance with Article 154 of the European Community Treaty (Treaty of Rome), the Union's goal is to establish and develop Trans-European Networks (TENs), in the sectors of transport, telecommunications and energy infrastructures. This will contribute to both the establishment of the internal market and economic and social cohesion.

Trans-European Transport Network: Network Approach

The efficiency of *transport systems* will play a critical role in safeguarding that the present expansion will succeed in integrating the new Member States, instead of turning them into burdensome territorial appendages.

The goal of the Union's transport policy is to ensure *sustainable mobility* for people and goods and create a coherent global transport system, which will produce the best possible returns, not only in terms of investment but also in securing safety and other environmental and social priorities. The overall objective of the Sustainable Mobility Policy is to reconcile economic growth and social demands for mobility with environmental impact and other costs of traffic movements. This policy focuses on transport systems and patterns and provides a means to meet economic, environmental and social needs, efficiently and equitably, while minimizing unnecessary adverse effects and their related costs, over relevant space and time scales (European Commission 2006a).

The European Union's search for a transport system which could contribute to the attainment of its major Community objectives, such as the smooth functioning of the internal market and the strengthening of economic and social cohesion, has resulted in the establishment of a *transport network* by means of a major infrastructure programme for a Trans-European Network in the field of transport (TEN-T) that actually comprises both infrastructure and traffic management systems.

The Maastricht Treaty also included the concept of the *trans-European network (TEN)*, which has enabled the Community to come up with a plan for transport infrastructure at the European level with the help of Community funding. The result of this process was Decision No. 1692/96/EC of the European Parliament and Council on the Community guidelines for the development of the trans-European transport network (TEN-T).

The significance of these guidelines lay in the fact that this was the first attempt to describe the vision of how an integrated network in the Union's territory should develop towards the horizon of 2010. A number of financial instruments have been set up at Community level and by the European Investment Bank in order to implement this policy, based on a specific budget for the funding of the TEN-T.

Throughout the years the TEN-T has been gradually established as one of the principal vehicles for the achievement of growth, competitiveness and employment, thus strengthening economic and social cohesion by reducing disparities between regions and linking peripheral regions with the central regions of the Union (PLANCO Consulting GmbH 2003).

Pan-European Transport Corridors: Corridors Approach

The idea of *Pan-European Transport Corridors* was first presented in 1989 to the Strasbourg summit. It was obvious that an effective transport policy could not just be limited to the construction of infrastructure on the TEN-T. The idea of corridors traversing Europe evolved in the 1980s, when it was already being acknowledged in Western Europe that the new dimensions of international traffic required planning in terms of transport corridors. The multimodal corridors conceived, at that time, can be regarded as a starting point for the idea of the TEN-T.

Later on, the question of East-West links in Europe gained significant momentum in the early 1990s, with the changes in the political and economic system of the former communist bloc. In 1994, the European Union (EU) took the decision to identify priority transport links to and within the Central and Eastern European countries (CEECs) which were applying at that time for EU membership. During the second Pan-European Transport Conference in Crete in 1994 and the third in Helsinki in 1997, the UN-ECE,[1] the ECMT,[2] the European Commission, and all European States supported the development of ten Pan-European Transport Corridors[3] and four Pan-European Transport Areas (PETrAs). The whole exercise was considered as an effort of the EU towards extending the trans-European network and improving East-West ties.

The central position of the CEECs, between the Western European countries and the Commonwealth of Independent States and between Nordic and Balkan countries, generated the necessity to create and exploit effective transport routes, adapted to EU standards. The aim of these countries to strengthen the links with the EU also pushed the development of this set of Corridors, combining infrastructure and services. This shows the sense of the exercise, which was simply to ensure the application of the EU standards for transport (technical and legal if possible) to critical transport links with new attractive markets. After the effective establishment and operation of the corridors, all the involved parties (the EU, the third countries, and the carriers) would benefit from this exercise. The Transport Corridors include cross-border road and rail traffic routes between the EU-15 and the Central and Eastern European countries, as well as airport, sea and river ports along the routes serving as intermodal nodes.

1 United Nations – Economic Commission for Europe.

2 European Conference of Ministers of Transport.

3 Also known as the 'Crete' or 'Helsinki Corridors'.

For all the Corridors and Areas, a Memorandum of Understanding has been signed between the Ministers of Transport of the respective governments and the European Commission. In addition, steering committees have been established for each Corridor and Area in order to monitor and promote the progress and performance and to coordinate required actions. Chairs and secretariats have been selected for each Corridor/Area to care for administrative matters.

The concept of the Pan-European Transport Corridors and Transport Areas has generally been accepted as an emerging priority regarding transport infrastructure development all over Europe. However, following the enlargement of the EU in 2004 and 2007, most of the Corridors are now part of the TEN-T network.

The establishment of the *Corridors* and *Areas* was part of the actions which formed *the Pan-European Transport Network Partnership* endorsed at the Third Pan-European Transport Conference at Helsinki in June 1997. This Partnership aimed to accomplish an infrastructure set-up all over the European Continent, which would allow transport services to follow the principle of sustainable mobility as set out in the CTP of the EU.[4]

TINA Network: Stepping from Corridors to a Network

The extension of the existing TEN-T in the accession countries was based on the results of the *Transport Infrastructure Needs Assessment* (TINA) process. TINA – led and funded by the EU – developed a multimodal transport network, which served as the main reference framework for the extension of the TEN-T in the enlarged EU (TINA 1999).

TINA was set up by the European Commission in order to supervise and coordinate the development of an integrated transport network in the countries that had applied for EU membership and would ensure coherence with the TEN-T within the enlarged EU. The idea was to upgrade the existing infrastructure or build new, in order to create a coherent network based on the existing transport corridors – which would be used as the '*backbone network*' – and thus maximize the potential of European trade. In this context, TINA followed the normal process of creating a network: first by defining a *backbone network* through a political process (common agreements at the highest level, in the two Pan-European Transport Conferences); and, second, by placing additional components (transport links) that transformed the corridors into a network.

The TINA Network was built on the following assumptions:

- The network should be in line with the criteria laid by the EU Guidelines for the Development of the TENs (Council Decision 1692/96/EC), according to the objectives described in Article 154 of the Treaty. In this respect, the TINA Network, like the TEN-T, should comply with provisions like promotion of

4 Third Pan-European Transport Conference, Helsinki, 23–25 June 1997: Helsinki Declaration: 'Towards a European wide transport policy, a set of common principles', 25 June 1997.

sustainable mobility of persons and goods, safety, environmental protection, optimal use of existing infrastructure and capacities, and so on.

- The technical standards of the future infrastructure should ensure consistency between the capacity of network components and their expected traffic. To achieve this, it was accepted that these standards should be in line with the recommendations of the UN/ECE Working Party on Transport Trends and Economics (WP.5) on the definition of transport infrastructure capacities (Trans/WP.5/R.60).[5]
- The time horizon for the realization of the network should be 2015 (in line with the provisions of Decision 1692/96/EC for the TEN guidelines).
- The costs of the network should be consistent with realistic forecasts of financial resources, so that the average costs would not exceed 1.5 per cent of each country's annual GDP over the period up to 2015. This assumption came from the fact that, in the 1990s, the EU Member States invested between slightly under 1 per cent and up to 2 per cent of GDP in Union-relevant infrastructure. This figure did not, however, concern Union-relevant infrastructure alone, but also infrastructure of solely national importance. The discussions also confirmed that the acceding countries needed to do somewhat more. In the EU most of the investments had already been made, while in the acceding countries major upgrading was required over the coming years. On the other hand, an overly high share of GDP would probably be considered unrealistic, since infrastructure investments were only one of the many investments the acceding countries had to undertake. So, TINA accepted, as an indicator for a realistic ceiling of planned infrastructure investments, that their cost should not, on average, exceed 1.5 per cent of the GDP in the coming years.

Approaching Transport Networks and Corridors

The European Commission has promoted the idea of the TEN-T, as an integrated transport system to cope with the needs of the internal market. The term *integrated transport network* was meant for a system incorporating compatible physical infrastructure and operational standards as well as similar institutional-legislative frameworks. In other words, this system would combine the planning, provision and operation of different modes of transport in such a way that journeys could be made as efficiently as possible.[6]

An integrated transport system implies:

- provision of maximum safety for the citizens using and operating the system;
- an environmentally sustainable transport system; and
- a strategic integration of transport and land use planning.

5 United Nations Economic Commission for Europe (see http://www.unece.org/trans/main/wp5/ /wp5.html).

6 For the case of the EU, the aim for sustainability has imposed the provision that this system should also minimize the use of the private car.

The *network concept* is used as the main tool for the spatial representation of a transport system. The integration of a network is a prerequisite for reaching the goals of the creation of the internal market and territorial cohesion by bringing together the peripheral, island or isolated areas with the central regions. In this context, a modern, interconnected and interoperable network enables the enhancement of the competitiveness of trade and the economy as a whole. At the same time, an efficient network management, based on the necessary infrastructure and the appropriate regulatory framework, would enhance the prospects of the internal market and the territorial cohesion of the Union.

The links and nodes are the elements of a network's physical infrastructure. The transport corridor concept is put into action when some appropriate routeings are selected, isolated from the rest of the physical infrastructure, in order to ensure certain efficient transport connections.

Operationalizing the Concept of Transport Corridors

The *concept of a transport corridor* in practice can become operational through some of its main characteristics:

- Canalization of transport;
- Concentration of transport volumes;
- '*Intermodality*', which refers to the combination of several transport modes (road, rail, sea and air) in a way that allows the use of at least two different transport modes for at least one single origin-destination trip. The concept presupposes the existence of a certain transport node, which allows the transfer between at least two different transport modes.
- '*Multimodality*', which refers to a characteristic of a transport network in which at least two modes compete/cooperate for taking trips in the *same corridor*. It implies the competition/cooperation between transport modes along the same corridor with the scope to achieve an optimal modal split. In recent years, the transport policy of the European Commission has been orientated towards the strengthening of the market position of modes other than road (railways, maritime, inland waterways). There are two reasons for this: the first concerns the practical problems of lack of capacity and bottlenecks, which many times can be overcome only with multimodal solutions; and the second concerns the better efficiency records of the other modes in terms of energy consumption and environmental impacts. The European programmes *Marco Polo* and *NAIADES* have been designed to promote innovative solutions in this field. The concept of the 'Motorways of the Sea' has been defined as a priority TEN-T project to promote maritime transport (European Commission 2006b).[7]
- '*Interoperability*' which refers to the quality of interaction of two or more transport systems. Better integration of transport modes into efficient

7 See also the Marco Polo Programme (http://ec.europa.eu/transport/marcopolo/index_en.htm) and Motorways of the Sea (http://ec.europa.eu/transport/intermodality/motorways_sea/index_en.htm).

logistics chains can be achieved through technical harmonization and improved interoperability across systems. It is the ability of transport systems to offer harmonized interfaces and an acceptable level of service by intermodal transport for the route, node or corridor under consideration and/or the use of the same mode services provided by different operations/ actors. *Interoperability* mainly deals with organizational issues (especially for terminals and transfer points), as well as the assessment and reduction of any kind of barriers (for example, institutional, financial, physical, technical, cultural, political and so on).

The *network approach*, in the context of transportation planning is a long-term approach, which aims to define a network consisting of both the physical infrastructure and the appropriate institutional rules in order to meet efficiently the traffic and market needs.

The *corridor-based approach*, on the other hand, is a short and medium-term approach, mainly market-oriented. Its aim is to satisfy the current traffic needs by improving the global services and using the upgraded infrastructure.

A study of a transport network should be enriched by a deep knowledge of the socio-economic profiles of the regions concerned. A main example that can be used is the general concept of Trans-European networks (TENs), which was developed as an instrument for achieving market integration, as well as economic and social cohesion. A network like an integrated transport system (TEN-T) may ensure that peripheral, remote or less developed regions can be fully absorbed into European economic life.

The *corridor concept* on the contrary, comes out of the necessity to serve specific transport needs, acting more as a tool than as a concept, without concentrating on the regional characteristics or level of performance of the regions concerned. Typical problems where the corridor concept can be used are optimization problems of carrying goods or people from point A to point B (TINA 1999). A knowledge of the socio-economic profile of regions or locations crossed by a corridor is, of course, important for the accuracy of any analysis results; however, the region's profile is of lesser importance for a corridor study than the corridor's own socio-economic characteristics.

Measuring the Performance of a Corridor

Any transport corridor (like any link of a network) is characterized by its technical characteristics[8], as well as its performance, incorporating traffic flows, capacity, safety, comfort, operational costs and so on. Obviously, there are common standards to measure the technical characteristics of transport networks and corridors, which incorporate both sections/links and nodes.

8 These technical characteristics can be the length, the geometrical standards, and other standards of the infrastructure such as the gauge of a railway line, the depth of an inland waterway and so on.

Apart from the technical characteristics of the infrastructure, in order to *measure the performance of a corridor*, several indicators have been used such as:

- transport volumes per time unit
- transport volume per distance
- modal split along the total (multimodal) corridor
- capacity and capacity used
- bottlenecks
- border delays
- transport prices per mode
- direct transport costs
- accident rates and
- accident costs.

Measuring the Performance of a Transport Network

In order to *measure the performance of a transport network*, several other indicators can also be used which describe the socio-economic and environmental considerations of the region.

The most important of these indicators, refer to:

- physical indicators, such as:
  - density, as a ratio of the network length to the region's area or population
- macro-economic indicators, such as:
  - gross domestic product (GDP)
  - GDP per capita
  - national income
  - consumption
  - investments
- socio-economic indicators, such as:
  - population
  - population density
  - workforce
  - employment, (jobs per capita)
  - unemployment
- environmental indicators, such as:
  - CO_2 per emitter
  - NO_x per emitter
  - CO per emitter
  - VOC per emitter
  - noise.

An Indicative Case

To give an example of how the above indicators can be used in the case of 'comparing' different transport networks, the case of the TINA Network versus the TEN-T is presented below:

The TINA Network should be consistent with the TEN-T. To compare the two networks, certain indicators were used which set out some interesting features that may characterize a transport network:

- The ratio of network length to surface area is an *indicator of the density* of the network.
- The ratio of network length to population gives an indication of the relative *availability of infrastructure* for the population.
- The ratios of construction cost to GDP, as well as of construction cost to population, are partial indicators of the *prospects for financing* the network.
- The ratio of construction cost to per capita GDP shows the *ability* of the country *to finance* the proposed projects.

Any assessment of the overall prospects for financing the network should take into account the balance of all the indicators related to GDP (TINA 1999).

Table 10.A1 in the Annex shows the comparison between the TEN-T and TINA Network, on the basis of the indicators used. This table gives some useful indices for the TINA Network of accession countries versus the TEN-T for the EU15, and sets out some interesting features of the TINA Network in comparison with the Union's TEN-T.

The ratio of Network length to surface area is an indicator of the density of the Network; this is generally significantly lower in the acceding countries than inside the EU, although the density of the Network in some TINA countries is very close to that of the TEN-T.

The ratio of Network length to population gives an indication of the relative availability of infrastructure for the population. Some states (mainly the Baltic states and Cyprus) are well served, compared with both the other candidates and with the Union, where the average is of a similar order to that of the candidates.

The ratios of construction cost to GDP and of construction cost to population are partial indicators of the prospects for financing the Network. Clearly, there will, in general, be fewer problems in financing the Network where these ratios are relatively low. This comment should, however, be qualified by an examination of the ratio of construction cost to per capita GDP. This will show, for example, that, although Slovenia has a very high ratio of construction cost to population, this is in part compensated by its relatively high level of per capita GDP, resulting in a correspondingly greater ability of the country to finance the proposed projects. The construction cost per GDP per capita expressed as a percentage of the population (last column of the table) means, for example, that in Latvia 3.2 per cent of the population should contribute, up to 2015, the equivalent of their 1995 GDP for the construction of the Network; the respective percentage in Hungary is only 1.7 per cent of the population, and in Cyprus 0.9 per cent.

Any assessment of the overall prospects for financing the network must therefore take into account the balance of all three above-mentioned indicators.

The Appraisal Framework

The purpose of this section is to provide some examples of significant methodological variations in the appraisal framework of transport investments (*projects*) when referring to transport corridors or networks.

Transport infrastructure investments may have significant impacts for the economy, the environment, and potentially for other aspects in the regions concerned. The principal aim of an appraisal framework is first to identify and assess the most important of these impacts, and second to report on the financial implications or financial non-viability of a project from the viewpoint of both the transport operators and the national governments.

In the global scheme of an appraisal process, there are four main groups of impacts likely to be considered, namely:

- effects on transport efficiency and safety;
- financial implications for transport providers;
- environmental impacts; and
- policy impacts beyond the transport system.

In each case, the impacts are defined as the differences between particular indicators in the do-something scenario (with the project) and the do-nothing (or do-minimum) scenario (without the project).

Different analytical methods are required for each of the above groups of impacts. The indicators chosen to represent the project impacts may vary according to the requirements of the particular *funding bodies*. It should be underlined, however, that the various funding or financing institutions have their own detailed requirements to document their investment decision (EIB et al. 1999).

The first two groups of impacts are essential for the assessment of transport corridors. For the assessment of transport networks, all four categories of impacts should be considered. This statement is explained in the following paragraphs.

Concerning the question about the need for environmental impacts assessment, one should distinguish between the impacts assessment of a project, and the strategic environmental assessment, as defined by the European Commission (European Commission and DG TREN 2005).

As it has been stated in the European Commission's SEA (Strategic Environmental Assessment) Manual (European Commission and DG TREN 2005, p. 28) 'In general … the experience [of the pilot studies focusing at] at corridor level has shown that most countries' planning framework does not include the concept of transport corridors as formalized planning units with explicit decision-making processes. This effectively means that – in general there are no "transport corridor" plans or programmes (or indeed policies), and as such, no decision into which the "SEA of a corridor" can feed.'

Furthermore, a persistent focus on 'corridors', 'infrastructure' and 'projects' is likely to conflict with the overall effort of promoting a more comprehensive, integrated approach to transport – essential in the pursuit of sustainable solutions, as widely illustrated in EU transport policy initiatives. The added value of a Strategic

Environmental Assessment (SEA) lies first and foremost in its ability to raise strategic questions; thus, it can be implied that the transport network is an ideal field for the application of complete SEAs.

In the context of transport planning, the aspect of finding a hierarchy of strategic transport initiatives (policies, plans and programmes) which leads in a linear manner to projects requires in-depth discussion. However, when we refer to individual projects, it is essential to mention that the same set of indicators, and the same environmental impact analysis should be applied for all the projects, regardless of whether they are examined as parts of a corridor or a network.

Effects on Transport Efficiency and Safety

Some of the direct effects of network projects will be on transport users (people and freight) and transport providers. The cost and time expended in getting from place to place will be reduced, both for personal travel and freight movement. This is at the heart of the concept of transport corridor efficiency. As travel time and cost decrease, travel will become more attractive and an increase in the overall transport use is likely to result. For the purpose of economic appraisal, the costs and benefits of this transport growth are also included within the analysis of transport corridor efficiency (European Commission 2001).

Transport *efficiency* effects together with *safety* effects (reduction of accidents) are expected to be included within a *social cost-benefit analysis (CBA)*.

Financial Implications

The financial analysis, required to meet the second aim of the appraisal framework, excludes non-market impacts for which social values are adopted in the social CBA (for example, time and safety) and instead limits itself to:

- financial investment costs;
- financial infrastructure maintenance and operating costs;
- vehicle operating costs (VOCs) met by operators;
- infrastructure and service operator revenues.

The financial analysis is concerned with the impact of these items on transport operators, infrastructure providers and governments, in cash-flow terms.

These two groups of impacts are usually enough in the assessment of a transport corridor. However, when assessing a network, investment decisions go further than simple decisions of how we spend money in a financially efficient manner, and are also influenced by a common understanding of the long-term social consequences of alternative actions. In this respect, two more groups of effects should be examined.

Environmental Impacts

It is essential that the cost of environmental damage is included in the costs of any individual project. However, transport system changes, and the resulting changes in transport use, affect not only participants within the transport system itself, but also those who are exposed to the system or its emissions without being directly involved. Environmental impacts occur at a local or regional level: for example, changes in exposure to noise and vibration, or to airborne pollutants. In this context, it is important to fully respect the provisions for *strategic environmental assessments* according to the European *acquis*.

Policy Impacts

The next group of impacts concerns those affecting broader public policy beyond the transport system. Governments (central and regional) typically invest in transport not only because of the expected national gain in economic efficiency and mobility but also because of the positive socio-economic effects that investment is expected to have on other policy areas of interest.

Such areas might be:

- regional/local economic development policies;
- land use policies;
- national or EU policies relating to other objectives.

In any *CBA*, elements that can be measured in monetary terms are usually the base of the assessment. A *Financial Analysis* is usually required by any financing institution. In addition to this, especially when we deal with integrated networks, there should be also an analysis of environmental and broader policy effects to be brought together with the CBA in a coherent framework. Multi-Criteria Analysis is also a valuable tool for an efficient overall assessment.

In the whole appraisal process, the initial stages include the definition and initial screening of candidate projects for financing. In the case of a corridor, the selection of projects is a relatively easy process, since all candidate investments aim at completing the same transport route. On the other hand, in cases where strategic routeing options do not exist, the full appraisal of various alternatives should be undertaken so as to demonstrate that the selected investment strategy is better than other available alternatives.

Thus, in the case of a network, screening should be included to ensure that priority projects are adequately justified, and in order to identify their broad performance relative to the main indicators, including transport system efficiency and safety, environmental impacts and wider policy effects. The analysis should be capable of allowing for the existence of many investment alternatives and should facilitate prioritization and ranking.[9]

9 In reality, with the existing serious budget limitations, the prioritization process tends to be the most important part of the analysis.

The Revision of the TEN-T

The focus of this section is on the factors that have led the European Union to re-orient the framework of its transport planning from the network approach to the corridors concept through the recent revision of the TEN-T.[10]

The TEN-T guidelines[11] were not meant to be either rigid or eternal, and should be revised in the light of any new conditions, such as traffic forecasts and changing economic, political and financial circumstances. In this context, in December 2003, the Council reached political agreement on a common position concerning the proposal for a Decision of the European Parliament and of the Council amending Decision No. 1692/96/EC[12] on Community guidelines for the development of the Trans-European Transport Network.

It is very important that the rationale of the amendment promotes the Corridors concept, stressing that the funding and management of the TEN-T needs to be better focused and more concentrated in the future. This shift in focus of efforts in the TEN-T will lead to a shift in emphasis towards the corridors, particularly those that connect the new Member States.

As the need to develop a more coherent transport system becomes pressing, a more reliable approach to assess priorities is needed. Some useful lessons from the experience of the Pan-European corridors over the past few years will be presented (de Palacio 2003):

- The Corridor concept has provided a useful instrument for monitoring the development of main *intermodal transport links* where various countries are involved.
- Better coordination of investments, and selective concentration of Community interventions on the major axes of the enlarged EU is needed.
- The way the Pan-European Corridors are managed may provide useful directions to the monitoring of the main links of the TEN. Such practices could allow for the regular monitoring of progress and a more transparent discussion on priorities and objectives.
- New forms of financing – using public private partnerships (PPPs) – should be adopted, as serious limitations to public funds are foreseen. The corridors are an ideal field to develop such financing mechanisms.

There is a strong relation between the network or corridor concept and the process of project prioritization. The term '*projects of common interest*' was introduced in the TEN guidelines. However, there was strong criticism about the lack of coherence, this being a general characteristic of the proposals for the projects of common interest.

10 Decision 884/2004 of the European Parliament and the Council of 29 April 2004, on Community Guidelines for the development of the Trans-European Transport Network.

11 TEN-T Guidelines: Decision 1692/96/EC of the European Parliament and the Council on the guidelines for the development of the Trans-European Network.

12 In May 2001, The European Parliament and the Council adopted Decision No. 1346/2001/EC, which amends the TEN-T guidelines as regards seaports, inland ports and intermodal terminals.

In order to reach consensus at the highest level, 14 large development projects were given priority by the European Summit in December 1994 in Essen concerning rail (mainly), road and airport infrastructure. Two years later, all these intentions were summarized and developed into more detailed guidelines (TEN-T Guidelines 1996). Besides the 14 priority projects of Essen, the problem remained open, as not all the countries respected the spirit or the objectives of the guidelines for the development of the TEN-T, and many additional projects of mere national interest were proposed as high priority for the Union. Thus, instead of initiating a real network concept, as this was the Commission's intention, the countries continued to plan in an outdated way.

The Union should therefore either impose network analyses for any project's initiation, or – at least – draw up precise guidelines for the definition of major transport corridors and routes. The first implies serious changes in the classic planning process;[13] the latter implies the definition of main corridors; it also implies fixing priorities for certain projects, especially in order to eliminate bottlenecks and construct the missing links. As a major consequence of the poor network management, the saturation of certain major routes and the lack of satisfactory connections with the peripheral regions were directly caused by delays in implementing the infrastructure of the Network. In the future, when the consequences of the EU enlargement will be more visible for transport, it is expected that the traffic will increase tremendously.[14] If the measures described in the 2001 White Paper for Common Transport Policy ('Time to Decide') (European Commission 2001) are not implemented, then the consequences will be very negative: most of the new traffic will come onto the road network, leading to serious capacity problems (bottlenecks) and very negative impacts for the environment. The optimistic case is that the European Union will succeed, and the future network will be characterized by a new modal split, where alternative modes such as railways or inland waterways have an enhanced role in the transport system.

Regarding the dilemma of choosing the network or corridor approach for the TEN-T, it should be mentioned that, recently, after the adoption of the last revision of the TEN-T guidelines (Decision 884/2004/EC), the Commission turned in the direction of promoting Corridors, even if its terminology continues to refer to a 'network'. In 884/2004/EC, the TEN-T was expanded to cover the new Member States, but the list of the 30 priorities define more a set of basic routes (Corridors) than a real network.

One of the new features of the recent expression of European transport policy is the different relationship established between Community subsidy and the candidate projects candidate for funding. In the case of assistance granted from the Structural

13 In a network design and assessment, measures taken in some of its parts might have benefits or negative impacts in its other parts, and this should be always examined and taken into account. Furthermore, any examination of a mode (for example, a road project) should always examine in parallel its impacts on other modes (for example, railways). This kind of assessment was rather rare in the case of TEN-T.

14 This will be a result both of the new economic, social and political conditions and of the generated (induced) traffic due to capacity expansion (European Commission 2006a).

Funds, projects were initiated by the Member States or other bodies eligible for funding.[15] Then, the European Commission had to assess whether the proposed projects were consistent with the common objectives, and whether these were justified according to other legal and economic requirements.

The revision of the TEN-T has changed the scene. Now, the European Commission is the body which sets the priorities and approves the implementation plans. The Community's position in this case is therefore substantially different: it now has a larger role to play than in the TEN guidelines of 1996. Its role is that of planning and establishing not simple guidelines but concrete priorities; this also includes identifying projects of common interest and puts the Community in a position which is not limited to that of a mere financing body, but it goes further, to become the body which decides and supervises the construction of infrastructure.

This approach includes the development of an action plan for each project in a form which aims to give the necessary political impetus for speeding up its implementation and financing.

In this context, in the current revision of the TEN-T, 30 major projects have been identified as high priorities, in a time horizon up to 2020.[16]

Conclusions

In transportation planning, the nature of a *network* has to be distinguished from that of a *corridor*, as it has a significantly different character and role. The network serves the integration of a region and sets up a harmonized transport system. The corridors set up trade connections, useful to link peripheral countries with the core centre (economic or political). The 'character' of a transport corridor is different from that of a transport network, as the former mostly focuses on the pure transport operations, while the latter focuses on the regional development of the area served by the network.

In reality, the TEN-T was designed to serve the internal needs of the Union, when integration was the main focus of the Common Transport Policy, together with the need to develop a harmonized transport system (harmonized in terms of both infrastructure and institutional standards) in place of the existing patchwork of national systems. The TEN-T was characterized by highly dense transport links within national boundaries and traffic with other states tending to be channelled through border crossing-points, the number of which was relatively limited compared with the national routes.

---

15 Such bodies were railway companies, public administrations, private enterprises and so on.

16 Vice President of the Commission and Commissioner for Energy and Transport Ms. Loyola de Palacio gave the mandate to a high-level group of experts from all existing and future Member States to identify these projects. The group, chaired by former Commissioner Mr. Karel Van Miert, worked from November 2002 to July 2003; its conclusions were ratified by the Council, in October 2003, in Luxembourg. Since 2004, they have been part of the European *acquis*: Decision No. 884/2004/EC regarding the revision of the guidelines for the development of the TEN-T.

On the other hand, a corridor can be an ideal tool for strengthening the links between peripheral and central regions, thus contributing towards economic cohesion. In the process of the extension of the TEN-T (during the Union's enlargement), the idea is that a first step before the creation of transport networks should be the establishment of transport corridors. The linear shape of the established multimodal transport routes (Helsinki Corridors) can attract investments and traffic, in a coordinated effort to quickly develop the necessary, suitable links to promote trade and social and political relations towards European regions that historically have remained rather isolated. The corridors are, however, just the first step, and will very likely be followed by future coordinated actions for network developments.

References

EIB, World Bank and EBRD (1999), 'Projects' Appraisal Framework', TINA Final Report.

European Commission (2001), 'White Paper 'European Transport Policy for 2010: Time to Decide'', (COM (2001) 370), Brussels.

—— (2006a), 'Communication from the Commission to the Council and the European Parliament – Keep Europe moving – Sustainable mobility for our continent', Mid-term review of the European Commission's 2001 Transport White Paper {SEC (2006) 768}, Brussels.

—— (2006b), 'NAIADES Action Programme: Communication on the Promotion of Inland Waterway Transport', COM (2006) 6 final.

European Commission and DG TREN (2005), 'The SEA Manual – A Sourcebook on Strategic Environmental Assessment of Transport Infrastructure Plans and Programmes', Brussels.

de Palacio, L. (2003), Opening address at the Symposium 'Towards the Integration of the Trans-European and Pan-European Transport Networks', organized by the Greek Presidency, Brussels, 6 May.

PLANCO Consulting GmbH (2003), 'T E N - I n v e s t/Transport Infrastructure costs and Investments between 1994 and 2010 on the Trans-European Transport Network and its connection to neighbouring regions including an inventory of the technical status of the Trans-European transport network for the year 2000', January–December 2002.

TINA (1999), 'Transport Infrastructure Needs Assessment'. Final Report, TINA Secretariat, November 1999.

Table 10.A1 Comparison of the TINA Network (in the accession countries) with the TEN-T (in the EU15)

	Surface area (in km²)	Populat-ion (in mil. inhabs)	Length (in km)		Km per thousand km²		Km per million Inhabitants		GDP (in billion euros)	GDP/c (in euros)	Construc-tion cost (in billion euros)	Construc-tion cost per GDP	Construc-tion cost per mil. inhabs (in billion euros)	Construction cost per GDP/c (in mil. inhabs)	
			Roads	Rails	Roads	Rails	Roads	Rails						in mil. inhabs	%
Bulgaria	110994	8.4	2025	2095	18.24	18.87	241.07	249.40	10.1	1200	5.3	0.5	0.63	4.4	2.9
Cyprus	9250	0.6	425	–	45.95	–	708.33	–	6.8	10570	1.1	0.2	1.76	0.1	0.9
Czech Republic	78860	10.3	1842	2341	23.36	29.69	178.83	227.28	35.9	3490	10.2	0.3	0.99	2.9	1.6
Estonia	45227	1.5	1000	657	22.11	14.53	666.67	438.00	2.8	1850	0.6	0.2	0.42	0.3	1.3
Hungary	93030	10.2	1448	2727	15.46	29.31	141.96	267.35	34.1	3340	10.2	0.3	1.00	3.0	1.7
Latvia	64589	2.5	1520	1343	23.53	20.79	608.00	537.20	3.4	1370	2.0	0.6	0.80	1.5	3.2
Lithuania	65300	3.7	1617	1100	24.76	16.85	437.03	297.30	4.6	1225	2.3	0.5	0.63	1.9	2.8
Poland	312685	38.6	4723	5529	15.10	17.68	122.36	143.24	91.1	2360	36.4	0.4	0.94	15.4	2.2
Romania	238391	22.7	2524	3163	10.59	13.27	111.19	139.34	27.2	1200	11.2	0.4	0.49	9.3	2.3
Slovakia	49036	5.4	949	1400	19.35	28.55	175.74	259.26	13.3	2470	6.5	0.5	1.21	2.6	2.7
Slovenia	20255	2	565	569	27.89	28.09	282.5	284.50	14.5	7240	5.8	0.4	2.89	0.8	2.2
TOTAL TINA	1087617	105.9	18638	20924	17.14	19.24	176.00	197.58	243.8	2302	91.6	0.4	0.86	39.8	2.1
EUROPEAN UNION 15 – TEN-T	2238700	372.1	75300	73900	33.64	33.01	202.36	198.60	6414.0	17237	–	–	–	–	–
Ratio CEEC/ EC	48.6 %	28.5 %	24.8 %	28.3 %	50.9 %	58.3 %	87.0 %	99.5 %	3.8 %	13.4 %	–	–	–	–	–

Source: TINA Final Report. October 1999: Main indices for the TINA Network versus the TEN-T.

Notes to Table 10.A1

Reference Network in the 11 Acceding Countries: the TINA Network.
Reference Network in the Member-States: the TEN-T.
The Construction Cost refers to all modes, and not only to road and rail.
GDP and population are figures taken from Agenda 2000 for the year 1995.
The ratio of Network length to surface area is an indicator of the density of the Network.
The ratio of Network length to population gives an indication of the relative availability of infrastructure for the population.
The ratios of construction cost to GDP, as well as of construction cost to population, are partial indicators of the prospects for financing the Network. Clearly, there will in general be fewer problems in financing the Network where these ratios are relatively low. This comment should, however, be qualified by an examination of the ratio of construction cost to per capita GDP. This will show, for example, that although Slovenia has a very high ratio of construction cost to population, this is in part compensated by its relatively high level of per capita GDP, resulting in a correspondingly greater ability of the country to finance the proposed projects. The construction cost per GDP per capita expressed as a percentage of the population (last column of the table) means, for example, that in Latvia 3.2 % of the population should contribute, up to 2015, the equivalent of their 1995 GDP for the construction of the Network; the respective percentage in Estonia is only 1.3 % of the population, and in Cyprus 0.9 %.
Any assessment of the overall prospects for financing the network must therefore take into account the balance of all three indicators.

Chapter 11

European Policy Aspects of Network Integration in the TEN Transport

Anastasia Stratigea, Maria Giaoutzi, Constantin Koutsopoulos

Introduction

During the enlargement process of recent years, 12 states have become full Members of the European Union. As Europe becomes an increasingly large territorial entity, the integration of countries with different cultural characteristics, economic potential, political orientation and so on, into a uniform market, signals new *patterns of interaction* taking place among different regions. At the same time, intensive trade relationships with neighbouring countries to the EU are contributing to the intensification of trade flows. Furthermore, the emergence of the information society has strengthened distant relationships and cooperation thus reinforcing the increasing mobility patterns of persons and goods. In such a context, the development of an infrastructure serving the steadily increasing patterns of interaction is of vital importance.

Transport networks, in this respect, are considered as *strategic elements*, since they constitute the necessary infrastructure along which flows of goods, persons and services can take place, thus facilitating the interaction between European regions and the achievement of the enlarged internal European market. Their significance is, moreover, widely recognized for the achievement of the goal of European enlargement and the establishment of external links of the European territory with the rest of the world. They also play a crucial role in the efforts of Europe to deal with regional disparities in economic performance across its territory. Such disparities are closely linked to *geographical location* and *accessibility*, in the sense that the more peripheral a region and the less accessible, the lower its development potential is likely to be.

While there are many other factors involved, it appears that, even in the age of the information society, transport infrastructure networks are often critical for *regional competitiveness* and *prosperity*, since they largely contribute to the goal of *social and economic cohesion* of the European territory. It is realized nowadays that the efficiency of a European-wide transport system is essential for the competitiveness of Europe, its growth and employment.

Along these lines, the development of the Trans-European Transport Network(TEN-T) constitutes a *strategic policy initiative* of the European Commission (EC), which is expected to meet the economic, social and environmental goals (Button et al. 1998). It places emphasis on the *integration* of national networks by either constructing new infrastructure (for example, in the case of missing links) and/or improving existing ones. The TEN-T is improving the accessibility for all parts of the European territory, both internally and to the rest of the world. At the same time, the TEN-T has to be consistent with the goal of *sustainable mobility*, which aims at reducing the pressure of transport on the environment by preventing congestion and maintaining the economic competitiveness of the regions.

In order to broaden the scope of sustainable mobility, a variety of issues have been incorporated, such as: safety and security in transport; increased transport options for the average EU citizen; building and financing of a solid infrastructure base; and ensuring that social and environmental considerations are designed to carry European transport into a robust future that can support the EU's enlargement and growth options. Hence, building such a sustainable transport network at the European scale should involve a holistic approach, taking into account a balanced analysis of all the costs and benefits of transport projects (EC 2006).

Intermodality, *Multimodality* and *Interoperability* (IMO) are characteristics of the transport infrastructure system that are considered to be at the *core* of the development of an integrated, efficient and effective Trans-European and Pan-European transport infrastructure system. As they have been defined in recent European research projects (EUROSIL 1999), *intermodality* is the feasibility to implement an integrated transport chain; *multimodality* is the competition between transport modes in the same corridors with the scope to achieve an optimal modal split; and *interoperability* mainly concerns organizational issues, as well as the elimination of any kind of barriers. The very nature of IMO projects may also have a potential influence on the development of the respective areas, with improvements in transport efficiency leading to improvements in accessibility and, in turn, to enhanced development perspectives of less developed regions

The present chapter focuses on the evolution of the *policy framework* elaborated by the EC towards the development of the TEN-T. In the next section the issues of intermodality, multimodality and interoperability are discussed as *cornerstones* in the TEN and Pan-European Network (PEN) transport policy. Then we elaborate on the *evolution of the TEN-T* in the European Union. After this, the most recent *developments* in the TEN-T policy framework are presented. Followed by an examination the *future policy guidelines* of the EU up to 2010. Finally, some conclusions are drawn and issues for further research are discussed.

The Role of Intermodality, Multimodality and Interoperability in the EU Policy

The EU's *Common Transport Policy* (CTP) calls for the establishment of transport systems capable of providing 'sustainable mobility', so that goods and people may be transported throughout the Community efficiently and safely, under the best possible social conditions, and fully respecting the environment. The direction of transport policy development for the years 1995–2000 is generally to proceed with the implementation of the basic policy measures already identified. The concept of 'sustainable mobility' remains at the core of the approach.

The well-defined objectives[1] of 'sustainable mobility', that is, to develop transport systems that can meet society's economic, social and environmental needs, remain valid. The goal is always to offer a high level of mobility to people and businesses throughout the Union, protect the environment, ensure energy security, promote minimum labour standards for the sector and protect the passenger and the citizen.

The fundamental objectives of CTP establish the link between the development of transport and economic activities in the various regions of the Member States. Transport infrastructure largely contributes to the strengthening of economic and social cohesion, the reduction of disparities among regions, and the linking of island, landlocked and peripheral regions with the central regions of the Community.

In the social system, the EU policy promotes employment quality improvement and better qualifications for European transport workers. EU policy also protects European citizens as users and providers of transport services, both as consumers and in terms of their safety and security.

Intermodality is one of the key issues in EU Transport Policy. Intermodal transport – including what the Commission calls 'combined transport' – is seen as a *policy tool* for encouraging freight to move from road to less environmentally intrusive modes, such as railways, waterways and short-sea shipping. EU policies referring to intermodality incorporate interoperability issues as well and, hence, the use of intermodality in this section should be considered to include interoperability (EUROSIL 1999).

Interoperability is seen as an important facilitator of good quality intermodal transport. The achievement of the *seamless journey* is dependent not only on interconnections between different modes, but also on consistency in technical standards, harmonization of regulatory frameworks (for access, operation, investment and safety) and exchange of information (Giaoutzi and Stratigea 2006). It is a goal of EU transport policy to have common standards of equipment and training to overcome the physical and institutional constraints of international transport operations (EUROSIL 1999).

Multimodality is also a key instrument in encouraging the shift from road transport to less environmentally intrusive transport modes. There have been a number of Commission initiatives to encourage a wider choice of transport mode, such as the 'Citizens' Network', 'Revitalising the Community's Railways', 'Fair and Efficient Pricing', and so on (EUROSIL 1999).

Until recently, European transport networks tended to be developed on *national grounds*, with emphasis more on nationally oriented modal networks than on integrated transport systems. The EU has placed increasing emphasis on the evolution of the Trans-European and Pan-European Transport Networks (TEN-T and PEN-T) during recent decades. A *key feature* of TEN-T/PEN-T development is the emphasis on multimodal, intermodal and interoperable networks.

---

1 From the transport White Paper of 1992, COM (92) 494: 'The future development of the Common Transport Policy' to the White Paper of 2001, COM (2001) 370: 'European transport policy for 2010: time to decide' and the Communication 'Keep Europe moving - Sustainable mobility for our continent' (COM (2006)314).

Intermodality, Multimodality and Interoperability (IMO) principles are integral to the development of the TEN-T and the Pan-European Corridors/TINA (Transport Infrastructure Needs Assessment) Network. The rationale behind this is that improvements in the IMO elements lead to the improved accessibility of the different regions of the Union, as well as between the Union and Central and Eastern Europe – which in turn, when applied in an appropriate manner, support more spatially balanced economic development and improved social cohesion (see Chapter 8 of this volume) .

The Evolution of the TEN-T

The concept of the Trans-European Transport Networks (TEN-T), originating at the end of the 1970s, aimed to *integrate the nationally oriented transport networks* of the Member States – national networks – to EC level and introduce major transportation links (land, inland waterways, air and sea) connecting the major transport centres of the European Union. Moreover, the TEN-T was intended to remove various types of *barriers* appearing in these networks (network barriers, institutional, linguistic, technical, cultural and so on) in the context of an integrated TEN-T (Suarez-Villa et al. 1992; Stratigea and Giaoutzi 2001).

In such a framework, the aim of the European Union has been to focus on the *integration* of national transport networks (bottom-up approach), aiming at the development of the interconnection and interoperability of these networks rather than at the development of a unique and uniform European transport system, which is not adaptable to real needs. This effort has been undertaken along the lines of the EU Common Transport Policy described in the following subsection.

The Need for a Common Transport Policy – Introduction of the TEN-T

The *completion of the Single European Market* in 1986 and the opening of borders for the free movement of people and goods have revealed many obstacles and inadequacies of the transport system, which have impeded this movement and restricted the development of the outermost regions. This fact pointed clearly to the necessity for a European-wide integrated transport network infrastructure, which would constitute the means towards the removal of these obstacles and inadequacies. Member States should cooperate and combine their efforts in order to conclude with a harmonized, interconnected and continuously developing integrated transport infrastructure, which would serve the EU's goal of free movement of people, goods and information.

Anticipating the future transportation problems in the EU and the importance of the TEN-T, an entire chapter of the *Maastricht Treaty* (1992) was devoted to these issues (Articles 129 B, 129 C and 129 D of Title XII), introducing the *legal framework of the TEN-T*. In this chapter it was stated that the 'Community should contribute to the establishment and development of Trans-European Networks in Transport, Telecommunications and Energy infrastructure sectors'. It further stated that 'in the framework of a system of open and competing markets, the Community aims at promoting interconnections and interoperability of national networks as well as access to those networks', to ensure the smooth functioning of the internal market, as well as to establish links between the central and the peripheral regions of the Union, thus serving the purposes of social and economic cohesion. However, the *political impulse* for the creation of the TEN-T came from the Copenhagen European Council (June 1993), which invited the Commission and the Council to speed up preparations in this field.

Following the signing of the Maastricht Treaty, President Delors's White Paper on 'Growth, Competitiveness and Employment' (EC 1992a) pointed out that large Trans-European Networks add to the efficiency of the Single Market and the EU's common policies, while large construction projects will also create jobs, thereby reducing the high European unemployment rates and helping the EU's troubled economy by boosting the competitiveness of the European markets (Club de Bruxelles 1990; EC 1995a, 1996).

The establishment and development of TEN-T as a European-wide integrated transport network would serve the *general objective* of the EU which is the achievement of the smooth *functioning of the internal market* and the *strengthening of economic and social cohesion.* Special care has been attached to the linking of islands, and landlocked and peripheral regions with the central regions of the Community.

This general objective should seek to meet the following dimensions: the *sustainable mobility* of persons and goods under the best possible social, environmental and safety conditions; *the integration of all modes of transport*, taking into account their comparative advantages; and the *creation of jobs*. These have been the directions which constitute *specific objectives* in the context of TEN-T in this period.

Along the lines of the *White Paper on the 'Future Development of the Common Transport Policy'* (EC 1992b), the EU has adopted policies encouraging the establishment of links between the Member States' national networks in order to ensure *network interoperability*, while at the same time taking environmental aspects into account. The overall aim of this TEN-T policy is to turn the 15 national transport networks into a single European transport network by eliminating *bottlenecks* and constructing the *missing links*.

More specifically, in this White Paper some serious *imbalances* of the European Transport System have been identified as follows (EC 1992b):

- an increasing *saturation of the transport network* due to the rapid growth of the transport market;
- an *unbalanced distribution of traffic burden among the different modes* of transport, which exhibit different degrees of saturation. Road transport is absorbing the heaviest burden, especially as a result of the remarkable increase of freight transport;
- *regional imbalances* due to poor connections of certain regions to the rest of the European transport network, and consequently to the main economic centres of the Union;
- *funding difficulties*, since investments on infrastructure are declining;

- *environmental deterioration aspects* attributed to rapidly increasing atmospheric pollution levels, with the transport sector being one of the main sources of CO_2 emissions;
- *safety aspects* with thousands of persons dying every year on EU roads; and
- *social aspects*.

To deal with the above-mentioned imbalances *seven objectives* have been posed in the context of the Common Transport Policy, related to (EC 1992b):

- the *establishment of the TEN-T*, which will enhance the potential of the transport network Europe-wide;
- the *integration of the transport systems* (different modes), which will enable a more efficient management of transport flows by shifting part of them to less polluting or underused transport modes: namely, rail, inland waterways or short sea shipping. The promotion of *combined transport* falls into this objective, where goods can be transported over long distances by rail and finally delivered by lorries. *Fair pricing* of the various transport modes consists of a specific policy tool, which will support the redistribution of traffic flows. In this respect, the full costs of infrastructure use are charged to the users, including external costs such as those generated by pollution, noise, congestion and accidents;
- the *protection of the environment* which calls for the implementation of policies with respect to polluting standards control, tax measures influencing travel behaviour, better use of public transport and so on;
- *safety* for individuals;
- *social safeguards* for all transport sectors, in the form of regulated admission to the occupation of transport operator and respective training requirements;
- *strengthening of the internal market* by promoting and properly implementing liberalization and harmonization measures, adopted as part of the Single Market; and
- *reinforcement of the external dimension of the Single Market*, which will allow the smooth interaction between the EU and other countries. This can take the form of bilateral agreements between EU and non-EU countries.

In order to implement these objectives, in 1995 the EC adopted the *Action Programme 1995–2000* supporting the *Common Transport Policy* (EC 1995b), which has provided a basis for action by all actors involved in the transport sector. This programme sets out *initiatives* in three fundamental areas:

- improving *quality* by developing integrated transport systems, particularly TEN-T based on *advanced technologies for traffic management*, which will also integrate environmental and safety objectives;
- improving the *functioning of the Single Market* in order to promote efficiency, choice and user-friendly provision of transport services, while safeguarding social standards;

- broadening the *external dimension* by improving transport links between the EU and third countries, and promoting the access of EU operators to transport markets in other parts of the world.

With regard to the *first objective*, the emphasis is put on developing the system. The various transport modes must be used more efficiently and in a more environmentally responsible way, enabling energy to be saved and providing more modal interconnections and greater interoperability.

With regard to the *second objective*, it clearly remains an important priority to monitor the application of the rules for creating a Single Market. Furthermore, in a liberalized market, the strict application of competition rules and State rules is of particular importance. The experience gained up to now shows that new legislation is required in certain areas. This is related in particular to the increased liberalization of rail transport, the allocation of slots at the airports, the gradual abolition of the queuing system for certain inland waterway markets, and the improved application of the responsibility rules and arrangements in the road haulage sector.

In a number of fields, the structural adjustment required by the passage from a national regulatory system to a Single Market implies that certain State aid is needed. The Commission must remain attentive to this. Lastly, in some cases the Commission will be called upon to help eliminate some structural capacity, such as the current overcapacity in inland waterway transport.

Moreover, being aware of the need to take action on infrastructure charges and the external costs of the various transport modes, the Commission has submitted a Green Paper on *fair pricing in transport*. This paper seeks to improve the balance between road and other modes of transport by considering the integration of costs arising from pollution, congestion and accidents into transport costs.

Nor has the *social dimension* of the market been overlooked. The Commission does its utmost to improve working conditions, in particular in the maritime and the air transport sectors.

Several tensions have also been noticed in a number of sectors during this period, in part resulting from the adaptations following liberalization. The Community's policies have been concentrated on specific issues such as working time, manning regimes particularly in the maritime sector and so on, which take account of the particular needs of the various transport activities concerned.

Finally, priorities for the *third objective* will be established once the added value of the Community's action has been clearly compared with the costs of inaction, given the present fragmentation of the market. At present, Member States are continuing to operate a system of reciprocal bilateral transport agreements with third countries, and this is likely to create substantial distortion. Initiatives in this field include the conclusion of the Council's work on mandates for establishing road and air transport links with the countries of Central and Eastern Europe, the follow-up to the mandate for air transport negotiations with the USA and action on shipping, in particular the finalization of the World Trade Organization (WTO) negotiations on liberalizing maritime transport. The Commission will also be taking action to give the EU a more prominent role within international transport organizations.

Table 11.1 European Union priorities for the TEN-T – 14 projects

Project	Influenced areas
1. High-speed train/ combined transport North-South	Berlin-Nuremberg/Munich-Verona
2. High-speed train PBKAL	Paris-Brussels/Brussels-Cologne-Amsterdam-London
3. High-speed train South	Madrid-Barcelona-Montpellier/Madrid-Vitoria-Dax
4. High-speed train East	Paris-Eastern France/Luxembourg-Southwest Germany
5. Conventional rail/ combined transport: Betuwe line	Rotterdam-Dutch/German Border (Rhine/Main)
6. High-speed train/ combined transport France-Italy	Lyon-Torino-Milano-Venezia-Trieste
7. Greek motorways	Greece
8. Motorway Lisbon-Valladolid	Portugal-Spain
9. Conventional rail link Cork-Dublin-Belfast-Larne-Stranraer	Cork-Dublin-Belfast-Larne-Stranraer
10. Milano-Malpensa Airport	Italy
11. Fixed rail/road link Denmark-Sweden	Oresund fixed link: Copenhagen- Malmö
12. Nordic triangle multimodal corridor	Stockholm-Swedish/Norwegian border, Stockholm-Malmö, Malmö-Goteborg-Swedish/Norwegian border, Stockholm-Turku-Helsinki-Finnish/Russian border
13. Ireland/United Kingdom/ Benelux road project	Ireland-United Kingdom
14. West coast main line (rail)	United Kingdom

At the *Essen Summit* in 1994, the European Council and the European Parliament recommended the Member States and the Community to support a list of *14 priority projects* (see Table 11.1) ranging from the development of high-speed railway links on international corridors to the construction of airports. Selected transport schemes aimed at the replacement of road transport: 80 per cent of the capital expenditure involved was for building railway lines and 9 per cent was for rail-road links.

Developing TEN-T

A Decision of the European Parliament and the Council was made on 23 July 1996 concerning the Community guidelines for the 'Development of the Trans-European Transport Network' (Decision No. 1692/96/EC). It included network maps for roads, rail, inland waterways and ports, airports and combined transport, as well as criteria and specifications for projects of common interest concerning roads, rail, inland waterways and inland ports, seaports, airports and combined transport, shipping information and management, air traffic management, positioning and navigation. These guidelines were intended to encourage the Member States and Community to carry out projects of *common interest*. 'Common interest', in such a framework, is associated with projects which promote *cohesion*, *interconnectivity* and *interoperability* aspects of the Trans-European Transport Network, as well as access to that network.

This Decision stresses the issues of multimodality, intermodality and interoperability as the *cornerstones* for the development of an integrated TEN-T, where *integration* is perceived in the context of both *different modes* (road, rail, and so on) and *different networks* (for example, missing links). At the same time, this Decision puts emphasis on the expansion of the TEN-T in order to fulfil social and economic cohesion purposes. All these issues have been reflected in the *objectives* in establishing the TEN-T, as they were articulated in this Decision, which include (Decision No. 1692/96/EC):

- *integrating land, sea and air transport infrastructure networks* throughout the Community gradually by 2010, thus contributing to the strengthening of the *social and economic cohesion*;
- ensuring the *sustainable mobility* of persons and goods within an area without internal frontiers under the best possible social and safety conditions;
- respecting the Community's objectives in relation to the *environment* and *competition*;
- including *all modes of transport*, taking account of their comparative advantages in terms of the optimal use of existing capacities;
- being *interoperable* within modes of transport and encouraging *intermodality* between the different modes of transport;
- *covering the whole territory* of the Community, so as to facilitate access in general, link islands and landlocked and peripheral regions to the central regions, and interlink without bottlenecks the major conurbations and regions of the Community;
- *assuring connections* to the networks of the *European Free Trade Association (EFTA) States*, the countries of *Central and Eastern Europe* (CEECs) and the *Mediterranean countries*, while at the same time promoting interoperability and access to these networks.

In summary, the main goal of this action can be defined as the establishment and development of the TEN-T, within a *system of open and competitive markets*, through the promotion of interconnectivity and interoperability of national networks and access thereto.

With this Decision, criteria and specifications for the identification of transport projects of common interest have been elaborated with respect to the road network, the rail network, the inland waterways network and inland ports, seaports, airports and the combined transport network.

On the basis of the Commission's Work Programme for 1999, the *priorities* in the transport sector and particularly in the TEN-T for the period 1999–2000, holding also for the period 2000–04, are focusing on the following aspects (COM 1998):

- improving *efficiency* and *competitiveness*;
- improving *quality*; and
- improving *external effectiveness*.

In relation to the *first priority* – efficiency and competitiveness – it seems that *five objectives* remain crucial to ensure that European Transport Systems realize their full potential in promoting competitiveness, growth, employment and environmental sustainability.

These are related to (EC 1998): liberalization of market access; enhancing of integrated transport systems across Europe, as well as effective traffic management systems for different modes; ensuring fair and efficient pricing within and between transport modes; strengthening the social dimension; and providing proper implementation of rules.

Transport systems are supposed to respond to the citizens' needs. In this context, the *second EU priority* is *quality* of transport, which is incorporating *safe*, *environmentally friendly* and customer friendly transport systems.

Safety is a permanent goal of the EC. In this respect, it puts effort into policies integrating technological advances in various transport modes in order to ensure safe travel. On the other hand, the development of *sustainable forms of transport* constitutes a *key priority* of the Commission. A set of policy initiatives applying to this issue have been proposed in its communication on 'A Strategy for Integrating Environment into EU Policies' (COM 1998), which underlines the importance of integrating the environmental dimension in sectoral policies, including transport.

Finally EU policies have been focused on the provision of *customer-friendly transport systems*, which are easily reachable and better integrated to provide door-to-door services, thus increasing the quality of transport services to citizens.

The *third priority* of the Commission's Work Programme for the period 1999-2000 was related to the improvement of the *external effectiveness* of the TEN-T. This has been mainly advocated by the enlargement perspectives of the EU towards the CEECs, as well as by the globalization trends of the new economy.

Despite the ambitious policy of EU on the TEN-T, *bottlenecks* are still present in many parts of the network especially in border areas. This strongly reflects the narrow national views on the network infrastructure of the countries involved (Stratigea and Giaoutzi 1999). In this context, the EC has proposed a *two-stage revision of the TEN-T guidelines*.

The *first stage in 2001* stresses the need to concentrate on the *elimination of bottlenecks* of the routes already identified in the Essen Summit. This includes the completion of the routes identified as EU priorities in the context of TEN-T in order to be able to absorb the traffic flows, which will be generated by the enlargement of the EU. Emphasis has been placed on frontier and outlying regions.

The *second stage in 2004* involves a *more extensive revision*, serving the purpose of strengthening the role of modes other than road. A specific aim of this stage in particular is the introduction of the concept of 'Motorways of the Sea' and the development of the *airport capacity* (EC 2001)+.

At the same time it places emphasis on the establishment of a *primary network* built up of the most important infrastructure, including sections of the Pan-European corridors. This is supported by the introduction of the concept of 'declaration of infrastructure of European interest', which is strategic infrastructure that ensures the smooth functioning of the internal market, as well as the reduction of congestion.

Latest Developments in the TEN-T Policy Framework

The enlargement of the European Union in 2004 has inevitably led to a reorientation of the efforts related to TEN-T by shifting the emphasis and focus onto those networks, which link up the EU with the new Member States. This reform has taken into account aspects such as financial restrictions, time-horizon of priority projects already set in Decision No. 1692/96, the objective to reduce bottlenecks in various European regions, the need to achieve an efficient network in the new Member States, and so on, which made it clear that the TEN-T needs to be more focused and more concentrated in the future, stressing the need to put more efforts into the key cross-border benefits in order to achieve adequate results within a more realistic time-horizon (De Palacio 2003).

Along these lines, a *major reform* of guidelines for the development of the TEN-T has been undertaken within the enlarged EU context, aiming at the *amendment of Decision No. 1692/96/EC*. The preparation of this reform has been worked out by a high-level group of representatives of Member States established by the Community.

Major aspects of this reform are (Decision No. 884/2004/EC):[2]

- the focus on the *corridor concept* instead of a network approach; and
- the shift of the *role of the EU* from a purely financing to a decision-making body in terms of planning the network and setting its priorities.

The *corridor concept* involves putting emphasis on the main transport routes and more specifically on those routes, which connect the EU and the new Member States. The reason for that lies mainly in:

- the need to connect corridors of the new Member States to the main body of the TEN-T in order to produce a *coherent network*;
- the necessity to *establish the EU's links beyond its borders*, for which purpose central-Eastern European new Member States are of vital importance in terms of connecting Western Europe to new opportunities opening up towards the East through the development of corridor mechanisms (De Palacio 2003).

Some useful experience has been gained from the development of Pan-European Corridors over the past years, which may provide valuable input in the new priorities-setting approach. This relates to (De Palacio 2003; and Chapter 10 of this volume):

- The corridor concept has been proven to be a useful instrument for monitoring the development of main intermodal transport links crossing various countries.

2 Decision No. 884/2004/EC, Amending Decision No. 1692/1996/EC on 'Community Guidelines for the Development of the Trans-European Transport Network', European Commission, Official Journal of the European Union, 30.4.2004, L167.

- It ensures a better coordination of investments and a concentration of community interventions on major transport routes.
- It provides useful information with respect to the monitoring of the main links of the TEN-T, which allow for a regular monitoring of the progress and a more coherent framework in relation to priorities and objectives.
- It consists of a proper field for attracting new forms of financing schemes by the use of public-private partnerships (PPPs).

As to the *role of the EU* with respect to the development of the TEN-T, a new approach has been set up, which shifts the responsibility of the EU from a purely financial body to a body playing a *decisive role in planning the network* and setting the priorities for future developments. Thus a different relationship has been established between the EU and its Member States. In such a context, the EU has the main task of planning the network based on concrete priorities, as well as supervising the construction of the relative infrastructure.

The revision of Decision No. 1692/96/EC was proposed by the Commission and approved by the Council and the Parliament on 21 April 2004 (Decision No. 884/2004/EC). According to the new decision, a revised list of priorities as well as projects was raised.

A basic comment on this point is that, although the construction of a TEN-T has been on the European agenda since 1990, the trends today do not seem very positive. A list of 14 priority projects were approved by the European Council in 1994, but the results of the constructions were rather poor. In 2004, the EU elaborated and approved a new list of 30 projects.

From a financial perspective, the figures show that from 1983 to 2006, €7.1 billion have been allocated to the TEN-T. The outcome of these investments is not promising:

In terms of the Essen projects:

- Only three out of the 14 Essen projects are completed.
- Most projects are considerably delayed: in 2004, the completion date for the 14 Essen projects was postponed from 2010 to 2020.

In terms of the new goals (Community of European Railway and Infrastructure Companies (CER) and Union of European Railway Industries (UNIFE) 2006)

- For the new budget period 2007–2013, the estimated investments required for the construction of the TEN-T network is €140 billion.
- The Commission proposal was to allocate €20 billion from the Union's budget.
- Finally, the European Council agreed in December 2005 to allocate approximately €7 billion.

It can be seen that, although the EU has raised high expectations, there are many difficulties to safeguard the necessary investments. To solve the problem, considerable efforts are being made to mobilize the private sector to invest in transport schemes. In this context, the profitability criteria for transport projects are continuously gaining importance.

In order to comply with the objectives of TEN-T, as laid down by the Decision No. 1692/96/EC, and to meet the transport challenges imposed by the enlargement, priorities have been revised, placing emphasis on intermodality, multimodality and mostly interoperability in various transport modes (for example, rail) and transport systems (for example, national networks). The revision of *priorities* implies (Article 5, Decision No. 884/2004/EC):

- the elimination of bottlenecks of the TEN-T by the establishment and development of *key links and interconnections*, fill in missing sections and complete the main routes, with particular emphasis on cross-border sections and across natural barriers, as well as on the improvement of interoperability on major routes;
- the development of the infrastructure which will establish the *interconnection* between national networks, so that the linkage among isolated or landlocked areas and central areas of the EU is promoted;
- the promotion of an *interoperable rail network* with routes adapted for freight transport;
- the promotion of *long-distance, short-sea, and inland shipping*;
- the promotion of the *integration of rail and air transport*;
- the optimization of the *capacity and efficiency* of existing and new infrastructure;
- the better *integration* of safety and environmental aspects in the context of the TEN-T;
- the *sustainable mobility* across the TEN-T.

According to these revised priorities of the TEN-T infrastructure, *priority projects*, that is, projects of common interest, are defined on the basis that (Article 19, Decision No. 884/2004/EC):

- They *eliminate bottlenecks* or *complete missing links* on major routes of TEN-T. Particular emphasis has been given to projects crossing borders, crossing natural barriers or having a cross-border section.
- Their scale of reference contributes to long-term planning in a European context.
- They have an impact on *socio-economic development* of the respective regions.
- They improve *sustainable mobility* as well as the *interoperability* of national networks.
- They support the *territorial cohesion* of the enlarged Europe and link peripheral to central European regions.

The list of priority projects as defined by Decision No. 1692/96/EC has been further revised, according to the suggestions of the High-Level Group. A list of 30 priority projects, which are thought to fulfil the above requirements, has emerged

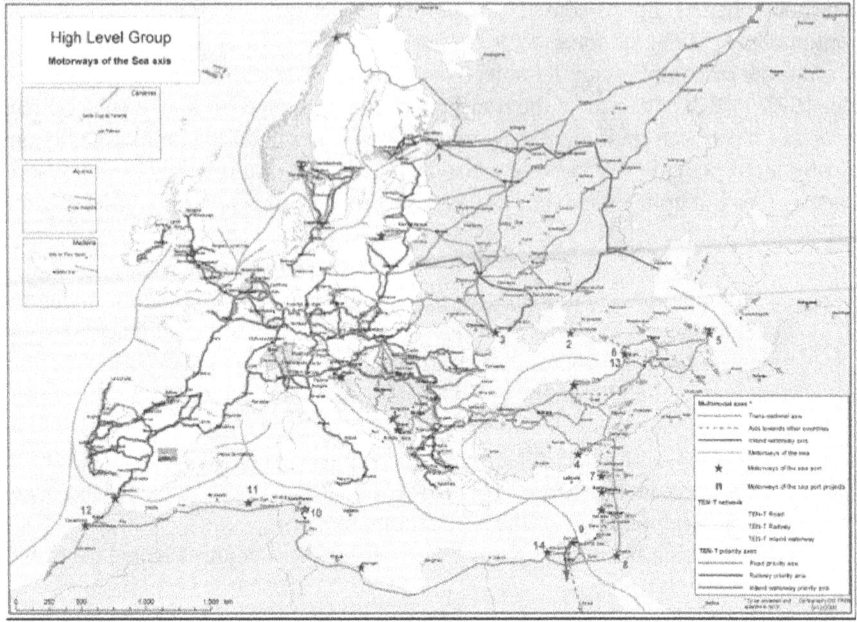

Figure 11.1 TEN-T priority projects in an enlarged Europe
Source: European Communities (2005).

from the evaluation process (Annex III, Decision No. 884/2004/EC). The revised priority projects are shown in Figure 11.1.

Future Trends in Policy Aspects

The Common Transport Policy (CTP) constitutes the cornerstone of bridging together different regions and people in a European context. It consists of a dynamic and continuously developing instrument of the EU designed to 'deliver an integrated transport system, even where the interests of different groups can pull in different directions' (COM 1998, p. 6).

Despite the efforts of the European Union towards the development of an effective integrated Trans-European Transport Network (TEN-T), this goal has not yet been fully realized. Moreover, some *trends* appearing during the last decade are going to negatively affect the effectiveness of the whole system. These trends have to be incorporated in a proper way in the EU's CTP.

A major issue in this respect is the *congestion* observed in certain areas and on certain routes as a result of imbalance amongst transport modes. This fact is having a progressively negative effect on the economic competitiveness of the areas involved. The White Paper on 'Growth, Competitiveness and Employment' in 1992 has placed emphasis on the issues of traffic jams, bottlenecks and missing links, as well as lack of interoperability between modes and transport systems. It stated that transport networks are 'the arteries of the single market and their malfunctions are reflected in lost opportunities and jobs' (EC 1992a).

Congestion is considered a major threat to the European transport system. Evidence shows that (EC 2001) almost 10 per cent of the road network (7,500 km) is affected daily by traffic jams; 20 per cent of the railway network (16,000 km) is classified as bottlenecks; in 16 of the main airports of the EU, delays of more than 15 minutes are recorded on approximately 30 per cent of their flights. All this congestion results in the consumption of an extra of 1.9 billion litres of fuel. Recent studies on this issue show that if nothing is done in this direction, road congestion will increase significantly by 2010.

Another very important aspect in this respect is the expected *growth in transport* due to the enlargement process of the European Union. This growth refers to both passenger and goods transport.

For *passenger transport*, the crucial sign supporting this prediction is the tremendous growth in car use in the last 30 years, during which time the number of cars has tripled, increasing by 3 million cars each year (EC 2001). Car ownership is expected to increase rapidly in all the candidate countries of the EU.

For *goods transport* as well, dramatic changes are expected. The relocation of many European industries towards production sites with low production costs is expected to increase the burden on road transportation. In addition, the accession of candidate countries in the EU is expected to bring heavier transport flows, in particular for road haulage traffic.

A major challenge of the CTP is its adaptation to the goal of *sustainable development*. This goal has already been introduced by the Amsterdam Treaty, but is still at the forefront of discussion because of the impact of the trends described above. This goal can only be achieved by the further integration of environmental considerations into all Community policies.

The scope of such an adaptation is on *shifting the balance between transport modes*, thus promoting transportation modes which are more environmentally friendly. In this context, it is essential to put efforts into the development of a Europe-wide *sustainable transport system*. This goal is of crucial importance, since the relevant studies are forecasting a considerable future increase of transport sector emissions: these are expected to increase by 50 per cent in 2010 if nothing is done (EC 2001).

The *future policy guidelines*, as they are expressed in the White Paper on the European CTP for the year 2010 are concentrated on the following issues (EC 2001):

- The shift of *the balance between transport modes* by favouring the more environmentally friendly modes. In this context, the EU is working out policies in the following directions: revitalizing the railways; improving quality in the road transport sector; improving road safety; promoting transport by sea and inland waterway; effective charging of different transport modes; striking a balance between growth in air transport and the protection of the environment; and putting research and technology at the service of clean, efficient transport.

- The elimination of *bottlenecks* in order to end up with integrated transport systems by introducing policies on: further developing their intermodality; and building the TEN-T.
- The placing of *users* of transport at the heart of transport policy by: raising the quality of transport services; recognizing the rights and obligations of users; and developing high quality of urban transport.
- The management of the *globalization of the European transport systems*, aimed at increasing the openness of the EU in a transport context.

Moreover, European transport policy has to take into account the evolving context, within which such a policy is decided and implemented. This context reveals that some *key issues* are coming to the fore (EC 2006), as described below.

A major issue in this respect relates to the *enlargement*, which gives the EU a *continental substance* within which there are two main subjects: the one is related to the *value of rail and waterborne transport* in the context of continental Europe; the second is associated with the increasing the *diversity of accessibility* of the Member States, which can be subject to differentiated solutions at the local, regional and national level along the lines of the European-wide transport market.

Another critical issue in the evolving context is that *technological advances* in the transport sector are becoming crucial for its further development. Transport is becoming a high-technology industry, where technological innovation, supported also by EU initiatives, contributes largely to increasing competitiveness, as well as facilitating adjustment to environmental and social goals. Efforts are dispersed in various fields, such as air traffic management, decongestion of transport corridors, urban mobility, intermodality and interoperability, safety and security of transport systems. Moreover, some very promising areas in progress are intelligent transport systems involving communication, navigation and automation, energy efficiency and use of alternative fuels, which will dramatically affect the potential for the effective management of transport systems as well as the fulfilment of energy and environmental goals.

These evolutions are crucial in the context of *global environmental commitments* and refer to the transport sector as well. Reducing harmful emissions, for which transport is one of the main sources, is a challenge at the global scale. This brings to the fore another very important issue, the *energy* consumed by the transport sector, which relates to both harmful emissions and scarcity of resources. Transport as a sector has to discover how to adjust to the goals set by the EU in the energy sector, which implies finding ways to improve energy efficiency and promote environmentally-friendly fuel alternatives.

Conclusions and Prospects

Mobility patterns are going to further increase continuously as a result of the economic growth and enlargement of the EU, as well as the globalization of the new economy. This has certain implications for the potential of the EU's Transport

Network System, such as congestion, environmental damage, safety and so on, which all lead to the reinforcement of bottlenecks of the system.

In order to be able to deal with such problems, the European Common Transport Policy should be subject to *continuous adaptations* based on new circumstances and always in conjunction with the main goals set by the Community. This will result in a Trans-European Transport Network which will make optimum use of the infrastructure, ensure the removal of bottlenecks, be environmentally friendly, provide access to peripheral and landlocked regions and contribute to the development of the European regions. However, transport policy alone is not sufficient to reach these objectives. A combined effort is needed to tie in with policies in other sectors, such as economic policies, urban and land use planning policies, social policies, local urban transport policies and so on.

Moreover, transport policy should take into consideration the evolving context within which decisions are made. New challenges are emerging in such a context, referring to the enlargement process and the new potential of specific European transport modes, the technological progress in the transport sector and its potential to deal with various transport inefficiencies at a European scale, the global environmental commitments and the way the transport sector makes its own contribution, as well as the issue of the energy consumption of the transport sector and its alignment with the energy goals set in the European context.

References

Button, K., Nijkamp, P. and Premius, H. (1998), *Transport Networks in Europe: Concepts, Analysis, and Policies* (Cheltenham and Northampton: Edward Elgar).

CER and UNIFE (2006), 'Fact Sheet: TEN-T budget for 2007-2013: Modern Infrastructure Is Key to European Competitiveness', <http://www.cer.be/files/ / TEN-T_Budget-093408A.pdf>.

Club de Bruxelles (1990), 'La Politique Européenne de l'Environnement' [The European Environmental Policy], Bruxelles.

COM (1998) 'A Strategy for Integrating Environment into EU Policies', 333final, 27.05.1998.

De Palacio, L. (2003), Opening speech, Symposium 'Towards the Integration of the Trans-European and Pan-European Transport Networks', Brussels, May.

European Commission (1992a), 'White Paper: Growth, Competitiveness and Employment', Office for Official Publications of the European Communities, Luxembourg.

—— (1992b), 'White Paper: Future Development of the Common Transport Policy', Office for Official Publications of the European Communities, Luxembourg.

—— (1995a), 'Trans-European Transport Network', Outline Plan.

—— (1995b), 'The Common Transport Policy – Sustainable Mobility: Perspectives for the Future', Action Programme 1995–2000, Commission Communication to the Council, European Parliament, Economic and Social Committee and Committee of the Regions (COM/95/302 final, 12.07.1995).

— (1996), 'Integrated Strategic Infrastructure Networks in Europe', COST 328 <http://cordis.europa.eu/cost-transport/src/cost-328.htm>.

— (1998), 'The Policy Priorities', The Commission's Work Programme for 1999–2000 (COM, 1998, 604, 28.10.98), Indicative List of Actions envisaged in 1999 in the Framework of the Commission's Work Programme (SEC, 1998, 1723, 5.11.98).

— (2001), 'White Paper: European Transport Policy for 2010 – Time to Decide', Office for Official Publications of the European Communities, Luxembourg.

— (2006), 'Keep Europe Moving: Sustainable Mobility for Our Continent', Mid-term Review of the EC's 2001 Transport White Paper, Communication from the Commission to the Council and the European Parliament, COM(2006)314 final, Brussels 22.06.2006 [SEC(2006)768].

European Communities (2005), 'High Level Group on the Extension of the Major Trans-European Transport Axes to the Neighbouring Countries and Regions', Final Report, November.

EUROSIL (1999), 'Guidelines for the Appraisal of the Interconnection between Intermodal Transport and Area Development, Deliverable 9/10, DG VII, Contract SC-1131, EUROSIL Consortium, Athens (Status: Restricted).

Giaoutzi, M. and Stratigea, A. (2006), 'Impact Assessment of Trans-European Networks on Area Development', in E.R. Alexander (ed.), *Evaluation in Planning: Evolution and Prospects* (Aldershot: Ashgate), pp. 221–36.

Stratigea, A. and Giaoutzi, M. (1999), 'The Role of TEN and PEN in Border Areas', *Rivista Geografica Italiana* 106:3–4, 361–78.

Stratigea, A. and Giaoutzi, M. (2001), 'Barriers Appearing in Large Scale Projects Evaluation: The EUROSIL Approach', NECTAR Workshop, Potenza, November.

Suarez-Villa, L., Giaoutzi, M. and Stratigea, A. (1992), 'Territorial and Border Barriers in Information and Communication Networks: A Conceptual Exploration', *Tijdschrift voor Economische en Sociale Geografie*, 83:2, 93–104.

Annex 11.A

List of priority projects according to Decision No. 884/2004/EC:

1. Railway axis Berlin–Verona/Milan–Bologna–Naples–Messina–Palermo.
2. High-speed railway axis Paris–Brussels–Cologne–Amsterdam–London.
3. High-speed railway axis of South-West Europe.
4. High-speed railway axis east.
5. Betuwe line.
6. Railway axis Lyons–Trieste–Divaca/Koper–Divaca–Ljubljana–Budapest–Ukrainian border.
7. Motorway axis Igoumenitsa/Patras–Athens–Sofia–Budapest.
8. Multimodal axis Portugal/Spain–rest of Europe.
9. Railway axis Cork–Dublin–Belfast–Stranraer.
10. Malpensa airport.
11. Øresund fixed link.
12. Nordic triangle railway/road axis.
13. United Kingdom/Ireland/Benelux road axis.
14. West coast main line, United Kingdom.
15. Galileo.
16. Freight railway axis Sines/Algeciras-Madrid-Paris.
17. Railway axis Paris–Strasbourg–Stuttgart–Vienna–Bratislava.
18. Rhine/Meuse–Main–Danube inland waterway axis.
19. High-speed rail interoperability on the Iberian peninsula.
20. Fehmarn belt railway axis.
21. Motorways of the sea.
22. Railway axis Athens–Sofia–Budapest–Vienna–Prague–Nuremberg/Dresden.
23. Railway axis Gdansk–Warsaw–Brno/Bratislava–Vienna.
24. Railway axis Lyons/Genoa–Basle–Duisburg–Rotterdam/Antwerp.
25. Motorway axis Gdansk–Brno/Bratislava–Vienna.
26. Railway/road axis Ireland/United Kingdom/continental Europe.
27. 'Rail Baltica' axis Warsaw–Kaunas–Riga–Tallinn–Helsinki.
28. 'Eurocaprail' on the Brussels–Luxembourg–Strasbourg railway axis.
29. Railway axis of the Ionian/Adriatic intermodal corridor.
30. Inland waterway Seine–Scheldt.

Chapter 12

European Integration Policy on Telecommunications Networks

Vassilios Vescoukis and Maria Giaoutzi

The Programme Context

In order to support the deployment of e-society applications and services, in 1997 the European Community launched the Trans-European Telecommunications Networks (TEN-Telecom) action, which since 2003 has been called 'eTEN'. TEN-Telecom aims to promote the use of global telecommunications networks in areas of high socio-economic value, by promoting across Europe new multimedia applications and generic services of common interest that such infrastructures make possible. It is biased towards applications from the users' point of view – not that of infrastructure deployment.

TEN-Telecom adopted the multi-network approach, addressing not single bearer telecommunications networks, but instead services which seamlessly integrate fixed and mobile network components. In one sense, TEN-Telecom helps participating parties to bridge the gap between technical evolution and market operations, accelerating the deployment of new services to the market, and encouraging investments in services of common interest that are not likely to obtain immediate funding.

TEN-Telecom is part of the TEN Community action. The TEN initiative is based on Articles 129b, c and d of the Maastricht Treaty, which considers that, in addition to transport and energy, telecommunications is a sector that should benefit from a European action favouring the interoperation of networks and bringing access to them for all. TEN-Telecom, as was the name of the Community Action at the time, has been legally based on the Financial Regulation of 18 September 1995.[1] This Regulation defines the conditions and procedures for the granting of Community financial aid in the field of trans-European networks for transport, energy and telecommunications, and was amended on 19 July 1999 in order to facilitate further access of the private sector to Community funding and to introduce a new form of Community aid: namely, risk capital participation for investment funds with a priority focus on TEN projects.

1 Council Regulation No. 2236/95 of 18 September 1995 concerning Community financial aid in the field of trans-European networks.

The particular thematic area of TEN-Telecom is defined in the TEN-Telecom Guidelines, adopted in June 1997 by the European Parliament and the Council.[2] These Guidelines cover the objectives and priorities of the action. Annex I of the Guidelines lists projects of common interest, eligible for receiving financial aid under the conditions set in the Financial Regulation. In January 1998, the Commission adopted the TEN-Telecom work programme, and, ever since, several revisions have been made[3] based on the experience and interim evaluations.[4]

TEN-Telecom Policy

The most important policy document of TEN-Telecom is the 'TEN-Telecom guidelines' (see Footnote 2), where the objectives and priorities of the action are defined. In Annex I of the same document and in its subsequent revisions, the thematic areas of projects within the scope of the action are defined. In the text below, the main policy concepts of TEN-Telecom will be presented.

Objectives and Priorities

The following objectives have been recognized as being of primary importance by TEN-Telecom:

> *Objective 1.* The facilitation of the transition towards the e-society for all. A special emphasis is put on the 'for all' phrase, which denotes that the needs of disabled and elderly persons should be considered, and that the social inclusion of all Community Members should be regarded as a priority, taking into account that e-exclusion is a phenomenon that can easily develop in the 'e' era;
> *Objective 2.* The improvement of the competitiveness of European enterprises, in particular small and medium-sized enterprises (SMEs), and the strengthening of the internal market;
> *Objective 3.* The support of economic and social cohesion in Europe, which can also be viewed as an outcome of the previous objectives;
> *Objective 4.* Last but not least, the stimulation of new activities that lead to job creation, in accordance with the EU policy for fighting unemployment.

In order to achieve these objectives, the guidelines of Annex I of Decision No. 1336/97/EC (hereafter 'Guidelines') identify the following priorities:

---

2 Decision No. 1336/97/EC of the European Parliament and of the Council of 17 June 1997 on a series of Guidelines for trans-European telecommunications networks. *Official Journal* No. L 183, 11/07/1997, pp. 12–20.

3 Report on the implementation of Decision no. 1336/97/EC on a series of guidelines for trans-European telecommunications networks. COM(2001) 742 final, 2001/0296 (COD), Brussels, 10.12.2001.

4 Court of auditors special report No. 9/2000 concerning trans-European-networks (TEN) – telecommunications, accompanied by the Commission's replies (2000/C 166/01).

Priority 1. The promotion of telecommunications network-based multimedia applications that are in line with the above objectives. Despite the common interest of the applications concerned, the uncertainty of short-term commercial viability, due to the innovative character of the application, as well as other both technical and non-technical factors, can discourage private initiatives in this domain. Therefore, the following categories of value-added service providers are relevant:

- ISPs (Internet Service Providers), resellers or wholesalers, who want to provide multimedia applications to both residential and business customers;
- infrastructure operators, including satellite and mobile operators, media players, and retail chains developing business models for multimedia services;
- administrations or public service organizations running services with significant socio-economic impact, under the condition that they are willing to offer their services at least at trans-European level.

Priority 2. The support of not end-customer, Internet-based generic services (such as value-added services and content services), which make possible the appearance of new applications on open service platforms and ensure their interoperability, reliability and security. This domain addresses innovative service providers bringing solutions for the e-society, such as e-commerce, e-education, e-health.

Priority 3. The interconnection of all types of communication networks (fixed, mobile, or satellite), and the interoperation of networking infrastructure through an application or generic service focus. The provision of transparent low-level network services will be made possible though actions that will focus on this priority.

The above-mentioned objectives and priorities of TEN-Telecom are shown in Table 12.1.

Table 12.1 TEN-Telecom objectives and priorities

What?	How?
• Promote transition to the e-society for all • Improve European competitiveness of SMEs • Support economic and social cohesion • Create new jobs	• Support the deployment of multimedia applications by ISPs, infrastructure operators, and the public sector • Support Internet-based middleware services for commerce, education and health. • Support network-level interconnection

A Three-Layer Model

TEN-Telecom will open up the Community market to the new applications and services which will form the basis for the development of the information society.

It is the Commission's position that such application and services are essential for supporting Community prosperity, creating jobs and strengthening economic and social cohesion.

In order to be able to obtain a holistic view of the projects that will be part of the action, a three-layer model has been proposed as a generally accepted framework for describing TEN-Telecom (Figure 12.1). The layers are:

- The Applications layer, through which users interact with generic services and basic networks in order to meet their professional, educational and social needs.
- The Generic Services layer, which is made up of compatible generic services and their management. These services supplement applications whilst aiding their interoperability through support of the applications' common requirements, and by providing common tools for the development and implementation of new applications.
- The Basic Networks layer, which provides the physical access, transport and switching elements of the networks, including their management and signalling. These elements ensure TEN interconnectivity in the liberalized telecommunications market context.

These three layers form a coherent structure, where applications are supported by the two lower layers: generic services and basic network. Applications, in particular, cannot be offered in the absence of one of the other two layers; however, each layer should be sufficiently open to support any element of the layer immediately above. In this context, projects of common interest should be identified on the basis of their operational capability to support the objectives discussed earlier. The following

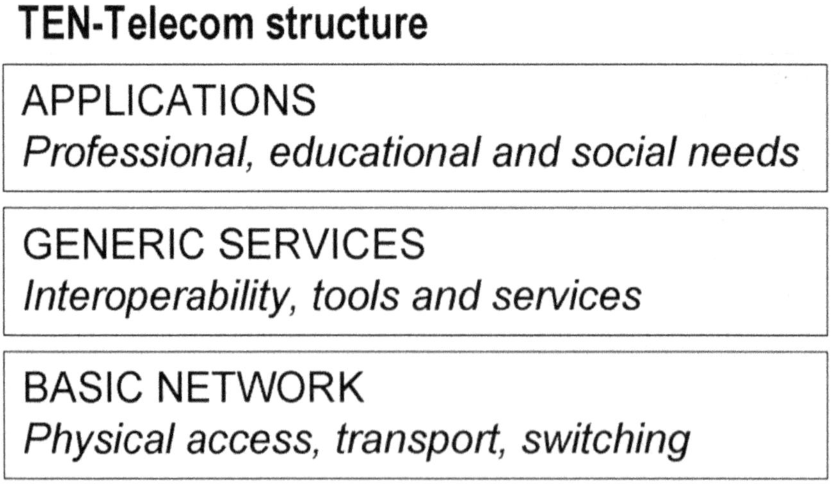

Figure 12.1 TEN-Telecom structure

sections identify projects of common interest in each layer of the early TEN-Telecom.

TEN-Telecom Projects

This chapter further specifies the projects of common interest already identified in Annex I of the Guidelines. Annex I identifies 12 application projects of common interest, three generic service projects, and four basic network projects, according to the three-layer model of the telecommunications sector.

TEN-Telecom Applications

TEN-Telecom for Education and Training

Education and training sectors increasingly use information and communication technologies, mainly based on the accelerating development of the Internet. This process, normally confined to a national basis, should also be encouraged on a European level by the creation of sustainable activities using integrated multimedia products and demonstrating their potential to generate revenues.

The objective in this sector is to stimulate the commercial deployment of multilingual, interactive and multimedia educational and training tools and services across the EU, and to promote Community-wide good practice for multimedia educational services. The European dimension of the education and training for all can be reinforced with:

- The commercial deployment of networked educational and training applications and services fostering student development and exposure to information technology through their networking curriculum. These services should have a demonstrated customer base, being part of the mainstream activities of key players in the field (for example, Schools, Universities, Corporations, SMEs, Business schools, Training centres).
- The development of brokerage services giving access to existing networked educational and training facilities. These services should offer facilities for the creation, access, broadcast, trading, and use of multilingual, multimedia educational and training materials. They should also provide appropriate copyright management environments, network access to end (individuals) and intermediate users (companies, educational institutions), and to all players concerned (for example, authors, teachers, learners, publishers, software developers, media players, telecom operators) at the broader European level.
- The promotion of existing (regional/national) networked education and training services as models of good practice. For example, identifying educational and training courses and services that can be extended to other areas within the EU and abroad (for example, for the promotion of digital literacy, for the long-term unemployed and job creation, for training on information technology, and for less developed countries).

TEN-Telecom for Access to Europe's Cultural Heritage

The objective is to create the conditions for the deployment of multimedia applications giving access to the European cultural heritage (collections and materials held in libraries, museums and audio-visual archives) and developing models that allow a fair return on investment, even if many content-holding organizations operate on a non-profit basis.

The successful deployment of these applications requires the creation of co-operation frameworks between media, cultural, legal and governmental actors and due consideration to legal constraints affecting the exploitation of applications (copyright, licensing). Applications should be interoperable and based on accepted standards and protocols for digitization and data exchange, thus allowing global, multilingual access to European cultural heritage.

Projects should be undertaken in the following areas:

> *Intermuseum virtual exhibitions*. Current technology allows the implementation of virtual exhibitions by bringing together multimedia information from various collections or from inside a museum. The objective is to develop new frameworks for virtual and thematic exhibitions enabling art trends to be known by a large public.
>
> Multimedia exhibitions adapted to tourism can also increase tourists' understanding and appreciation of Europe's cultural heritage. Furthermore, they can support the dissemination and promotion of Europe's cultural wealth, as well as contribute to preserving cultural awareness.
>
> *Cultural heritage and education*. The use of multimedia by teachers is becoming increasingly routine as an aid to teaching students. The integration of these products into school and university courses has to be addressed from both the museum and the educational perspective. The objective is to stimulate the deployment of educational multimedia products which exploit Europe's cultural heritage.

TEN-Telecom Applications for SMEs

In the light of increasing global competition, many SMEs are looking to international markets and adopting innovative marketing strategies. Multilingual capabilities for marketing, trade and doing business should be offered in a trans-European framework which will ensure interoperable, secure and confidential transactions for SMEs willing to operate on the network.

Projects in this area will support the use of TEN-Telecom applications and services by SMEs, with links to public authorities, trade associations, consumers and suppliers. Priority is given to projects focusing on:

- information retrieval and push services which include applications supplying geographic information data or other public sector data (for example, customs and taxation information, and trade statistics) over the network for use in local business, and users' customized transactional services provided over the Internet;

- electronic tendering with an emphasis on projects that create an electronic marketplace, which brings together buyers and suppliers of products and services for public procurement purposes, providing seamless services on a European scale;
- electronic commerce with a focus on projects facilitating the creation and development of SMEs through the streamlining and automation of business processes, increasingly using Internet facilities;
- management networks for SME support organizations providing assistance to local companies on the access to qualified resources and solutions and the automation of cross-border business processes. Public Administration as well as Territorial Coordination Agencies are organizations that can play an important role.

TEN-Telecom for Transport and Mobility

The transportation sector is heavily reliant on telecommunications services. The sector's requirements are complex and often trans-European in character, making it a target customer base for new telecommunications applications.

Full advantage has to be taken of TEN-Telecom to provide user-oriented and value-added services, which will considerably enhance the quality of life and the optimizing of transport, travelling and leisure activities considering, in particular:

- the logistical support for transport industries, freight and fleet management for the different transport modes;
- innovative security (for example, anti-theft services), emergency and maintenance support services, such as the mobile emergency calls all over Europe (for example, the Europe-wide emergency phone number 112);
- new traffic data acquisition and integration services (for example, based on the floating car data concept);
- the provision of seamless, dynamic and location-based travel, tourism and traffic information, such as trip planning and route guidance enriched with general or personalized information services to the mobile citizen on a real-time basis.

In addition, networked transport services in urban and rural areas will be covered, taking into account standardization and interoperability requirements and the potential for integration with other sectoral applications (for example, tourist information, air quality monitoring and control, geographic information and global positioning).

The deployment of all these services will be based on global/regional telecommunications infrastructure networks of fixed, mobile, and satellite components and should exploit synergies between communication components and global positioning and navigation systems. It should also satisfy, wherever applicable, the necessary complementarity and interoperability with the trans-European transport network (TEN-T).

TEN-Telecom for Environment and Emergency Management

TEN-Telecom can significantly contribute to the monitoring and management of the environment, including emergency management. There is a growing demand for environmental information services (covering, for example, air, water, soil and bio-diversity) on the part of environmental managers in both public administrations and private enterprises in order to enable them to comply with existing and emerging national and European policies and directives (for example, to guarantee air quality in our cities). On the other hand, trans-European value-added services based on harmonized environmental public data will create new markets with more user-oriented products and services (for example, weather information, pollution, ultraviolet and pollen levels), as well as contribute to better environment management decisions.

For global emergency management and public safety systems (such as, but not limited to, flood management for urban areas, rivers and estuaries, forest fires and industrial risk management), reliable communication systems and networks (mobile, satellite, radio and so on) and accurate positioning of resources form the essential backbone for real-time decision aid in crisis situations. These systems are also important to forecast natural disasters and estimate the damage they cause. Remote sensing applications utilizing space-borne and terrestrial data will play a more prominent role in environmental emergency management and monitoring (for example, sea ice and coastal water monitoring and supervision of unattended points).

TEN-Telecom for Health

The health care sector is subject to continuous transformation. Following progress in medical science, new methods of diagnosis and treatment have been introduced. Meanwhile, life expectancy is rising, and people have higher requirements concerning the health care services provided. Computerization and the use of telecommunications-based applications provide opportunities for innovation in the health care delivery process, setting the goals of allocating resources more efficiently, increasing the quality of care and the quality of life of the citizens.

The objective in this sector is to support and facilitate a structured information flow between all actors of the health care chain (for example, hospitals, either public or private, practitioners, pharmacies, social insurance) for the benefit of the patient. Projects in this area have a process-oriented approach, describing how the proposed services affect and improve the existing health care delivery process and what are the new business models that can be derived. The real-environment hurdles for launching the service should be addressed, taking account of the various conditions of the different Member States. Three mainstream activities are mentioned as typical examples:

Public health information services. Information services for authorities, health care institutions, professionals and citizens are to be deployed, in order to increase access to and quality of data related to the prevention, diagnosis and therapy of diseases (for example, Internet-based health portals). The opportunities should be improved through e-health centres where citizens can make informed choices of health care service providers, self-care, healthy living, health and safety at work, and disease prevention.

Continuity of care and health care management. By developing applications and integrated networks for complementary health centres, shared access to transparent patient-related information should be promoted, respecting rights of access. The matching of patients' needs to available and relevant health care resources is another subject for exploitation.

Telemedicine. The purpose is to link points of care and to develop supporting services. This involves teleconsultation and telediagnosis, including diagnosis and supervision of patients and citizens with special needs at home or living in small, remote and isolated communities.

The objectives are to develop services that permit remote consultation between professionals in specialized centres, peripheral hospitals and other points of care, and that provide citizens with effective health care in their homes, in isolated places and in cases of emergency. By this, the interoperability of different and multilingual health care systems in Europe should be emphasized and improved.

TEN-Telecom for City and Regional Information

A significant awareness has been established amongst local and regional authorities about the importance of telecommunications networks for the economic and social life in their area and the advantages their use can offer. There is substantial scope for improving and extending communication on both a local and a regional basis between public authorities and administrations, business and the citizens, which would constitute a significant step towards the e-society. The improvements affect the democratic and bureaucratic processes, equality of information access, cost efficiency and the quality of life for the citizens.

User-friendly applications at a 'single point of entry' should be set up, demonstrating the possibilities of affordable and sustainable access for the citizens to an integrated set of services of collective interest. As typical examples, we mention here the development of systems and services for public administrations to improve cost-effective mobile and fixed access to regulatory information and legal documents, as well as the facilitation of exchanges between administrations, citizens and business.

A suitable user platform should be developed on existing networks and infrastructure, addressing bottlenecks and interoperability, and using appropriate standards and generic services in order to run the applications (for example, via Intranet/Internet solutions and smart cards). The acceptance and use of secure electronic documents could be an important component for these purposes. The involvement of, and liaison with, political and business decision makers encourages the integration of IT strategy with socio-economic and commercial objectives, and brings a stronger deployment orientation to the projects.

As for the required trans-European dimension, simultaneous implementation in different cities, and common testing and market validation of new services will allow cross-fertilization and exchange of best practice between the project promoters. Thereby, possibilities will also be provided for interconnectivity between different cities and regions in the next step.

TEN-Telecom for New Work Methods and Services for the Job Market

Businesses require employees with higher flexibility and mobility. While some factors may increase the need for mobility, such as companies having various geographical locations, or the general tendency to leave city centres for cleaner suburban areas, mobility is limited because of increased demand for transport systems. New ways of working should propose sustainable and intelligent workplaces and working conditions that bring better life for citizens, economic growth and job creation.

The objective in this sector is to promote the deployment of applications and services that integrate interdisciplinary work methods and tools for a large number of business situations, within a sound technical and operational framework. This framework should take into account the impact of telework on production and innovation costs, on flexible individual solutions at work (for example, for the disabled), as well as on energy consumption and the environment.

Applications should be able to manage large amounts of data, collaboration amongst work groups, organizational change processes, and virtual and modular organizations resulting in reduced physical workplace presence and the integration of work, learning and leisure. These applications are expected to support new work organizations for companies with a dispersed and mobile workforce, with appropriate provisions for data protection, technical and information assistance, invoicing, social security, and health and safety at work.

Network services such as job information databases should be deployed to support the changing labour market in the Community and to help tackle unemployment and the social integration of disabled people.

TEN-Telecom for Research

Electronic networks are an indispensable instrument in a researcher's daily work, for the easy access, exchange and dissemination of information. The objectives in this sector are to demonstrate the commercial viability of trans-European broadband interconnections between national and industrial research and education networks providing multimedia training, collaborative research and access to resources, as well as to support peering agreements between the research networks, beneficial for industry, university and research centres.

Generic TEN-Telecom Services

Generic services are those network-based services which facilitate the operation and management of applications supporting commercial and societal activity in the e-society. They can form building blocks in the provision of specialized services, or may be of widespread use in broad sectors of industry, commerce or the general life of the citizen. The provision of standardized generic services, which can be employed and managed by a wide range of applications, reduces both the development cost and time to market for more specific network-based applications.

Generic services should be widely accessible and extend the availability of the e-society, in particular to the private individual and to SMEs. They should therefore be based upon existing and available network services; this implies that they should be focused on the Internet. Particular attention should be paid to mobility, as this has both great potential for commercial growth and increasingly wide application in daily life.

Some projects of common interest in the area of generic services are discussed below in the following sections.

Generic Information Services

These are services permitting the handling of large volumes of data associated with multimedia. They include the management and transmission of complex information such as video. Services should include storage, indexing, retrieval and interaction.

Services should be broadly accessible in the context of a multilingual society. Attention should therefore be paid to interfaces offering multilingual support and accessibility in the user's mother tongue through, for example, use of automatic translation. Services should be useable by persons without specialized technical knowledge, requiring a wide range of multimedia information access services through user-friendly interfaces.

Services are required to enable users to identify and procure network offerings in a cost-effective manner, through unified directory services, associated with content-checking services, and intelligent tariff information services. Services should also be able to conduct transactions over networks in a secure manner both by protecting exchanges of sensitive information and the integrity of data, and by authentication services (electronic identification and digital signature, certified e-mail operators, or alternative approaches tailored to the specific needs of different classes of customers).

Services for Industry and Commerce

These services enable users, particularly in SMEs, to use network services in their commercial and trading operations. These include all aspects of:

- secure dealing over networks (secure transmission, trust services, fraud prevention, online dispute settlement and alternative redress procedures, authentication and non-repudiation);

- services enabling users to publicize their products and obtain information about available goods and services, as well as infomediaries, combining and integrating services and content;
- services designed to protect intellectual property rights, copyrights, patents and origin of products, and to gather income arising from licences and other forms of intellectual property.

Information services to support decisions in the procurement of other network services are also required.

New Access Methods, Mobility and Convergence

These are services which enable users to access higher capacity network services either through existing installations or by means of newly commercially available equipment (for example, satellite, mobile and so on). Convergence of services is an important topic as the barriers between voice and data services erode. Projects should address the provision of multiple services over networks.

Mobile services are of great importance. Projects should address issues of personal mobility and number portability. Projects are also invited to address methods of accessing services from portable devices (for example, handheld next-generation mobile phones) and in mobile situations (for example, in a car or train).

TEN-Telecom Basic Networks

New requirements emerging from the e-society demand high performance access and backbone infrastructures. TEN-Telecom will encourage and lay the basis for the deployment of high-quality multimedia services for end-users and support the gradual introduction and interoperation of integrated broadband communication (IBC) network solutions.

Euro-Integrated Services Digital Network (EURO - ISDN)

The Community intervention in support of Euro-ISDN (Integrated Services Digital Network) has fulfilled its aims. ISDN is now an active niche market well supported by European industry. TEN-Telecom will not prioritize ISDN-based applications, it will rather adopt, as far as possible, a technology-neutral, multi-network approach integrating, when necessary, the fixed and mobile networks.

Interoperation of Broadband Networks

Satellite-based infrastructures increasingly meet key user requirements such as personal mobility, access to high speed Internet for multimedia services, and global connectivity. TEN-Telecom supports the deployment and scalability, over heterogeneous network infrastructure, of satellite-based multimedia and interactive services and applications, hopefully involving a range of actors in the value chain (application developers, content providers, service and network providers, terminal equipment suppliers and financing bodies) who are committed to becoming global players. The objective is to demonstrate the commercial viability of such services and applications when supported by a mix of interworking satellite and terrestrial infrastructures. It includes (but is not restricted to) the validation of traffic and revenue models, the operational viability of the management of satellite and terrestrial networks and the commercial adequacy of administrative services ensuring the provision of an adequate quality service, tariff structure and integrated billing.

The rapid penetration of *mobile telecommunications* offers new opportunities for multimedia services and applications. TEN-Telecom supports the deployment of advanced and innovative mobile services (for example, very high data communications, Internet/Intranet applications and mobile video) based on mobile networks. These services should be delivered on a range of wireless access technologies and networks based on open standards and will demonstrate the market viability of terrestrial wireless systems to provide feature-rich applications in a mass market or in a public service context.

Interconnection and interoperation of the existing networking technologies is required to create an effective trans-European networking infrastructure. Fixed, satellite and mobile networking technologies supporting a range of protocols at different levels (for example, ATM, xDSL and IP) are becoming available and it is unlikely that a single solution will emerge as the universal one. The objective in this area is to commercially validate and encourage end-to-end seamless service provision in an operational context supported by a heterogeneous yet inter-networked infrastructure, including core and access networks, with Internet protocols (IP) as one important network technologies integrator

Implementation Experience

TEN-Telecom has been funding projects since 1998. Since then, an external contractor as well as the European Court of Auditors have reported on the implementation of the action.[4,5] The discussion on the social impact of TEN-Telecom is based on anticipated results shown in the annual project reviews. However, a qualitative approach to the assessment of the programme can be safely taken.

The principal objective of the programme is facilitating the transition towards the Information Society for all. Given that the term 'Information Society' has no clear and straightforward definition, there is no single, global perception of what the Information Society is. However, sticking to the generic definition of information society in the EU context, the weight is put on the 'for all' buzzword. In this respect, increased access to the Information Society is expected from generic services that provide low-cost telecommunication services over existing infrastructure.

At the applications layer, TEN-Telecom has funded several projects focusing on providing new modes of interaction between local and regional administrations

5 Intermediate Evaluation of the TEN-Telecom Programme, Final Report, PLS RAMBØLL Management A/S <http://europa.eu.int/information_society/programmes/evaluation/ index_en.htm#tentelecom>.

and citizens. In line with the objective of improving the competitiveness of SMEs, TEN-Telecom-supported satellite services can potentially lower the entry barrier to broadband services. Projects on the applications layer are likely to provide benefits to specific areas and to the digital market as well.

Contributions to social inclusion are anticipated from projects providing interactive and communicative services to citizens, such as easing communication with local administrations (E-GAP) and developing demand-responsive public transport.

Several technical and policy remarks have been made by auditors. The most important one concerning the TEN-Telecom policy relates to the deployment of TEN-Telecom along with other community initiatives. Provided that there are several Community actions in the area of telecommunications, a need to avoid overlap with the Research Framework programme and other sources of Community funding is identified. It has also been made clear that:

- more deployment projects (not studies) should be promoted;
- the public sector should be more heavily involved;
- the interconnection and interoperability of networks, as well as the coordination with other players in the structure of Figure 12.1, should be given further weight.

In the technical field, some minor observations concerning the programme administration and monitoring, as well as the communication of the programme profile itself to the interested parties, have also been recorded by the auditors.

Current Status and Future Prospects

Since 2003, TEN-Telecom has been called eTEN. The latest Call for Proposals for eTEN was issued in 2006, under the generic title 'Deploying Trans-European e-Services for All'.[6] It implements a transition towards the new Commission initiative for the Information Society, i2010, as a key element of the Lisbon Agenda. Beyond 2006, the instruments proposed to support these policies will be part of the Competitiveness and Innovation Framework Programme (CIP) to be implemented under the Financial Perspectives 2007-2013. eTEN has already funded a significant number of projects that address 'caring for people' with pilot services that broaden inclusion, that provide home care for the sick and elderly, and that provide easy access to Europe's cultural heritage and diversity, as shown in Table 12.2.

The i2010 initiative envisages an inclusive EU Information Society with high-quality ICT-enabled public services that support growth and employment.[7] It foresees an Information Society that benefits all citizens, provides better, more efficient and

6 eTEN Work Programme 2006, DG Information Society and Media, Trans-European Telecommunications Networks <http://ec.europa.eu/information_society/activities/eten/library/ /reference/workprog2006_en.pdf>.

7 i2010 – A European Information Society for growth and employment <http://ec.europa.eu/information_society/eeurope/i2010/index_en.htm>.

Table 12.2 Projects funded

Area	Number of projects
e-Government	57
e-Health	56
e-Inclusion	15
Learning and Culture	36
SMEs	84
Trust and Security	15

more accessible public services and improves the quality of life. In 2006 eTEN will place particular emphasis on proposals that address the third pillar of i2010, an inclusive society,[8] as demonstrated in the objectives outlined below.

- The principles of an inclusive information society should be considered in all services supported by eTEN.
- They provide support for new challenges and innovative approaches to services in the public interest, including organizational aspects and actions that increase their effectiveness and enhance their deployment and impact.
- They provide support for communicating project results including 'showcasing' pilot services and demonstrating concrete achievements at the national, regional and local level.

TEN-Telecom which has evolved into eTEN was an early EU initiative towards the integration of telecommunication infrastructures and services across Europe. eTEN has funded a significant number of projects with participation from all 25 EU members, which have resulted in useful cases of services that broaden inclusion, that provide home care for the less privileged, and that provide easy access to Europe's cultural heritage and diversity. Several of these projects may contribute to the 'flagship initiatives' of i2010, which is the next EU major initiative for the development of Information Society actions up to 2010 in the EU-25 social, economic and cultural arena.

8 Inclusion, better public services and quality of life <http://ec.europa.eu/information_society/eeurope/ /i2010/inclusion/index_en.htm>.

PART 3
Assessment of Network Integration Effects

Chapter 13

The Network Effect of Aircraft and High-Speed Train Substitution under Airline and Railway Integration

Moshe Givoni

Introduction

Traditionally aircraft and railways have competed with each other on routes where they provided comparable travel times. However, with the increase in congestion and in environmental concern regarding aircraft operation, the Air Transport Industry (ATI) has begun to see an opportunity in the railways, making the way for cooperation between the modes. Predominantly, such cooperation takes the form of using the railways as an access mode to airports, but recently it has also taken the form of using the railway to replace the aircraft without replacing the service provider (the airline). Substitution of the aircraft by railway services is usually confined to routes where the railway service is by High-Speed Train (HST).

Since airlines probably have no interest, at least in the near future, to operate railway services and add trains to their aircraft fleet, substitution of the aircraft by HST can take place only if the Train Operating Company (TOC) operates the railway services provided by the airlines.[1] For such cooperation to work, the level of cooperation between the airline and the TOC must lead to a full integration of the aircraft and the railway service. This type of cooperation is the focus of this chapter.

In order to clarify the differences between airline and railway competition, cooperation and integration, the following definitions are given:

> *Competition* between the airlines and the railways takes place on routes where both offer (separate) services. Under competition, the railway service starts at the city centre and not at the airport.
>
> *Cooperation* between the airlines and the railways takes place on routes where the railway and the aircraft journeys complement each other, and one mode does not substitute the other. In this case, the railway service starts at the airport.

1 'It also appears that airlines themselves are not the best providers of rail services' (Buchanan and Partners 1995, p. 10-4).

> *Integration* between the airlines and the railways takes place on routes where the railway and the aircraft journeys complement each other, but the train substitutes the aircraft on the railway segment of the journey. Again, in this case, the railway service starts at the airport. Integration means, and requires, seamless transfer between the plane and train to match the transfer characteristics between two flights operated by the same airline.

Aircraft and HST substitution is increasingly recognized as one way to confront the congestion and the environmental problems faced by the ATI. Yet, if such substitution leads to competition between the modes, it might even exacerbate these problems, because of the airlines' reaction to the competition, which might 'create more pressure for slots at airports as the airlines remaining on the routes would have to compete with railway on frequency' (Caves and Gosling 1999, p. 59). Therefore, it is assumed that, by avoiding direct competition between the airlines and the railways following mode substitution, the ATI can enjoy the benefits provided by substitution. If airlines will not benefit from mode substitution it will not take place, thus preventing benefits to passengers and society from materializing. On the basis of this assumption, the research adopts airline and railway integration as the preferred model for aircraft and HST substitution.

This chapter focuses on one aspect of airline and railway integration: its potential to free up runway capacity by using the HST to substitute for the aircraft on airlines' services. In addition, the effect of mode substitution on these routes is analysed, and extra network benefits, provided through connecting the air and rail networks, are discussed.

The model for integration on a specific route is illustrated in Figure 13.1. It is assumed that passengers who would benefit from aircraft and HST substitution are passengers who arrive at the airport in order to transfer between flights, or passengers who prefer, for different reasons, the airport over the city-centre HST station as the origin of their journey. The air journey begins with the passenger arriving at a hub airport by a flight, and then transferring to another flight that takes him to the destination city airport. At the destination airport the passenger transfers again, this time to a surface mode, to complete the journey to the city centre. In the case of airline and railway integration, the railway option, the passenger would arrive at the hub airport by a flight and then smoothly transfer to an HST service that takes him directly to the city centre of the destination city.

London Heathrow airport, one of the world's biggest airports[2], is used as the case study. For airline and railway integration to take place the airport must be connected to the HST network, the airport should be equipped to provide a seamless transfer between the plane and the train for both passengers and their baggage, and HST lines should be available on routes identified as suitable for mode substitution. In this chapter all these requirements are assumed.

2 In 2002, the airport handled 63.47 million passengers, third in the world, and 466,554 Air Transport Movements (ATMs), sixth in the world (ACI 2004).

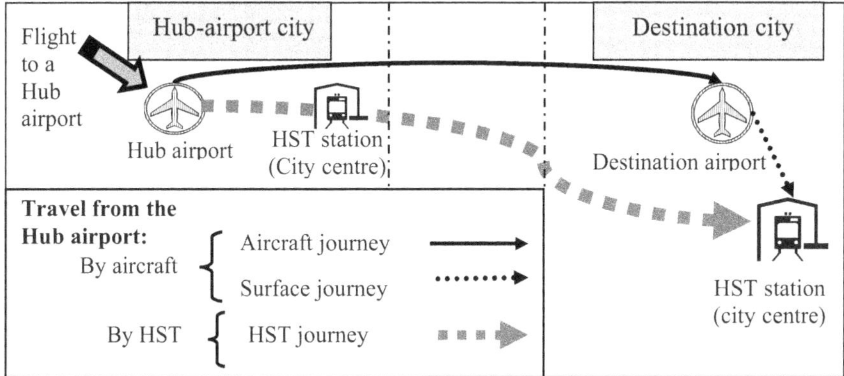

Figure 13.1 Schematic illustration of the aircraft and HST journeys

Airline and Railway Integration at Heathrow – The Potential to Free Up Runway Capacity

Several routes served from Heathrow are suitable for airline and railway integration and these routes, as well as the current capacity they consume, must be identified so that policy makers and airport owners can decide whether to provide the infrastructure for integration to take place. By evaluating the capacity which can be freed through mode substitution, the potential to relieve congestion at airports can be estimated.

Routes Suitable for Aircraft and HST Substitution at Heathrow

The criteria for aircraft and HST substitution are usually based on route distance. There is agreement that the HST ceases to be a competitive substitute for the aircraft on distances greater than 1,000 km (for example, Pavaux 1994; Janić 2003). Based on that and the routes currently served from Heathrow (BAA 2004), the following routes can be considered for aircraft and HST substitution (Table 13.1). In practice, even aircraft do not always follow the great circle distance between origin and destination because of Air Traffic Control (ATC) restrictions, and therefore 10 per cent was added to estimate the aircraft route distance. Rail services rarely follow the shortest route from origin to destination, and this must be accounted for when considering mode substitution. The route distance for the HST journey from Heathrow to the destinations considered for mode substitution was calculated based on the road distance from London and accounting for the proposed European HST network (UIC 2003, Figure 13.2). Considering the great circle distance, all routes in Table 13.1 seem good candidates for mode substitution, certainly the ones up to 800 km, which include all the routes from Leeds/Bradford to Zurich. Yet, a different picture emerges when taking into account the HST route distance. Then the routes to Aberdeen, Frankfurt, Hanover, Hamburg, Geneva, Stuttgart, Lyon and Zurich no longer seem good candidates for mode substitution. Furthermore, it is important to considerthe ratio between the aircraft and the HST route distance, since this can

Table 13.1 Routes from Heathrow suitable for consideration of aircraft and HST substitution (based on route distance)

	Great circle distance (in km)[a]	Aircraft route (in km)[b]	HST route (in km)[c]	Distance ratio (HST/aircraft)
Leeds/Bradford	227.4	250	313	1.25
Manchester	242	266	321	1.21
Paris	347.7	382	510	1.33
Brussels	350.3	385	398	1.03
Amsterdam	369.9	407	612	1.50
Newcastle	404	444	446	1.00
Düsseldorf	500.9	551	650	1.18
Edinburgh	532.5	586	641	1.09
Cologne	533.9	587	610	1.04
Glasgow	553.7	609	648	1.06
Aberdeen	646.2	711	832	1.17
Frankfurt	654.4	720	802	1.11
Hanover	702.5	773	904	1.17
Hamburg	744.6	819	1,063	1.30
Geneva	754.2	830	1,051	1.27
Stuttgart	756	832	1,006	1.21
Lyon	757.4	833	974	1.17
Zurich	788.3	867	1,204	1.39
Bilbao	927	1,020	1,425	1.40
Munich	941.2	1,035	1,229	1.19
Berlin	961	1,057	1,187	1.12
Nice	1,040.3	1,144	1,493	1.30
Barcelona	1,147.7	1,262	1,633	1.29
Madrid	1,254.9	1,380	1,811	1.31

Notes:

[a] *Source*: Landings.com (2003).

[b] Great circle distance plus 10 per cent.

[c] *Source*: Route Planner (http://www.europe.opel.com) based on the forecasted European High-Speed Network (UIC 2003).

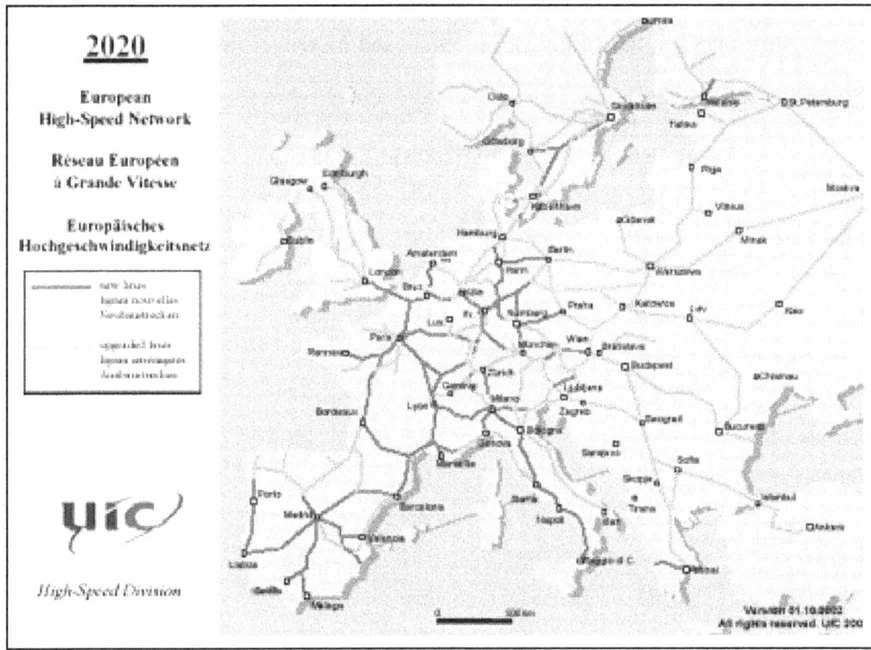

Figure 13.2 European High-Speed Network 2020 (UIC 2003)

show substitution is probably not worthwhile even if the HST route distance is within the range considered favourable for substitution. Thus, the route from Heathrow to Amsterdam is probably not suitable for mode substitution, as it would appear from considering the aircraft route distance, and even the HST route distance.

In practice, however, the potential for aircraft and HST substitution depends on the travel time more than on the distance, although the latter can provide an indication of travel time advantage. Assuming that the transfer time between two aircraft and between aircraft and HST at Heathrow is the same, then the HST journey travel time consists of the railway journey, while the aircraft journey travel time consists of the duration of the flight, the egress journey from the destination airport to the city centre, and the transfer between them (see Figure 13.1). Assuming the egress journey, on average, takes 20 minutes, and the transfer at the destination airport 60 minutes, the aircraft journey based on the airlines' advertised schedule (Innovata Flight Schedules 2004)[3] is shown in Table 13.2. The HST journey time depends on the average speed, and this is estimated at 200 kph for a good HST service, and 250 kph for an outstanding service.[4] Based on the average speeds, the HST journey travel

3 Different flights on the same routes have a different scheduled time, depending on expected levels of congestion at different airports and at different times of the day. The flight time presented in Table 13.2 is the most common low flight time (that is, not necessarily the fastest advertised time).

4 The benchmark for average travel speed could be the TGV Méditerranée line. According to the timetable a journey from Paris to Marseilles (750 km) runs at 250 kph

Table 13.2 Routes from Heathrow suitable to be considered for aircraft and HST substitution (minutes, based on travel time)

	Aircraft		HST time (avg. speed)		Travel time adv. HST	
	Flight	Journey*	(200 kph)	(250 kph)	(200)	(250)
Leeds/Bradford	55	135	94	75	41	60
Manchester	60	140	96	77	44	63
Paris	65	145	153	122	–8	23
Brussels	70	150	119	96	31	54
Amsterdam	70	150	184	147	–34	3
Newcastle	65	145	134	107	11	38
Düsseldorf	75	155	195	156	–40	–1
Edinburgh	75	155	192	154	–37	1
Cologne	75	155	183	146	–28	9
Glasgow	80	160	194	156	–34	4
Aberdeen	85	165	250	200	–85	–35
Frankfurt	95	175	241	192	–66	–17

* Flight time plus 20 minutes egress journey, plus 60 minutes transfer time between flight and egress journey.

time is presented in Table 13.2, and the advantage to travelling by HST over travelling by aircraft is derived. Under the assumptions described above, there seems to be a clear advantage to the HST, and hence a case for airline and railway integration on the following routes from Heathrow to: Leeds/Bradford, Manchester, Brussels, and Newcastle, even when the HST average speed is only 200 kph. But if the HST can obtain an average speed of 250 kph then there is also a case for mode substitution on the following routes from Heathrow: Paris, Cologne, Glasgow, Amsterdam and Edinburgh. On the routes to Düsseldorf, Frankfurt and Aberdeen, there is no case for mode substitution, at least from the travel time savings aspect.

This analysis indicates that, while route distance can provide a rough estimate as to whether airline and railway integration is a favourable option on a certain route, it is not accurate enough when the route distance is less than 1,000 km and probably when it is above 300 km. In this case the specific HST route distance needs to be accounted for, together with the ratio between the aircraft and the HST routes' distance. However, the route distance is only an estimate for the travel time, which depends mainly on the HST average speed. The travel time of the aircraft journey compared with the HST journey is the recommended criterion when identifying routes suitable for airline and railway integration.

(Perren 2001). For comparison, the current average speed between London and Paris (495 km) is about 165 kph, but should increase to around 210 kph upon completion of the Channel Tunnel Rail Link (CTRL).

The extent of the capacity that these routes consume at the airport depends on the level of service on each of them. This is examined next.

The Level of Service on the Routes Suitable for Aircraft and HST Substitution at Heathrow

Using data from the airlines' schedules for the summer of 2004 (Innovata Flight Schedules 2004), the number of services on the routes identified as suitable for substitution were calculated based on the number of one-way services from Heathrow on a Wednesday weekday.[5] Then, the daily level of the one-way service was multiplied by 364 days of the year, and then by 2 to account for the return journey as well. Table 13.3 shows the number of yearly ATMs, or slots, at Heathrow and the capacity they represent of Heathrow's total ATM capacity, which was in 2002 (the latest figure available) 466,554 ATMs, for each route and in total.[6]

Table 13.3 The scope for freed capacity at Heathrow following airline and railway integration (ranked by travel time advantage according to Table 13.2)

	Daily services (one way)			Yearly services (two way)		
	Total	By BA	By BMI	ATM	% of Heathrow	Aggregate
1. Manchester	15	8	7	10,920	2.3 %	2.3 %
2. Leeds/Bradford	4	–	4	2,912	0.6 %	3.0 %
3. Brussels	13	6	7	9,464	2.0 %	5.0 %
4. Newcastle	4	4	–	2,912	0.6 %	5.6 %
5. Paris	27*	9	5	19,656	4.2 %	9.8 %
6. Cologne	6	3	3	4,368	0.9 %	10.8 %
7. Glasgow	18	10	8	13,104	2.8 %	13.6 %
8. Amsterdam	23*	6	8	16,744	3.6 %	17.2 %
9. Edinburgh	16	8	8	11,648	2.5 %	19.7 %
10. Düsseldorf	9	3	5	5,824	1.2 %	20.9 %
Total	*134*	*57*	*55*	*97,552*		*20.9 %*

* The other services are by AF on the route to Paris, and KLM on the route to Amsterdam.

5 Wednesday is considered to represent an average level of service throughout the week. It is expected that the level of service will be lower at the weekend and higher at the start and end of the working week. Since only the 2004 timetable was available for routes other than the case-study route, this was also applied, in this section, to the case-study route.

6 Although 2004 traffic is expected to be higher than that of 2002, and the percentage of capacity would be an overestimate, CODA (2003) traffic figures for Paris, Amsterdam, and Brussels of 21,034, 17,919, and 15,064 respectively, during 2002 suggest that the figures in Table 13.3 are more likely to be an underestimate.

Table 13.3 shows that, if on all routes considered suitable for mode substitution airline and railway integration would take place, over 20 per cent of Heathrow capacity, one of the largest airports in the world, would be freed. Even if airline and railway integration only took place on the routes on which there are clear favourable conditions for substitution (routes 1–5 in Table 13.3), about 10 per cent of Heathrow capacity will be freed.[7]

However, not all airlines operating from Heathrow on the routes favourable for substitution are expected to adopt airline and railway integration. Airline and railway integration is particularly attractive for airlines operating hub-and-spoke (H&S) systems. At Heathrow this is British Airways (BA), and to a lesser extent British Midland International (BMI) through its alliance with Lufthansa (LH) and other Star alliance airlines. BA's level of service on the routes found suitable for substitution represent 8.9 per cent of Heathrow ATM capacity. Most of the capacity freed will come from BA's routes to Glasgow (1.6 per cent of Heathrow's capacity), Paris (1.4 per cent), Manchester (1.2 per cent), and Edinburgh (1.2 per cent).

Benefits from Airline and Railway Integration at the Airport Level

In general, aircraft routes which are suitable to be served by HST under airline and railway integration have similar characteristics in terms of route distance, travel time by each mode, HST average speed and so on. Therefore, the results of the evaluation of benefits from airline and railway integration on the route Heathrow to Paris (Givoni 2004) can indicate whether benefits can be expected on other routes as well.

The evaluation of operating costs (OCs) savings to airlines following airline and railway integration on the Heathrow-Paris route showed that in seat-km units the HST OCs are lower (Givoni 2003) but on the Heathrow-Paris route the longer distance covered by the HST results in the HST being more expensive to operate. Considering that the ratio between the HST and the aircraft OCs per seat-km is 1.33,[8] it can be expected that airlines will benefit from 'integration' on the routes in Table 13.3, except the routes to Paris and Amsterdam. Table 13.2 suggests that passengers almost certainly will enjoy travel time savings on routes 1 to 4 in Table 13.3, while on the other routes benefits can be expected if a very fast HST service is available. The scale of the environmental benefits from airline and railway integration found on the Heathrow-Paris route suggests that, on all the routes examined, reduction in local air pollution and in impact on climate change can be expected (IPCC 1999). Accordingly, net benefits from airline and railway integration on routes 1 to 4 in Table 13.3 can be expected, and also on the other routes if the HST service is provided at an average speed of 250 kph or more.

7 The scope for substitution is often considered to be 10 per cent of European internal passengers (Sharp 2002), 10 per cent of European scheduled airline capacity (Caves and Gosling 1999), or 10 per cent of all air passengers at the airport concerned (Buchanan and Partners 1995).

8 The OCs were €0.076 per seat-km for the aircraft and €0.057 per seat-km for the HST, a ratio of 1.33.

Additional Network Benefits

Airlines operating from congested airports cannot expand their services and add routes to their network. Airline and railway integration, however, can offer an alternative to the runway and allows airlines to expand their network and services even at congested airports. This holds for any airline, but is of much more importance for airlines operating a network of services, hub-and-spoke (H&S) operation, rather than a collection of point-to-point services with no direct connection between them. Hence, airline and railway integration allows airlines to increase the number of destinations in their network by offering rail (HST) services to destinations not served before because of lack of capacity. In the UK, there are no BA services from Leeds to Heathrow, which leads passengers to use other European airlines and airports for their long-haul journeys instead of using (British) airlines based at Heathrow.

For BA, the lack of capacity at Heathrow has meant that it has had to scale down its H&S operation by either shifting some services to Gatwick or withdrawing from some routes and, instead, concentrating on long-haul routes, mainly across the Atlantic;[9] in the meantime, Lufthansa (LH) and Air France (AF) are moving in the opposite direction and emphasizing their H&S operation (Doganis 2002). Lack of capacity has also meant that BA could not schedule 'waves' of incoming and outgoing flights next to each other, which is one of the main elements of H&S operation, in order to allow passengers a range of connecting opportunities at the hub airport in a relatively short time. 'Air France schedules 52 departures in 55 min at CDG, KLM 63 departures in 75 min at AMS and Lufthansa 68 departures in 105 min at FRA. In contrast BA schedules about 18 departures in every hour at Heathrow' (Doganis 2002, p. 2). HST services at Heathrow, like the services at Frankfurt, would allow BA to substantially increase its hourly departures at Heathrow. Yet, a prerequisite for that is, as outlined earlier, a fully integrated railway service and a Minimum Connecting Time of 45 minutes for transfer between a flight and a rail service.

The lack of capacity and, consequently, the obstacles in operating a H&S system, make it harder for BA to compete against its main rivals Air France and KLM, and Lufthansa[10] (see Doganis 2002), and likewise for any airline based at Heathrow. The concentration of BA on the long-haul (trans-Atlantic) routes comes at the expense of serving the local market, which in turn is captured by airlines such as KLM which enjoy geographical proximity to the UK. Airline and railway integration could allow BA to serve the local market without giving up long-haul services, and could be used to feed traffic into the long-haul route network. For an airline like BA, two apparent benefits seem to exist. First, it should be easier for a company like BA to gain the market share from other network carriers in the domestic market rather than competing with them for markets outside the UK.[11] Second, BA would be able to consolidate its services from Gatwick to Heathrow, for example on the

9 However, these changes are also associated with the changes in the market and the changes in demand that BA had to face.

10 Even harder after the merger of AF and KLM.

11 Assuming home airlines enjoy some market power over foreign airlines (Doganis 1992).

London-Paris route, enjoying the network benefits of higher frequency (for example, higher frequency of service to Paris would mean lower average connecting time for passengers). In addition, the benefits of BA's H&S network would also be enjoyed by UK passengers, and not only by overseas passengers, if many destinations in the UK gained access to BA's range of destinations from Heathrow.

The Finnish flag carrier Finnair's agreement with the Swiss Federal Railway SBB illustrates some of the benefits described above. Recently, Finnair added four new destinations to its scheduled network: Bern, Basel, Lausanne, and Lucerne. These destinations are served by a Helsinki-Zürich flight and then a rail journey. For Finnair passengers, the integration of the aircraft and rail service is complete. The passenger experiences similar conditions as if the journey is by two connecting flights (international plus domestic). This means the luggage is checked in and collected at both ends of the journey; the reservation and sale of tickets for both segments of the journey is made at the same time (through Finnair); one ticket is used for both segments of the journey; seat allocation for both segments is done by the airline; passengers can earn airmiles on the rail journey; and the rail service is designated a flight number (with the Finnair code, AY) and the destinations receive an 'airport' code (for example, ZDH for Basel) (Finnair 2002). In this example of integration, the train service does not *replace* the aircraft service, certainly not Finnair's aircraft service, but *complements* it. Thus, mode substitution does not take place. A railway station at Heathrow integrated with the UK (future HST) rail network would allow airlines to add destinations like Bristol, Birmingham, Sheffield and so on to their network of routes from Heathrow.

Heathrow provided aircraft services to 191 destinations in 2004 (BAA 2004) but lack of capacity means that this cannot be increased and might decrease if, in order to increase services to high demand (international) routes, airlines stop services to low demand (regional) routes, a pattern that was observed by Graham and Guyer (2000). Currently, Birmingham, one of the biggest cities in the UK, has no direct access to Heathrow, and despite the size of Birmingham airport[12] it cannot match the level of service and range of destinations offered from Heathrow. A railway station at Heathrow with a good airline-railway interchange can solve this. The following illustrates the potential for Heathrow, and some airlines: 'If you look up a journey between Birmingham and Hong Kong, you are usually offered just one connection in OAG, the airline timetable. This is either via Amsterdam or Paris … If however you travel via London, you are given the choice of six flights by three carriers (two of them British) – but the timetable doesn't give you the option of going via London because there are no flights from Birmingham to London' (Sharp 2002, p. 2).

Airports do not benefit only from airline and railway integration per se. They also benefit from improved means of access. These benefits can be summarized as follows: 'intermodality is seen by many major airports … as a way to increase their catchment area. Indeed railways can reach populations in areas where air services are not available or where frequencies are not good ... Intermodality is also a way to

12 In 2003, the airport was ranked fifth in the UK in terms of ATMs (116,040) and passengers (9.07 million) (CAA 2004).

offer more city linked services to a given airport versus another one, and strengthen competitive edge' (IATA 2003, p. 99).

Conclusions

To identify the routes suitable for airline and railway integration, that is, for aircraft and HST substitution, the route distance was used as one criterion. However, this was found to be not sufficient and the HST route distance proved more suitable. However, this criterion was considered inferior to estimate the travel time by each mode. Furthermore, it was found that, because the HST route can be much longer than the aircraft route, the ratio between the modes' route distance should be considered when the HST route distance falls within the range considered suitable for mode substitution (assumed here to be under 800km).

Following the analysis presented above, better criteria for mode substitution can be defined as follows: *On HST routes of under 800 km in length, where the HST route is not more than 30 per cent longer than the aircraft route, and the average HST speed along the route is at least 200 kph, aircraft and HST substitution under airline and railway integration is beneficial to airlines, passengers and society.*

This definition is more robust than the ones usually cited in the literature based on route (aerial) distance. However, the analysis above also suggests that, overall, operating airline and railway integration depends on whether travel time benefits could be achieved, and this is a more appropriate criterion to determine the scope for airline and railway integration.

Ten aircraft routes from Heathrow were found to be suitable for airline and railway integration if, on these routes, the HST service could achieve an average speed of at least 250 kph. These routes consume about 20 per cent of Heathrow annual capacity. Five routes were identified to be suitable for airline and railway integration, even if the HST service only achieves an average speed of only 200 kph. These routes, to Manchester, Leeds/Bradford, Brussels, Newcastle and Paris consume almost 10 per cent of Heathrow's annual capacity. On the first four of these routes, benefits are expected even at lower average speeds.

The discussion above indicates that airline and railway integration is also likely to be beneficial to the TOC,[13] and that the railways will substantially benefit from the network effects of connection to a (major) international airport.

References

ACI (Airport Council International) (2004), 'ACI Passenger Traffic Reports'. ACI <http://www.airports.org> (28/04/2004).

BAA (2004), 'Terminal Information', BAA web site <http://www.baa.com/main/ /airlinedb/heathrow/any_any_page.html> (09/03/2004).

13 The focus on the airlines is because it is principally their decision whether to operate and adopt airline and railway integration.

Buchanan and Partners (1995), 'Optimising Rail/Air Intermodality in Europe'. European Commission – DG VII, London, November.

CAA (Civil Aviation Authority) (2004), 'UK Airport Statistics: 2003 – Annual'. CAA, Economic Regulation Group <http://www.caa.co.uk/erg/erg_stats/ /sgl.asp?sglid=3&fld=2003Annual> (18/05/2004)].

Caves, E.R. and Gosling, D.G. (1999), *Strategic Airport Planning* (Amsterdam: Pergamon).

CODA (Central Office for Delay Analysis) (2003), 'Delays to Air Transport in Europe Annual 2002'. Eurocontrol/ECAC.

Doganis, R. (1992), *The Airport Business* (London: Routledge).

—— (2002), 'The Future of Hubbing in London'. Rigas Doganis & Associates, October <http://www.baa.co.uk/pdf/HubbinginLondon.pdf> (23/5/03).

Finnair (2002), 'Blue Wing'. Finnair, December.

Givoni, M. (2003), 'Evaluating Aircraft and HST Operating Costs', in K. Cederlund and S. Ulf (eds), *New Trends in the European Air Traffic*: NECTAR Cluster 1 Workshop Networks Land Use and Space', Department of Social and Economic Geography, Lund University, Sweden.

—— (2004), 'Aircraft and High Speed Train Substitution: The Case for Airline and Railway Integration'. PhD. Thesis, University of London.

Graham, B. and Guyer, C. (2000), 'The Role of Regional Airports and Air Services in the United Kingdom', *Journal of Transport Geography* 8, 249–62.

IATA (International Air Transport Association) (2003), 'Air/Rail Intermodality Study'. IATA, February.

Innovata Flight Schedules (2004) <http://www.quicktrip.com/flightbase/schedules> (12/03/2004).

IPCC (Intergovernmental Panel on Climate Change) (1999), *Aviation and the Global Atmosphere* (Cambridge: Cambridge University Press).

Janić, M. (2003), 'The Potential for Modal Substitution', in P. Upham, J. Maughan, D. Raper and C. Thomas (eds), *Towards Sustainable Aviation* (London: Earthscan).

Pavaux, J. (1994), 'Air-Rail Complementarity in Europe: Lessons for Intermodalism in America'. Paper presented at the Transportation Research Board (TRB) 73rd Annual Meeting, Washington, DC, January 9–13.

Perren, B. (2001), 'Special Report: Paris-Marseille in Three Hours', *Modern Railways* 10 July.

Sharp, A. (2002), 'Rail-Air Replacement: Problems and Opportunities'. Air-rail transfer seminar, Manchester Metropolitan University, 21 November.

UIC (International Union of Railways) (2003) <http://www.uic.asso.fr> (21/7/2003).

Chapter 14

Airports as Nodes in Global Knowledge Networks

Marina van Geenhuizen and Holmer Doornbos

A Focus on Global Knowledge

It is almost commonplace to state that the capability to apply new knowledge is the most important asset of developed economies in their attempts to innovate and remain competitive in a global world. Whereas knowledge in neoclassical growth theory in the 1960s was perceived as an important external factor that could not be explained by labour and capital input, since the 1980s and increasingly today it has been assigned an endogenous role that can be steered and decided upon by companies (for example, Smith 2002; Foray 2004). An additional explanation for the current attention on the application of new knowledge is the increased mobility of capital and labour in a world in which modern traffic, communication technology and the lifting of particular borders have led to increased competition in attracting production with the highest value added (OECD 1996).

In recent studies on innovation in metropolitan economies, there has been much interest in knowledge as an important component of agglomeration advantages. Cities provide advantages of knowledge spillover effects and an abundant availability of knowledge workers in the labour market (Audretsch 1998; Acs 2002). Spatial concentration of similar and/or dissimilar activities increases the opportunities for interaction and knowledge transfer, and the resulting spillover effects reduce the cost of searching and obtaining new knowledge. In addition, knowledge workers preferably interact with each other in agglomerated environments to reduce interaction costs, and they are more productive in such environments (Florida 2002). Particularly the facilitation of face-to-face contact and opportunity to establish trust through spatial and social proximity, and the limits imposed by some kind of geographical borders of, for example, a daily activity system or central business district, have urged many authors to place an emphasis on knowledge transfer and learning primarily on a *local* basis (for example, Storper and Scott 1995; Maskell and Malmberg 1999; Rosenthal and Strange 2001).

By contrast, much less attention has been paid to the *global* nature of some knowledge – as in biotechnology and nanotechnology. The need to benefit from high competence levels and specialization in distant places abroad may encourage knowledge workers to travel by air to participate in the knowledge networks concerned by engaging in face-to-face meetings. In addition, if a firm is a subsidiary of a foreign mother company, this may also cause frequent participation by its staff in knowledge networks abroad. The attention for achieving tacit knowledge at *distant* places, through personal visits or sending employees, is relatively new (for example, Rutten and Boekema 2004; Doornbos 2005) and points to another advantage of large metropolitan areas: namely, access to an international airport (Simmie and Sennett 1999). More recently, an increasing number of authors active in research on clusters have abandoned the stance of predominantly local (regional) learning and now recognize the co-existence of local and global knowledge networks (for example, Bathelt et al. 2004). This also holds true for one of the most clustered high-technology sectors, biotechnology, in which interactive innovative relations are employed on very different spatial scales, often as dual local-global patterns (Zeller 2001; Coenen et al. 2004; Gertler and Levitte 2005).

From the perspective of impacts on the regional economy, the importance of air transportation for high-technology business has been recognized since the late 1990s (for example, Button et al. 1998). In US studies it appears that employees in high-technology sectors fly more often than those in traditional industries (1.6 times more), leading to the assumption that close proximity of regions to an airport contributes to the presence of high-technology employment. Indeed, the location of a hub-airport, amongst other factors, provides a positive statistically significant explanation for part of the variation in high-technology employment (Button and Lall 1999). The studies in the US have also revealed a rough estimation of the job stimulus associated with being close to a hub-airport, that is, an increase of a region's high-technology employment by over 12,000, as an average across a number of hub cities.

More specifically, international airports may contribute to business practice in high-technology companies in three different ways. First, as alluded above, international airports act as nodes enabling knowledge workers to conveniently access knowledge centres around the globe in personal visits, thereby creating the opportunity to utilize the highest quality level of knowledge. Secondly, international airports may act as places where global knowledge workers see colleagues from abroad who are in transit, or meet them in a conference organized at the airport itself, or at a close distance from the airport. Thirdly, international airports enable innovative companies from abroad to establish a subsidiary and still be able to utilize specialized manpower from the mother company abroad in the starting period, or even longer. We may summarize these direct functions as: the connecting function, the meeting function, and the attraction function.

The importance of international airports as nodes in global knowledge networks may, however, decline as a result of the use of advanced ICT. There may be a partial substitution of face-to-face meetings by videoconferencing, teleconferencing, simultaneous design sessions, electronic data mining, and so on. In this respect, we observe the rise of what are called 'virtual teams' (Martins et al. 2004). Virtual teams are a response to deal with the increased travel time and travel expenses, and coordination costs associated with bringing geographically and functionally dispersed employees to work together on a common task. The support of ICTs in such tasks and other business operations is based on various unique and far-reaching advantages (for example, Kenney and Curry 2001; Cairncross 2002). First, the Internet provides increased access to many places around the globe. Addresses connected via the Internet are almost unlimited and almost equally accessible. Accordingly, knowledge workers are able to search for the best knowledge sources across the globe, and they can make themselves known as searching for particular knowledge. Also, the speed of transmission and the complexity of the knowledge concerned have significantly increased. The second important feature is interactivity, such as in information feedback loops between producers in the composition of a product, or in monitoring by suppliers of product use at customers' sites (remote diagnostics). A third and probably the most powerful feature is intelligence. Intelligence is the ability to collect information across the network, to store and process the information, and utilize the results in one node or redistribute the results across the network between the members of a virtual team. These activities include search functions, selection functions (decision support) and data analysis including data mining and monitoring, and so on. At the same time, it is said that the use of ICTs is still limited because of various practical problems. There only seems to be substitution between physical and virtual if the communication and connected economic activity are non-material and sufficiently standardized, and if there is sufficient trust between the interacting partners. If the interactions are concerned with negotiation and unique problem solving or with risk taking, electronic communication tends to be hampered by various basic shortcomings and too high costs (for example, Bolisani and Scarso 2000).

Given the above contradictory trends, the aim of this chapter is to explore the relevance of global knowledge networks and the role of international airports in enhancing the knowledge economy of metropolitan areas, particularly concerning young, innovative companies. Accordingly, the following questions will be addressed:

1. Which factors determine the use of global knowledge networks by young innovative companies in metropolitan areas?
2. Which factors determine the importance of international airports in such networks? What is the role of telecommunication in substitution of long-distance travelling?
3. Which change profiles can be observed with regard to the importance of international airports in knowledge networking?

The structure of the chapter is as follows. We first address notions from resource dependence theory to gain a better understanding of the needs of young, innovative companies, and pay attention to the theoretical aspects of knowledge networks and communication modes. This is followed by a brief discussion of the research design, particularly the way of sampling and the use of rough set analysis. Next is the interpretation of the empirical results concerning the global dimension of knowledge networks and concerning access to Amsterdam Schiphol Airport. A brief exploration of changes in the importance of such access follows. The chapter concludes with a summary and evaluation of the results, and indicates a few future research paths.

Knowledge as a Resource

New knowledge is one of companies' most important resources in achieving competitive edge, particularly if the knowledge is unique and not easily transferable to other companies. New knowledge in business practice not only refers to new products and processes, but also to new markets and management models. Accordingly, it is not only about know-what and know-how, but also know-who and know-why (for example, Lundvall and Johnson 1994). This classification of subject matter is often associated with different types of knowledge in terms of codification and transferability (for example, Gorman 2002). Thus, know-what and know-why mainly refer to codified knowledge found in manuals, databases, conference proceedings and so on, whereas know-who and know-how are mainly connected with practical experience, learning by doing and social interaction, including the transfer of tacit knowledge.

Unlike codified knowledge, the creation and use of tacit knowledge is strongly dependent upon the social and organizational context. This context determines – through shared beliefs, perceptions and experiences – the interpretation of tacit knowledge and the communication requirements for transfer. One may mention, on the basis of Bolisani and Scarso (2000), real-time interaction, open communication – not hindered by predetermined structures – and use of a large variety of communication means to increase understanding, all of which demonstrate why face-to-face contact is the most efficient way of communication. Note that the various types of knowledge do not work separately, but interact. For example, tacit knowledge is necessary to understand codified knowledge. What the previous notions do not clarify is the spatial layout of the knowledge networks concerned. Although tacit knowledge is often associated with proximity and regional or local scales, it may also work on a global scale provided that there is similarity of social (cultural) context and a sufficient number of personal (face-to-face) meetings.

Resource dependence theory is help for understanding differences in opportunities between companies and in the sets of resources available to them in realizing these opportunities (for example, Barney 1991; Reid and Garnsey 1998; Lockett and Thompson 2001; Vohora et al. 2003). In modern versions of resource dependence theory, the competitive edge of companies is seen as residing in the particular resource configurations that managers build. The more unique, immobile and irreplaceable resources are, the higher the potential of competitive edge may be. Companies make use, on a temporary basis, of various combinations of resources (bundles), such as knowledge, capital, employees and networks. The growth of companies depends on their capability to match the development of their own resources or the disclosure of external resources with the business opportunities they perceive (for example, Barney 1991). Growth is constrained if there is a shortage or weakness in companies' capabilities to generate internal resources, and if companies lack the capability to disclose external resources.

With regard to the growth of particularly young high-tech companies, Reid and Garnsey (1998) distinguish between three different stages, running from achieving access to resources, and mobilization of resources, to their own generation of resources. Knowledge is a critical resource in all three stages, including knowledge about commercial opportunities, the business start-up process, and business development, marketing and finance (Teece 2000). The use of the right combination of resources at the right time enables high-tech companies to undertake a jump in their growth (next development stage). Alternatively, if they are not able to perceive the right opportunities in time and fail in organizing access to the necessary resources, they may fall back. More recently, an emphasis has been placed on the *heterogeneity* between, on the one hand, high-technology start-ups at their start in terms of perceived opportunities and resource requirements, for example, following from a different position of the entrepreneurs (different sets of experiences), different corporate positions (resources from the mother company or company of origin, or absence of such resources) and, on the other hand, the different resource requirements of the main activities, such as contract R&D, technical services or in-house basic research (for example, Druilhe and Garnsey 2004; Heirman and Clarysse 2004).

Networks can be seen as a specific external source of resources that may be utilized to exploit particular opportunities (for example, Brush et al. 2001). The analysis of entrepreneurship from a network perspective has become quite fashionable since the mid-1980s in organizational and economic studies (for example, Håkansson 1987; Kamann 1989; Hoang and Antoncic 2002; Borgatti and Foster 2003). Networks comprise sets of actors, such as persons, teams and companies, connected by a set of ties (Granovetter 1973). Companies establish new networks or decide to participate in existing ones if the perceived benefits outweigh the costs concerned. To achieve the best knowledge may be the main purpose of a network (for example, a strategic alliance on R&D) or may be a positive side-effect of other networks, like those with suppliers and customers. Networks, including knowledge networks, face many dimensions aside from the geographical layout (for example, Kamann 1989), including directed or non-directed ties such as personal advice to a network member or shared facilities for all members due to physical proximity; different structures such as ego-centred networks and multiple focal networks; and different stability as in temporary networks and (semi) permanent networks.

An underlying assumption of this study is that young, innovative companies are operating in global knowledge networks if they have a strong global orientation via their customers and/or supplier relations and are supported by strong capabilities stemming from previous experience of entrepreneurs and/or strong resources from a current mother company (if they are a subsidiary), or simply by their own capabilities developed at a later stage in their development. We also assume that high levels of innovativeness require highly specialized knowledge, available in just a few spots on earth. These ideas will be addressed in the discussion of the research design in the next section.

Which mode of communication will be employed in the knowledge networks and whether there is substitution between air transport and telecommunication can be understood on the basis of the potential advantages and disadvantages of different modes of communication. To this purpose, these modes are divided in a two-by-two matrix, on the basis of (1) simultaneous contact or contact at different times; and (2) located in one place or distributed (Table 14.1) (Mitchell 1999; Van Geenhuizen and Nijkamp 2004). For example, e-mail qualifies as not simultaneous and distributed, whereas telephone (including imaged), teleconferencing and videoconferencing

Table 14.1 Nature of modes of communication

	Simultaneous	**Not simultaneous**
Local	– Requires coordination – Enables high intensity and personal impact – Enables protection of new information – *Overall: very high costs*[a]	– No coordination – *Overall: low costs*
Distributed	– Requires coordination – Enables a limited level of personal contact – *Overall: less costs than above*[b]	– No coordination – Superficial and non-personal contact – *Overall: very low costs*

Notes:

[a] including costs of absence at workplace.

[b] possibly high initial costs, for example, of video equipment and a videoconferencing room.

Source: Adapted from van Geenhuizen and Nijkamp (2004).

qualify as simultaneous and distributed. The combination of local and simultaneous represents the mode of face-to-face meetings, while the combination of local and non-simultaneous represents less common situations in which a letter or note is left in the place of a person while he/she is not there. With the rise of digital networks, there has been a shift along the diagonal to not simultaneous and distributed, but the question still remains to what extent face-to-face contact on the spot can be replaced by other modes. It seems that companies use a diversified set of modes, including various combinations, based on a continuous evaluation of advantages and disadvantages with regard to the required quality of communication, in terms of speed, depth (richness) of content and personal touch, and costs (for example, Mitchell 1999). Virtual (distributed) teams, for example, rely heavily on various telecommunication modes, but, at critical points while tasks are running, key personal meetings need to be inserted.

Research Design and Methodology

The design of this study implies a particular selection of urban economic activity: namely, of producers of intermediate goods/services (not consumer goods/services), of young companies engaged in such activity, and of sectors representing various growth dynamics in the current development of large cities in the Netherlands (Van Geenhuizen 2005). This study has employed an inductive approach by using a limited number of carefully selected case studies. The case-study design permits a logic in the sense of 'replication', allowing the case analysis to be treated as a series of independent experiments (Yin 1994). This approach works through a close correspondence between theory and data, a process in which the emergent theory is grounded in the data. Accordingly, we utilized a detailed field study of 21 young, innovative companies in the city regions in the Northern Randstad in the Western part of the Netherlands and in the region of Eindhoven in the adjacent Southeast, all within approximately 1.5 hours' travel time between a company site and Amsterdam Schiphol Airport (AMS). The companies were selected from biotechnology, ICT services and mechatronics (optronics), as important drivers of urban economic growth in recent times. Data were mainly derived from in-depth interviews with corporate managers. The research design implied that companies were selected to contain a substantial degree of variation on the dimensions considered relevant from a resource-dependence perspective, for example, company size, position (corporate origin) and degree of innovativeness.

Information from semi-structured interviews was used to develop a case-study database as a matrix that constitutes a concise representation of the underlying field information: the information table. An information table contains the features of the companies used in the 'explanation' (condition attributes), as well as the attribute of the companies that needs to be 'explained' (decision attribute). Two such tables were developed, that is, one to serve as a basis for a systematic analysis of the spatial layout of knowledge networks (first step), and a second to serve such an analysis concerning the importance of AMS (second step). Conventional statistical analysis – such as multiple regression analysis or discrete choice modelling – could not be applied in the current study because of the low level of measurement of some (categorical) variables, the sometimes fuzzy nature of the data, and the small sample. Therefore, another technique was used that has become popular in recent years as a pattern recognition and classification technique, that is, rough set analysis (for example, Pawlak 1991; for details, see Polkowski and Skowron 1998). An additional advantage is that in rough set analysis – unlike more conventional methods – only one assumption needs to be made about the data, that is, that the value of the determining factors can be categorized. Rough set analysis works as follows. If a distinction is made between stimuli (condition attributes) and response (decision attributes), rough set analysis is able to identify causal linkages between classified conditions and a decision variable. In our analysis we focus on the decision algorithms produced by a stepwise scanning of the data matrix. These contain conditional statements of an 'if ..., then ...' nature. Accordingly, we can identify which conditions (combinations of attributes of the conditional variables) lead – in a logical deterministic way – to a particular state of the decision variable. The decision variables in our study are the spatial layout of knowledge networks (first step) and the relevance of AMS in employing global knowledge networks (second step).

Knowledge networks were measured as 'relations dealing with knowledge', for example, concerning the personal networks of the manager (chief executive officer (CEO)), customers, suppliers, knowledge institutes, alliance partners, head office and so on (van Geenhuizen 2005). In our approach to emphasizing interpersonal networks, tacit knowledge forms a substantial part of the knowledge concerned. In addition, the more stable relations were included. The knowledge relations identified could cover ego-centred networks (as between the company and a university research group in a dedicated project), as well as multiple focus networks (as between the company and customers with multiple customer relations). The spatial pattern of knowledge networks was measured on the basis of *stated* preference, that is, the most appreciated knowledge sources and the location of the relationships involved.

The condition attributes used in the current study refer to different knowledge requirements and a different access to external knowledge, as, for instance, caused by the position of the company (spin-off, subsidiary, independent and so on) and its age. With regard to the latter, the aim was to avoid young companies still vulnerable to some starting failures and focus on slightly older ones that had survived this stage (but not older than approximately 12 years). Size of the company, main activity (manufacturing or services) and development time of innovations were also included. The reason for using the latter was to capture different levels of innovativeness and related needs for highly specialized knowledge. Furthermore, a proxy for the generic spatial orientation was added, on the basis of the company's dominant supplier and customer relationships. The second step included the following condition attributes: importance of global knowledge networks; use of advanced telecommunication as indicated by the use of videoconferencing; size of the company, because of the assumption that a larger size corresponds with larger travel budgets; and accessibility of AMS for the company, measured as travel time between the company and AMS (door-to-door). In the latter respect, the assumption was that a certain travel time is critical (that is, beyond which the importance of AMS will decrease), given the specific geographical situation in the Netherlands where a longer distance from AMS often implies a shorter distance to other international airports, like Zaventem (Belgium), Dűsseldorf and Niederrhein (both in Germany). The condition attributes used in the first and second step of the analysis can be summarized as follows: 1.1) position; 1.2) age; 1.3) company size; 1.4) main activity; 1.5) development time of innovations; 1.6) general spatial orientation; 2.1) relevance of global networks; 2.2) use of advanced telecommunication; 2.3) company size; and 2.4) travel time to AMS.

Each rough set analysis produces a number of decision rules and the concomitant coverage of each of these rules (in percentages). The coverage is an indicator of the strength of the rule and gives the percentage of all cases sharing a similar score on the decision variable for which the rule is true. For example, if a decision rule covers five companies facing similar condition attributes that lead to global knowledge networks, whereas the total number of companies employing global networks is twelve, the coverage of this rule is 5/12 (41.7 per cent). Another indicator in the interpretation of rough set results is the frequency in which a condition attribute is included in the set of rules, referring to the strength of this attribute. We mention that the interpretation of the results of the rough set analysis is valid to the extent to which the case studies selected provide a fair representation of young, innovative companies in the intermediate sector located in major city regions in the Netherlands.

The rough set procedure also produces indicators of the quality of the data (condition attributes and decision attribute) in the information table. First, each estimation produces a distinction between 'core attributes' and other condition attributes. If all condition attributes belong to the core, then the conclusion can be drawn that all these attributes contribute to an explanation, and no attribute gives redundant information. In the first estimation all but one condition attribute belong to the core, in the second estimation all condition attributes belong to the core. In addition, the quality of the core is different. In the analysis in the first step the quality is 1.0, meaning that the reliability of the classification for the decision attribute (dependent variable) and the overall quality of the information table are at their maximum. In the second step, the quality is 0.62, pointing to a limited inconsistency in the data content.

In various studies employing rough set analysis, the *prediction accuracy* of the decision rules was tested on the basis of new samples and this has revealed some satisfactory outcomes. The prediction accuracy, measured as the percentage share of correctly classified companies, ranges between 75.0 per cent and 68.3 per cent (company acquisition: see Słowiński et al. 1997), between 98.8 per cent and 50.0 per cent (company failure, see Dimitras et al. 1999), and between 80.6 per cent and 65.9 per cent (failure as a result of insolvency of insurance companies: see Sanchis et al. 2006). The average prediction accuracy for these three factors is 71.7 per cent, 74.4 per cent and 74.6 per cent respectively. We may learn from the previous ranges and averages that the decision rules derived from our analysis tend to be quite robust.

How Global Networks Are Being Shaped

In this section the results are presented from the comparative analysis of the spatial lay-out of the knowledge networks. The application of the rough set methodology has led to a set of 12 decision rules (see Annex 14.A). These rules fall into two parts, that is, they refer to mainly regional networks and mainly global networks. The discussion in this chapter is limited to four rules concerning global networks that are solid and not affected by highly individual circumstances (Table 14.2).

Table 14.2 Rough set results on mainly global knowledge networks

Condition attributes in the decision rules	Coverage (%)[a]	Rules and conditions; additional information in italics[b]
Position and spatial orientation	25.0 % (3)	*Rule 1*: Independent or subsidiary, and no specific orientation. *ICT services and engineering services.*
Position and main activity type	8.3 % (1)	*Rule 2*: Corporate spin-off, providing services. *Advanced biotechnology services, network adopted from multinational (origin).*
Age and duration of innovation projects	41.7 % (5)	*Rule 3*: Older age and (very) long-lasting innovation projects. *Biotechnology research and optronics development and manufacturing*
Age and spatial orientation	16.7 % (2)	*Rule 4*: Young age and global orientation. *Biotechnology and ICT services.*

Notes:

[a] The numbers of supporting cases are in brackets.

[b] Not mentioned in this table is a rule that constitutes a subcategory within Rule 3.

Source: Adapted from van Geenhuizen (2005).

The trends can be summarized as follows (Table 14.2 and Annex 14.A):

1. On the basis of frequency of appearance of condition attributes in the rules, two factors exert a strong influence on the spatial layout of knowledge networks, that is, the position of the company and the company's general spatial orientation (Annex 14.A). The remaining factors, derived from resource dependence theory, like age and size, tend to be less important.
2. With regard to *global* networks, the strongest factor is the position of the company (Annex 14.A). This result suggests an important influence of organizational capabilities through past experience of the management or through current resources provided by the mother company.
3. Rule 3 is the strongest rule in terms of coverage, that is, it is supported by five cases from a total of 12 cases employing global networks (5/12 = 41.7 per cent) (Table 14.2). What makes this rule even stronger is that it includes different high-tech sectors (biotechnology and optronics). The rule says that, if a company is somewhat older and is involved in (very) long-lasting innovation projects, then it employs global knowledge networks. It suggests that companies in R&D and manufacturing develop global knowledge networks if they have overcome difficulties in the early years and if they are highly innovative, thereby utilizing highly specialized knowledge only available outside the Netherlands.

By focusing on the scope (type of partners) and spatial reach of the global networks of companies in the largest class (according to Rule 3), it appears that companies subject to the same shaping factors of globalization have broadly similar patterns (Table 14.3). The scope tends to be either 40 or 60 per cent and tends to include a focus on personal relations. The spatial reach of the networks tends to cover 50 or 75 per cent, with a focus on the US and the parts of Europe beyond driving distance from the Netherlands (for example, central France and Switzerland). Knowledge networks in the US serve the development and testing of new medicines (clinical trials) as well as the development of application knowledge concerning specialized optic devices (by customers). It appears that the company having the broadest scope in terms of types of partners and the largest spatial reach is a former spin-off of a multinational and is currently a subsidiary of an American company. This illustrates the influence of position.

The preliminary conclusion is that amongst young innovators, predominantly global networks tend to be shaped by network configurations of the mother company and/or company of origin (adding to the experience and capabilities of the entrepreneurs), an older age and a high level of innovativeness; all this tends to go together with a dominance of personal networks and of relations in the US and in those parts of Europe beyond driving distance from the Netherlands.

It is appealing to assume that if companies employ mainly global knowledge, AMS is an important node in their knowledge networks. However, the more routine parts of knowledge interaction can also be performed using telecommunication, for example, videoconferencing or teleconferencing and e-mail. The reasons why knowledge workers and managers would prefer face-to-face contact on the spot

Table 14.3 Features of global knowledge networks (five case studies covered by Rule 3)

Case study characteristics through Rule 3 (in italics) and other characteristics (2005)	Scope of global networks (%)[a]	Spatial reach of global networks (%)[b]
– *age of 12 years and innovation projects of some years* – advanced optronics company serving global customers (job size of 55)	40 % (personal networks and customers)	75 % (Europe I, Asia and US/Canada)
– *age of 12 years and long-lasting innovation projects* – biotechnology research company (job size of 25)	40 % (personal networks, universities)	75 % (Europe I and II and US/Canada)
– *age of 7 years and long-lasting innovation projects* – biotechnology research company (job size of 35)	60 % (personal networks, customers, universities)	50 % (Europe II and US/Canada)
– *age of 8 years and long-lasting innovation projects* – advanced optronics company, corporate spin-off and subsidiary (job size of 450)	60 % (personal networks, customers and alliance partners or HQ/sister company)	75 % (Europe I, Asia and US/Canada)
– *age of 10 years and innovation projects of some years* – biotechnology research company (job size of 95)	40 % (customers and alliance partners)	50 % (Europe II and US/Canada)

Notes:

[a] The maximum level is 100 per cent, meaning a highly differentiated globalization through: (1) personal networks; (2) customers; (3) suppliers; (4) universities or other knowledge institutes; and (5) alliance partners or headquarters/sister company.

[b] The maximum level is 100 per cent, meaning the largest reach including: (1) Europe I (driving distance by car from the Netherlands); (2) Europe II (remaining Europe); (3) Asia; and (4) United States/Canada.

Source: Adapted from van Geenhuizen (2007).

can be found in Table 14.4. The advantages can be divided into those affecting the processes between the interacting partners and those affecting the content of the communication (though both cannot be fully separated). The results from our in-depth interviews suggest a greater perceived importance of the processes compared with content. Some of the advantages listed are certainly not new and are often addressed in the literature, such as the possibility to establish trust and to deepen it; other advantages have received much less attention, such as the possibility to maintain momentum or increase speed in interaction in teams, and easier means of protection against the undesired spread of new knowledge.

As previously indicated, the choice of communication modes should be seen as the result of an evaluation of advantages and disadvantages of all modes available, including the electronic ones. This may lead to a combination of communication modes in the same relationship. The relevance of Amsterdam Schiphol Airport in this context will be discussed in the next section.

Table 14.4 Advantages of face-to-face contact in knowledge relations

Dimension		Number of opinions
Processes (relation) between partners	• Enables trust to be established and trust to be deepened • Enables team spirit to be created • Puts an emphasis on the importance of the partner as a person • Puts an emphasis on personal honour for the partner • Increases efficiency in problem solving • Enables direct feedback to increase understanding • Enables momentum to be maintained or speed of processes to increase • Protects against unwanted spread of new knowledge	10
Content	• Enables the analysis to be deepened and the content to be enriched • Enables something new to be created • Allows for truth to be checked	6
Totals		16

Source: Adapted from van Geenhuizen (2005).

Importance of Amsterdam Schiphol Airport

Application of the rough set methodology using the 21 case studies has led to a set of nine decision rules. Five rules are presented here in detail because these refer to the great importance of AMS for the participation of innovative companies in global knowledge networks (Table 14.5 and Annex 14.A).

The trends can be summarized as follows:

1. Three factors have the same strength based on the frequency of condition attributes in the rules, and one factor is slightly weaker (Annex 14.A). Thus, the relevance of global knowledge networks, use of advanced telecommunication and company size tend to be equally important, whereas travel time from the company site to AMS is slightly weaker in importance. In terms of travel time the limit seems to be one hour, in the sense that for travel times over one hour alternative airports (in Belgium and Germany's borderland) gain in importance (Doornbos 2005).
2. The strongest rule is Rule 1, which has a coverage of four companies (33.3 per cent). The rule says that, if a company employs a moderate use of telecommunication, then AMS is important in global knowledge networking. This can be understood as follows: limited telecommunication use goes

Table 14.5 Rough set results on a high relevance of AMS

Condition attributes in the decision rules	Coverage (%)*	Rules and conditions
Use of advanced telecommunication	33.3 % (4)	*Rule 1*: A moderate use of advanced telecommunication.
Size, global knowledge networks and use of advanced telecommunication	16.7 % (2)	*Rule 2*: Medium-sized, highly important global networks, frequent use of advanced telecommunication.
Global knowledge networks, travel time to AMS	16.7 % (2)	*Rule 3*: Highly important global networks, short travel time to AMS.
Size and use of advanced telecommunication	8.3 % (1)	*Rule 4*: Small size and frequent use of advanced telecommunication.

* The numbers of supporting cases are in brackets.

Source: Adapted from Doornbos (2005).

together with (or does not substitute for) a strong importance of transport by air. If we also consider Rule 2, it appears that both a relatively moderate as well as a frequent use of advanced telecommunication go together with high importance of AMS. Most probably, the conditions concerning advanced telecommunication use do not refer to direct 'causal' relations, but refer to an underlying causal structure, connected to different needs for diversified communication modes.

A strong demand for a diversified use of communication modes can be illustrated with the following case study: a small IT company employing a rich combination of all communication modes, including videoconferencing and the Internet (to develop and share creative ideas), as well as frequent international flights to meet customers (decision making) and to participate in creative processes (co-design). This situation suggests a certain complementarity between telecommunication and face-to-face meetings.

To conclude, the factors determining the importance of access to AMS are relevance of global knowledge networks, use of advanced telecommunication and size. According to the strength of the decision rules, the use of advanced telecommunication is most important. However, the rules concerned may not be interpreted as 'causal' links but as reflections of underlying patterns of the communication preferences of high-technology companies.

The Role of Amsterdam Schiphol Airport in Perspective

To identify relevant changes in the role of AMS, companies were selected that satisfied the following criteria: (1) AMS is relevant according to the importance assigned to AMS (stated preference) and according to the location decisions taken in the past; (2) AMS is relevant according to revealed preference: namely, the actual frequency of flights taken in order to participate in global meetings. It appeared that a division of the companies according to their corporate position makes sense because this gives information about the way in which the global networks came into being, as discussed in a previous section. The main change profiles and connections with the functions of an international airport can be summarized as follows:

1. *Independent start-ups.* In the first stages, AMS is not important but, after the companies start growing and increase their level of specialization in innovation, the knowledge networks become global as a result of global customer and/or suppliers' markets and global research alliances. This holds true for IT services (knowledge from customers) and for optronics (knowledge from customers, suppliers and alliances). Actual AMS use shows a quite broad range, between 1 or 2 and 12–48 times per year, depending on where the company stands in the above development. This profile illustrates an increasing connection function.
2. *Foreign subsidiaries.* At the start, AMS is highly important because it enables a company to fly in personnel from the head office, but this importance decreases when sufficient employees are found in the Netherlands and the national (regional) markets are successfully conquered (profile 2.1). In this case, there is also a trend for substitution of international flights by videoconferencing. Alternatively (Profile 2.2), AMS remains important because the subsidiary in the Netherlands serves various countries from the Netherlands and/or important knowledge relations with the headquarters abroad remain. This all concerns services (IT and engineering). Flights tend to be more often than for Profile 1, that is between 3–6 and 12–48 times per year. These profiles illustrate that the attraction function may decrease or may remain strong and stable.
3. *Corporate spin-offs* (from multinationals). These adopt the customer base abroad from the mother company and are engaged in high-level innovative services, requiring highly specialized knowledge and presence at customers' sites (as in advanced biotechnology services). AMS is an important node from the start and the frequency of flights tends to be higher than for profile 1 and profile 2, that is between 12 and 48 times per year. This profile illustrates a strong and stable connection function.

The above change profiles indicate that the selected segments of innovative companies may follow many possible changes concerning importance of access to AMS, that is, an increase, a decrease or a consolidation at a high level.

Discussion

This chapter has addressed the relevance of global knowledge networks for young, innovative companies in metropolitan areas and the connected role of international airports as a node in such networks. Various city regions in the Netherlands served as an example of large metropolitan areas experiencing particular economic growth and Amsterdam Schiphol Airport (AMS) served as an example of international airports. Because of the mostly qualitative and sometimes fuzzy data, and the small sample, rough set data analysis was used as a straightforward technique to identify logical deterministic relations, presented as decision rules. The rough set estimation concerning the occurrence of global knowledge networks revealed three factors as most determining: corporate position, age and a high level of innovativeness. This result is in line with our previous assumptions, except that a general global strategy concerning supplier or customer relations does not play an important role. The second estimation – concerning the importance of access to AMS for participation in global knowledge networks – has brought to light three main influencing factors, that is, the relevance of global knowledge networks, use of advanced telecommunication and size of the company, of which the first is most important. It appeared that both a relatively moderate as well as strong use of telecommunication (videoconferencing) goes together with frequent flights to destinations abroad. This pattern is difficult to understand and may reflect underlying different preferences for use of diverse combinations of communication modes, subject to as yet unknown impacts of complementarity and substitution. The study has also brought to light that the importance of AMS for young high-technology companies may change over time, including a decrease, an increase and high-level consolidation.

This study has various limitations as a result of the selections made in the research design, that is, high-technology industry and services limited to intermediate production, and relatively young companies. The outcomes may be different for other segments. For example, in the upper (creative) segments of consumer industries, such as fashion, multi-media and entertainment, other factors may be shaping global knowledge networks and access to an international airport may be more important. In addition, this study has employed a research design that identifies relationships that can be generalized only on the basis of selected theoretical positions. On the basis of similar studies using rough set analysis, however, we may assume a relatively high prediction accuracy of the rough set rules derived in this study, albeit more for the rules on global networks than those on the importance of AMS. Accordingly, it would be interesting to test the prediction accuracy of our outcomes in a future step of this study, using another sample of similar companies. In terms of current research results that call for further clarification, the issue of substitution versus complementarity can be put forward. Such clarification is needed to understand preferences for different combinations of communication modes, and the conditions under which these preferences change in favour of complementarity or substitution.

Acknowledgements

This study is part of the Centre of 'Sustainable Urban Areas' of TU Delft, and has benefited from support from the Netherlands Ministry of Economic Affairs.

References

Acs, Z.J. (2002), *Innovation and the Growth of Cities* (Cheltenham and Northampton: Edward Elgar).

Audretsch, D.B. (1998), 'Agglomeration and the Location of Innovative Activity', *Oxford Review of Economic Policy* 14:2, 18–29.

Barney, J. (1991), 'Firm Resources and Sustained Competitive Advantage', *Journal of Management* 17:1, 99–120.

Bathelt, H., Malmberg, A. and Maskell, P. (2004), 'Clusters and Knowledge: Local Buzz, Global Pipelines and the Process of Knowledge Creation', *Progress in Human Geography* 28:1, 31–56.

Bolisani, E. and Scarso, E. (2000), 'Electronic Communication and Knowledge Transfer', *International Journal of Technology Management* 20:1/2, 116–33.

Borgatti, S.P. and Foster, P. (2003), 'The Network Paradigm in Organizational Research: A Review and Typology', *Journal of Management* 29:6, 991–1013.

Brush, C.G., Greene, P.G. and Hart, M.M. (2001), 'From Initial Idea to Unique Advantage: The Entrepreneurial Challenge of Constructing a Resource Base', *Engineering Management Review, IEEE* 30:1, 86.

Button, K.J. and Lall, S. (1999), 'The Economics of Being an Airline Hub City', *Research in Transport Economics* 5:1, 75–106.

Button, K.J., Haynes, K.E. and Stough, R. (1998), *Flying into the Future: Air Transport Policy in the European Union* (Cheltenham: Edward Elgar).

Cairncross, F. (2002), *The Company of the Future. How the Communications Revolution is Changing Management* (Boston: Harvard Business School Press).

Coenen, L., Moodyson, J. and Asheim, B.T. (2004), 'Nodes, Networks and Proximities: On the Knowledge Dynamics of the Medicon Valley Biotech Cluster', *European Planning Studies* 12:7, 1003–18.

Dimitras A.I, Słowiński, R., Susmaga, R. and Zapounidis, C. (1999), 'Business Failure Prediction Using Rough Sets', *European Journal of Operational Research* 114:2, 263–80.

Doornbos, H. (2005), 'Mainport/Brainport? An Explorative Study of Schiphol Airport as a Link between Knowledge-Based Companies and Global Networks', MSc thesis (in Dutch), Faculty of Technology Policy and Management, TU Delft and Ministry of Economic Affairs, Delft and The Hague.

Druilhe, C. and Garnsey, E. (2004), 'Do Academic Spin-Outs Differ and Does it Matter?' *Journal of Technology Transfer* 29:3/4, 269–85.

Florida, R. (2002), *The Rise of the Creative Class, and How It Is Transforming Work, Leisure, Community and Everyday Life* (New York: Basic Books).

Foray, D. (2004), *The Economics of Knowledge* (Cambridge: The MIT Press).

van Geenhuizen, M. (2005), *ICT, Location Dynamics and the Future of Cities* (Delft and The Hague: Delft University of Technology, Faculty of Technology, Policy and Management/Netherlands Science Foundation), <http://www.nwo.nl/ /nwohome.nsf>.

—— (2007), 'Knowledge Networks of Young Innovators in the Urban Economy: Biotechnology as a Case Study', *Entrepreneurship and Regional Development* (forthcoming).

van Geenhuizen, M. and Nijkamp, P. (2004), 'In Search of Urban Futures in the E-Economy', in M. Beuthe, V. Himanen, A. Reggiani and L. Zamparini (eds), *Transport Developments and Innovations in an Evolving World* (Berlin, Heidelberg and New York: Springer-Verlag), pp 69–83.

Gertler, M. and Levitte, Y.M. (2005), 'Local Nodes in Global Networks: The Geography of Knowledge Flows in Biotechnology Innovation', *Industry and Innovation* 12:4, 487–507.

Gorman, M.E. (2002), 'Types of Knowledge and their Roles in Technology Transfer', *Journal of Technology Transfer* 27:3, 219–31.

Granovetter, M. (1973), 'The Strength of Weak Ties', *The American Journal of Sociology* 78:6, 1360–80.

Håkansson, H. (1987), *Industrial Technological Development. A Network Approach* (London: Croom Helm).

Heirman, A. and Clarysse, B. (2004), 'How and Why Do Research-Based Start-Ups Differ at Founding? A Resource-Based Configurational Perspective', *Journal of Technology Transfer* 29:3–4, 247–68.

Hoang, H. and Antoncic, B. (2002), 'Network-Based Research in Entrepreneurship: A Critical Review', *Journal of Business Venturing* 17:1, 1–23.

Kamann, D.J. (1989), 'Actors in Networks', in F.W.M. Boekema and D.J. Kamann (eds), *Socio-Economic Networks* (in Dutch) (Groningen: Wolters-Noordhoff), pp. 29–84.

Kenney, M. and Curry, J. (2001), 'Beyond Transaction Costs: E-Commerce and the Power of the Internet Dataspace', in T.R. Leinbach and S.D. Brunn (eds), *Worlds of E-Commerce: Economic, Geographic and Social Dimensions* (Chichester: Wiley), pp. 45–65.

Lockett, A. and Thompson, S. (2001), 'Resource-Based View and Economics', *Journal of Management* 27:6, 723–55.

Lundvall, B.A. and Johnson, B. (1994), 'The Learning Economy', *Journal of Industry Studies* 1:2, 23–42.

Martins, L.L., Gilson, L.L. and Maynard, M.T. (2004), 'Virtual Teams: What Do We Know and Where Do We Go From Here?', *Journal of Management* 30:6, 805–35.

Maskell, P. and Malmberg, A. (1999), 'Localised Learning and Industrial Competitiveness', *Cambridge Journal of Economics* 23:2, 167–85.

Mitchell, W.J. (1999), *E-Topia "Urban Life, Jim – But not as We Know It"* (Cambridge: The MIT Press).

Organistion for Economic Co-operation and Development (OECD) (1996), *The Knowledge-Based Economy* (Paris: OECD).

Pawlak, Z. (1991), *Rough Sets* (Dordrecht: Kluwer).

Polkowski, L. and Skowron, A. (eds) (1998), *Rough Set in Knowledge Discovery* (Heidelberg and New York: Physica-Verlag).

Reid, S. and Garnsey, E. (1998), 'Incubation Policy and Resource Provision: Meeting the Needs of Young Innovative Firms', in R. Oakey and W. Dunning (eds), *New Technology Based Firms in the 1990s*, Vol V (London: Paul Chapman), pp.67–80.

Rosenthal, S.S. and Strange, W.C. (2001), 'The Determinants of Agglomeration', *Journal of Urban Economics* 50:2, 191–229.

Rutten, R. and Boekema, F. (2004), 'A Knowledge-Based View on Innovation in Regional Networks: The Case of the KIC Project', in H.L.F. de Groot, P. Nijkamp and R. Stough (eds), *Entrepreneurship and Regional Economic Development. A Spatial Perspective* (Cheltenham: Edward Elgar), pp. 175–97.

Sanchis, A., Segovia, M.J., Gil, J.A., Heras, A. and Vilar, J.L. (2006), 'Rough Sets and the Role of the Monetary Policy in Financial Stability (Macroeconomic Problem) and the Prediction of Insolvency in the Insurance Sector (Microeconomic Problem)', *European Journal of Operational Research* (forthcoming).

Simmie, J. and Sennett, J. (1999), 'Innovative Clusters: Global or Local Linkages?' *National Institute Economic Review* 170:1, 87–98.

Słowiński, R., Zopounidis, C. and Dimitras, A.I. (1997), 'Prediction of Company Acquisition in Greece by means of the Rough Set Approach', *European Journal of Operational Research* 100:1, 1–15.

Smith, K. (2002), 'What is the Knowledge Economy? Knowledge Intensity and Distributed Knowledge Bases', Discussion Paper 06, The United Nations University.

Storper, M. and Scott, A. (1995), 'The Wealth of Regions: Market Forces and Policy Imperatives in Local and Global Context', *Futures* 27:5, 505–26.

Teece, D. (2000), 'Strategies for Managing Knowledge Assets: The Role of Firm Structure and Industrial Context', *Long Range Planning* 33:1, 35–54.

Vohora, A., Wright, M. and Lockett, A. (2003), 'Critical Junctures in the Development of University High-Tech Spinout Companies', *Research Policy* 33:1, 147–75.

Yin, R.K. (1994), *Case Study Research: Design and Methods* (Thousand Oaks: Sage).

Zeller, C. (2001), 'Clustering Biotech: A Recipe for Success? Spatial Patterns of Growth of Biotechnology in Munich, Rhineland and Hamburg', *Small Business Economics* 17:1–2, 123–41.

Chapter 14 Appendix

Table 14.A1 Results of the first step of rough set analysis

Condition attribute	Local/regional or global knowledge networks	Global knowledge networks
	Frequency (in 12 decision rules)	*Frequency (in 6 decision rules)*
1.1 Position	5	3
1.2 Age	3	2
1.3 Size	3	1
1.4 Main activity (industry, services)	2	1
1.5 Duration of innovation projects	3	2
1.6 General spatial orientation	6	2
STRENGTH OF INFORMATION TABLE		
NUMBER OF CORE ATTRIBUTES	5 out of 6 condition attributes	–
QUALITY OF THE CORE	1.00	–

Table 14.A2 Results of the second step rough set analysis

Condition attribute	Importance of AMS
	Frequency (in 9 decision rules)
2.1 Relevance of global networks	5
2.2 Use of advanced telecommunication	5
2.3 Size	5
2.4 Travel time to AMS	4
STRENGTH OF INFORMATION TABLE	
NUMBER OF CORE ATTRIBUTES	5 out of 5 condition attributes
QUALITY OF THE CORE	0.62

Chapter 15

Ports in International Networks: A Multicriteria Evaluation of Policy Options for Accommodating Growth

Ron Vreeker, Peter Nijkamp, Ingeborg van Ansem and Ekko van Ierland

Water as a Strategic Network Element

More than 70 per cent of the earth's surface consists of water. Consequently, a significant part of the physical transport of people and goods uses waterways, such as deep-sea shipping, short-sea shipping or inland navigation. Water is apparently not only something that separates areas, but also connects areas. For example, the Mediterranean is historically a water surface that was essential for the integration of the regions around it.

Given the specific nature of waterways, transport over water is especially suitable for mass transport, in particular for goods. Water is one of the dominant modes for the physical transport of commodities, for example, bulk transport and container transport. Scale advantages have led to the emergence of large hubs in international waterway transport, with fierce competition between major ports such as Rotterdam, Hamburg, Marseilles, New York, Singapore, Hong Kong and so on. Hub-and-spoke systems have become a natural organizational constellation in a competitive market.

The proper functioning of international networks for waterway transport evidently depends on the use of sophisticated logistic systems, as well as on the capacity of the ports. Ports are therefore attempting to improve their competitive position by investing in modern logistics and port expansion in order to accommodate the ever- increasing flows of physical goods, often shipped by vessels ever increasing in size. When international ports are seen as major hubs in a global network, it is plausible that policy-making authorities will tend to invest in the improvement and extension of port operations, even though such investments are large scale in nature and costly. In many cases, however, port expansion is not so easy, as land is scarce and environmental regulations have become stricter.

As a maritime nation, the Netherlands has a strong historical position in the transport of goods. It is thus no wonder that port expansion has always been looked at favourably. This has, for example, led to the major offshore expansion of the Port of Rotterdam in the 1970s, by reclaiming a new port area from the North Sea, the Maasvlakte. This area is gradually reaching its capacity limits, and consequently an intensive debate has started on the question whether yet another large area might be reclaimed from the sea. The present chapter maps out the strategic issues involved, the various choice alternatives and the relevant policy criteria for evaluating this large-scale project. Then, a multicriteria analysis is deployed to identify the best possible choice.

Capacity problems in the Port of Rotterdam

The Port of Rotterdam is one of the two main ports of the Netherlands and acts as a 'gateway to Europe'. Direct employment in the port amounts to 65,000 jobs, and indirect employment to 200,000 jobs (van Ansem 2000).

The growth of the Port of Rotterdam seems to be unlimited. Several calculations of the Rotterdam Port Authority (GHR 1998, 1999) and The Netherlands Bureau for Economic Policy Analysis (CPB 1997, 1998) show that, by the year 2020, there will be a shortage of space. It is assumed that this shortage will hamper the economic growth and affect the competitive position of Rotterdam.

The main port of Rotterdam consists of a variety of locations and connections. The shortage of space will predominantly manifest itself in the port and industrial areas of Rotterdam. This lack of space is caused by:

1. autonomous developments in the maritime and related industries;
2. the upsurge of new economic activities and production;
3. the westward expansion of the city of Rotterdam;
4. citizens' decreasing tolerance of the negative environmental effects of industry and traffic.

In order to gain insight into the uncertainties surrounding the above-mentioned developments, the GHR and CPB conducted various scenario analyses. These investigations revealed that future developments in the transhipment, storage and processing of goods would lead to a demand for more space. When this demand for more space is confronted with the stock of available port and industrial land, the (future) shortage of space becomes immediately apparent.

The scenario studies are all based on three CPB scenarios regarding the development of the European and Dutch economy: namely, the Global Competition scenario (GC), the European Coordination Scenario (EC), and the Divided Europe scenario (DE).

The findings of the studies of the GHR and CPB are summarized in Table 15.1. This table depicts the extremes in the shortages in the year 2020. Depending on the economic growth rates of the Netherlands, the resulting space shortage varies between 1260 and 0 hectares. In the GC-scenario especially the petrochemical industry, deep-sea shipping and related distribution activities determine the demand for space. This also holds good for the EC-scenario. However, the petrochemical industry is growing less, resulting in a lower demand for space.

According to both studies, a lack of space will occur in the port area as early as 2004. Both forecasts show a comparable growth in transhipment activities in the North-Western European Seaports. However, the CPB estimates a lower demand for

Table 15.1 Space shortages in the Rotterdam Port Area, projections for 2020

	GC-scenario	EC-scenario	DE-scenario
Extrapolations for Rotterdam Port Authority 2020	1260 ha	750 ha	0
CPB	610 ha	370 ha	0

Source: PMR (1999).

space than the GHR, especially in the GC-scenario. The variation in shortages can be explained by differences in the assumptions underlying both studies:

- The CPB assumes a weaker market position of Rotterdam in the transhipment of goods than the GHR does.
- The CPB assumes that possibilities to improve the efficiency of land use by the container sector are less apparent than the GHR does.
- According to the CPB, possibilities for re-using vacant parcels are limited.

The Dutch government considers the prognoses of the GHR as the upper limit of the forecasted space shortages. These prognoses indicate the maximum space shortage expected in the year 2020. If economic growth is below expectations, it might well be possible that no space shortage will occur in the Port of Rotterdam. This is deemed the lower limit by the Dutch government.

When investigating possible solutions for the spatial shortages, the Dutch government stated that solutions should lead to: 'a quality improvement for the economy as well as the social environment'. This quality improvement should be achieved 'by strengthening the competitive position of the port, achieving sustainable land use, improving the social climate, decreasing the environmental pressures, and improving the quality of the adjacent nature areas (Voordelta)'. To realize these goals the Dutch government designed an integrated package of measures aimed at (PMR 2001):

- improving the competitive position of the Port of Rotterdam by reducing the lack of space in the port and related industrial locations;
- improving the quality of the (urban) environment in the Rijnmond area.

Both objectives are deemed equally important. This implies that measures aimed at reducing the lack of space should also result in an improved quality of the (urban) environment.

Description of the Selected Port Development Alternatives

The *Rotterdam Mainport Development* Project (PMR) was initiated in 1997 to tackle the problem of lack of space in the Rotterdam port area and to improve the spatial quality of the Rijnmond area. Within this project (organization) the following authorities are collaborating: the Ministry of Transport, Public Works and Water Management, the Ministry of Economic Affairs, the Ministry of Finance, the Ministry of Housing, Spatial Planning and the Environment, the Ministry of Agriculture, Nature and Food Quality, the Municipality of Rotterdam, Rotterdam City Region, the Rotterdam Port Authority and the Provinces of South-Holland and Zeeland.

In order to fulfil these two objectives, a set of relevant alternatives is defined by the project organization. One of these alternatives is the seaward expansion of the port (*Tweede Maasvlakte*, otherwise known as Maasvlakte II). This expansion consists of a land reclamation of around 1,000 hectares in the North Sea. Other solutions are to intensify the current land use situation, which should result in more efficient forms of land use, and to relocate activities to other ports such as Moerdijk, Flushing and Terneuzen. Even locations more remote from the current port were considered. These include the 'Eemsmond', Arnhem-Nijmegen, Amsterdam and Venlo. Eemsmond and Amsterdam were considered important opportunities to accommodate short-sea shipping activities, Venlo and Arnhem-Nijmegen were deemed important in the light of intermodal inland shipping activities. The locations considered all have great potential for the expansion of container-related distribution activities. However, with regard to petrochemical activities the locations were deemed unsuitable. The constraints imposed by the development of the petrochemical activities resulted in the definition of three global strategies:

- the expansion of the Port of Rotterdam with land reclamation of 1,000 hectares (Tweede Maasvlakte);
- intensifying land use in the current port area;
- the relocation of activities to alternative port and industrial sites in the Southwest part of the Netherlands (Moerdijk, Flushing and Terneuzen).

The possibilities of three port areas in the Southwest part of the Netherlands were investigated. A study was conducted to investigate the possibility of relocating parts of the deep-sea container sector and petrochemical sector. The study indicated that opportunities for the relocation of deep-sea container activities are limited. The ports of Moerdijk, Terneuzen and Flushing are too shallow to accommodate deep-sea container vessels. Furthermore, both ports do not offer the high quality of hinterland connections, economic scale benefits and level of logistical services required by the container sector.

Various chemical companies are already present in the three ports. Space available in these ports will, however, be used to accommodate the autonomous growth of these existing companies. Moreover, any further expansion of chemical activities at Moerdijk would result in a deterioration of the spatial quality.

As a consequence of the limited possibilities in the ports described above, the third strategy has been replaced by an alternative strategy. This combines the first two strategies: *intensifying land use and land reclamation.*

Alternative A: Seaward Land Reclamation

One alternative solution for the possible space shortage is a land reclamation project known as *Tweede Maasvlakte* (or Maasvlakte II). Three variants for land reclamation have been designed which differ in terms of access to the sea, positioning of port and industrial activities and nature and recreational areas. Furthermore, the variants differ with respect to possibilities for future expansion and phasing of construction.

Any damage caused by a certain type of land reclamation to the main Dutch ecological structure should be compensated. This will be done by means of the development of a nature and recreational area of 750 hectares adjacent to the land reclamation. In this situation an integral development, which specifically improves the existing nature values in the 'Voordelta', is preferred. The considered types of land reclamation are the following:

Land reclamation Variant A1 This design for land reclamation offers space for the future development of the port and related industries, as well as for nature and recreation. This option would provide a safe and direct access to the port for bigger ships, and high-quality industrial sites. In addition, if necessary, the land reclamation could be expanded towards the sea. This would be possible without expensive adjustments to constructions such as piers, quays, flood barriers, dikes and the port opening. The land reclamation would also accommodate a nature and recreational area of 750 hectares.

Land reclamation Variant A2 This option adjoins the current and first seaward port extension and the present nature and recreational areas. The new land reclamation would proceed towards the sea. The current nature and recreational areas at the south side of the port would be expanded westwards by means of a broad strip of land consisting of dunes and a beach within the land reclamation. Furthermore, the important influence of the sea on the dunes of Voorne and Goeree would be maximized by means of direct access to the sea. However, without expensive adjustments to the port entrance and dike, the newly reclaimed land could not be easily expanded to create additional space for future port and industrial activities.

Land reclamation Variant A3 This option for land reclamation offers 1,000 hectares of newly created land. However, expansion of this reclaimed land for future activities is only possible at very high financial cost and is deemed impossible. The existing port entrance would not change. The reclaimed land would be protected by a dike 8 kilometres long. This land reclamation would not contain a nature area.

Alternative B: Urban Infill and Intensifying Land Use

A second strategy to solve the space shortage is to intensify current land use in the port area. By utilizing the existing space in a more efficient way space could be created to accommodate new port and industrial activities. Three variants have been developed to pursue intensive land use in the port area.

Intensive land use Variant B1 This variant comprises the efficient and sustainable use of land in the existing port and industrial areas. Emphasis is placed on the use of currently vacant parcels in the area which were once used by industries now in decline (infill). On the basis of various suitability criteria, land use functions would be assigned to parcels. This would result in the clustering of industrial activities, such as the petrochemical industry, dry bulk activities, transhipment and roll on/roll off distribution activities.

To realize land use efficiency gains, measures would be taken to expand the supply of land in the port area and to reduce the demand for space by certain industries. These measures include:

- The relocation of port and industrial activities leading to the spatial clustering of activities. It is aimed to create a situation of increasing returns to scale.
- Increasing the supply of land by small land reclamations (*dempingen*) within the port area.
- Technological innovations which would lead to more intensive land use in the storage of dry and wet bulk, and distribution activities.
- Reuse of parcels which are not in use as a result of excess capacity in certain sectors (for example, tankage). Because of high land detoxification or demolition costs, some parcels of land are currently left idle.
- Relocation of activities outside the current port area. This concerns the activities of sectors which do not need access to port facilities such as cranes, quays or docks or port-related industries.

To summarize, Variant B1 would involve the reuse of currently underused parcels. The relocation of firms would be avoided as much as possible. However, the land use claims by the chemical sector and the container sector would be accommodated in less strategically important parcels in the port area. An area of 60 hectares would be withdrawn from existing and future land reserves. However, Variant B1 calls for a number of expensive spatial investments, such as small land reclamations, land detoxification, and the restructuring of industrial sites and infrastructure.

Intensive land use Variant B2 In this variant, measures would be taken which are designed to lead to an improvement of the spatial quality in the port area and the city of Rotterdam. This variant is based upon Variant B1 presented above. First, it was investigated whether measures could be taken to mitigate the negative aspects of the first variant. It became clear that mitigating measures would invariably result in a redesign of the spatial plan of the port area. Major adjustments could be avoided. On the one hand, these measures concern the limitation of port activities which cause severe nuisance. On the other hand, they concern incremental functional changes resulting in port locations being used for urban non-port activities. Both types of measures would result in less space being available for the port activities.

Intensive land use Variant B3 Within this variant, two options can be discerned; Variant B3a and Variant B3b. These options are specifically related to the border area between port and city which has changed over time. The expanding city needs more space and nowadays a number of former port areas are used for non-port activities. As a result the border areas exhibit a mixture of activities. The two options presented here illustrate the political choice between, on the one hand, maintaining the current border between city and port and, on the other hand, altering the border at the cost of the port.

Intensive land use Variant B3a This option provides a considerable amount of room for port expansion. Up to the year 2020, no major functional changes would take place in the spatial plans for the city of Rotterdam. As a result of more stringent environmental policy measures, activities that cause nuisance would be relocated. There would be major plans for the development of Schiedam and Vlaardingen. In Schiedam, the city would gain more space at the cost of the port. In Vlaardingen, the focus in the spatial plans would be on mixing functions. The measures described in this option are predominantly focused on solving the spatial shortage. In 2020 this would result in at least 4,995 ha of the available area of 5,330 ha being in use. The calculated shortages would decline from 1,260 ha to 590 ha, but by then the reserves would be almost fully used up.

Intensive land use Variant B3b In this variant, especially functional changes in favour of the city would take place in the border area. This variant relates to Variant B1. It would, however, take more space from the port on behalf of urban activities. At the same time, this option would provide for the green buffers between port and city to grow to a substantial size in the former port and industrial areas. This option differs from option B3a specifically regarding the design of the Waalhaven-East and Fruitport. The areas next to the river Meuse would also be utilized for housing. In 2020, of the 5,150 ha available area, at least 5,005 ha would be in use. The calculated space shortage would decrease from 1,260 to 780 ha; however, there would scarcely be enough space to accommodate the demand in the growth sectors and instead of the envisioned strategic reserves, in 2020, only 145 ha at the most would be available.

Alternative C: Intensifying Land Use and Land Reclamation

This last alternative comprises two variants. Variant C1 is a combination of Variant B1 and Variant A3. This option focuses on the first policy objective: improving the competitive position of the port by reducing the lack of space. The second, Variant C2, is a combination of Variant B2 and Variant A1. The focus in this variant is on the second objective: the improvement of the spatial quality in the port area.

Combination Variant C1 This variant aims to use the space within the existing port area and the new land reclamation as efficiently and sustainably as possible. The development of the land reclamation (*Tweede Maasvlakte*) of a total of 1,000 ha would be divided into three separate phases. The total investment costs for the land reclamation would amount to €1,970 million in this variant. The other measures that would be taken within this variant are listed in Table 15.2.

From the table and the data related to the costs of the land reclamation, the following conclusions can be drawn. In Variant C1, the various measures would be taken before new land is reclaimed. In 2020 this would result in a shortage of approximately 10 ha.

Combination Variant C2 This variant focuses on the second policy objective. Within the existing port area and on the newly reclaimed land, space is reserved for nature and recreational areas. The total investment costs for the land reclamation would amount to €3,100 million, just as in alternative A1: Space for port development. Table 15.3 lists the costs associated with the other measures taken in this variant and in Variant B2.

Table 15.2 Cost overview for the measures in Combination Variant C1 (in million euros)

Measure	Ha	Total costs (min.)	Total costs (max.)	Total costs (avg.)	Costs/ha
Reuse of vacant parcels	135	114	128	121	1
Land reclamation within the port area	115	188	188	188	1.6
Underground bulk storage	70	250	333	291	4.1
Underground oil storage	150	613	729	671	4.5
Intensifying land use in the container sector	75	302	662	482	6.3
Intensifying land use in the distribution sector	80	836	836	836	10.4
Infrastructure	–	253	330	292	–
Total	625	–	–	2,880	–

Source: SBRG (1999).

Table 15.3 Cost overview of the measures in Combination Variant C2 (in million euros)

Measure	Ha	Min. Costs	Max. costs	Total costs (avg.)	Costs/ha
Reuse of vacant parcels	135	114	128	121	1
Land reclamation within the port area	115	188	188	188	1.6
Underground bulk storage	70	250	333	291	4.1
Underground oil storage	150	613	729	671	4.5
Intensifying land use in the container sector	75	302	662	482	6.3
Intensifying land use in the distribution sector	80	836	836	836	10.4
Loss of space Vierhaven/ Fruitport	–70	268	402	335	–4.8
Total	555	–	–	2,924	–

Source: SBRG (1999).

From the table above, it can be concluded that the sum of the average total costs of all measures, including the costs for loss of space from the Vierhaven/Fruitport, are lower that the investment costs for the land reclamation. A total of 555 ha of land would be gained within the current harbour area and 1,000 ha of new land reclamation. In this option, the spatial shortage (1,260 ha) would have reversed into a spatial surplus of 295 ha in the year 2020. Table 15.4 gives a detailed overview of the different alternatives and options, including the accompanying measures and costs.

We will now proceed to discuss the spatial-economic effects of the different alternatives and make the final assessment of these alternatives.

Multicriteria Analysis

In the literature on evaluation, various types of evaluation techniques are distinguished. A major distinction is often made between *monetary* and *non-monetary* techniques. Monetary techniques are characterized by an attempt to measure all effects in monetary units, whereas non-monetary evaluation utilizes a wide variety of measurement units to assess the effects. Cost benefit analysis (CBA) belongs to the first category of techniques while multicriteria analysis (MCA) belongs to the latter.

The most important difference between the CBA and MCA is the decision rule used in the evaluation process. CBA uses prices to make efficiency attributes compatible, whereas MCA is characterized by a weighting system implicitly or explicitly involving the relative priorities of decision makers. Another difference concerns the type of criteria used in the evaluation. In contrast to CBA, which is focused on efficiency and therefore on Benefit-Cost ratios or Net Present Values, MCA does not impose any limits on the number and nature of criteria. This causes, however, a serious risk of 'double counting' the effects. Furthermore, to apply CBA, prices need to be known. CBA therefore requires that effects on efficiency attributes are measured in quantitative terms. MCA does not impose such strict requirements regarding the measurement of effects (for a detailed discussion, see Munda et al. 1995; Nijkamp et Vreeker 2005).

MCA methods aim to take into account heterogeneous and conflicting dimensions of decision problems. Rather than a specific appraisal method, MCA is a family of methods. This family comprises a collection of around 100 techniques that share some basic principles, but differ in other, mainly technical aspects. MCA methods can be classified according to the aggregation method they use. Three types of aggregation are distinguished: *complete aggregation*, *partial aggregation* and *iterative aggregation*. Methods that apply complete aggregation include all (sub-)criteria in the evaluation process. Furthermore, it is assumed that all (sub-)criteria are mutually comparable and full compensation between criteria scores is allowed. This means that a bad score on one criterion can be compensated by a good score on another (Vincke 1992). Complete aggregation is present in methods belonging to Multi-attribute Utility Theory (MAUT) such as *weighted summation*, *the reference point method*, *goal programming* and *the indifference method*. Although the Analytic

Table 15.4 Synthesis of alternatives, options and accompanying characteristics

Strategy	**Land reclamation**			**Intensive land use**				**Combination**	
Alternative	Land reclamation Variant A1	Land reclamation Variant (A2)	Land reclamation Variant A3	Intensive land use Variant B1	Intensive land use Variant B2	Intensive land use Variant (B3a)	Intensive land use Variant (B3b)	Combination Variant C1	Combination Variant C2
Characteristics									
Investment costs	€3.2 billion	€2.8 billion	€2 billion	€2.95 billion	€3.2 billion	€3 billion	€3.6 billion	€2.3 billion	€6.35 billion
Extra ha	1,000 ha	1,000 ha	1,000 ha	695 ha	600 ha	670 ha	480 ha	1,650 ha	1,730 ha
Expansion	Good expansion possibilities towards the sea	Expansion possible with substantive investments	Expansion possible with major investments	n.a.	n.a.	n.a.	n.a.	Possible (high investment costs)	Possible (high investment costs)
Recreational activities	Intensive recreational activities	Intensive, attraction-focused recreational activities	Extensive recreational activities	No extra recreational services possible	Intensive recreational activities	Intensive recreational activities	Intensive recreational activities	Extensive recreational activities	Extremely intensive recreational activities
Nature	750 ha new nature areas on MV II (intertidal area)	Keeping open Haringvliet-mond dunes, Goeree and Voorne	No new nature area is realized, this is done somewhere else	No extra room for nature areas	By maintaining the borders of usage	No extra room for nature areas	Green buffer zones between harbour and city area	No extra room for nature areas	750 ha new nature area (intertidal area)
Harbour access	Direct access	Harbour access less far into the sea	Maintaining harbour entrance	n.a.	n.a.	n.a.	n.a.	Maintaining harbour entrance	Direct harbour access
Measures taken	n.a.	n.a.	n.a.	Land reclamation, intensifying, relocation of non-harbour related activities, reuse of abandoned parcels.	B1 + relocation of Vierhavens/ Fruit-port	B1 + Withdrawal from Waalhaven-East, Vierhavens, parts of Schiedam and Vlaardingen	See B3a	Land reclamation and reuse of parcels, development of 1st phase MVII	Land reclamation, reuse of parcels, development of 1st phase MVII

Source: SBRG (1999) and SM2V (1999).

Hierarchy Process method does not belong to MAUT, it allows the complete aggregation of criteria.

Partial aggregation is used by what are called 'outranking methods' such as concordance analysis (Roy 1968, 1972; Roy and Jacquet-Lagreze 1977). Methods applying partial aggregation do not use all (sub-)criteria to determine a final ranking. Methods based on partial aggregation try to build an outranking relationship by means of pairwise comparisons between alternatives. The underlying idea of outranking techniques is that it is better to accept a result less information-rich than one yielded by MAUT, by avoiding the introduction of mathematical hypotheses, which are too strict, and by not asking the decision maker questions which are too complicated. Well-known outranking methods are ELECTRE, PROMETHEE, NAIADE and REGIME.

Iterative aggregation is used by what are referred to as 'continuous multicriteria evaluation methods'. The application of such methods requires that the alternatives considered are formulated continuously, as a consequence of which the number of alternatives is infinitely large. MCA methods based on iterative aggregation are called 'multicriteria objective programming methods'. These methods maximize or minimize several objectives, subject to a set of constraints. Because of conflicts between the various objectives included in the analysis, a simultaneous optimization of these objectives is impossible. Instead of optimization, the aim of the decision maker is to achieve a set of predefined goals as closely as possible (satisfactory behaviour). Multiobjective programming is aimed at finding a set of efficient solutions and divides the feasible set of solutions into a set of efficient solutions and a subset of inefficient solutions (Ballestero and Romero 1998). The final or best solution is the solution preferred by the decision maker above all other efficient solutions (for a detailed discussion, see Korhonen 2005).

Since the evaluation of the PMR alternatives entails the ranking of a restricted number of discrete project alternatives, the application of partial aggregation methods is not suitable. In our case study we will apply a method stemming from the outranking class: namely, Regime Analysis.

Regime Analysis is a discrete MCA method. The strength of Regime Analysis is that it is able to cope with binary, ordinal, categorical and cardinal data. This applies to both the effects and the weights in the evaluation of alternatives.

Regime Analysis uses two kinds of input data: an impact matrix and a set of weights (Hinloopen et al. 1983; Nijkamp et al. 1990). The impact matrix is composed of elements that measure the effect of each considered alternative in relation to each policy-relevant criterion. The set of weights incorporates information concerning the relative importance of the criteria in the evaluation.

Regime Analysis is a generalized form of concordance analysis. In order to gain a better understanding of Regime Analysis, let us explain the principles of concordance analysis. The basic idea in concordance analysis is to rank a set of alternatives by means of their pairwise comparisons in relation to the chosen criteria. We consider a choice problem where we have a set of alternatives i and a set of criteria k. We begin our analysis by comparing alternative i with alternative k in relation to all criteria. After this, we select all criteria for which alternative i performs better than, or equal to, alternative k. This class of criteria we call a 'concordance set'. Similarly, we define the class of criteria for which alternative *i* performs worse than, or equal to, alternative *k*. This set of criteria is called a 'discordance set'.

In order to rank the alternatives, we introduce the concordance index. The concordance index is the sum of the weights that are related to the criteria for which *i* performs better than *k*. We call this sum C_{ik}. Then we calculate the concordance index for the same alternatives, but by considering the criteria for which *k* is better than *i*, that is, C_{ki}.

After having calculated these two sums, we subtract these two values in order to obtain the net concordance index $\mu_{ik} = C_{ik} - C_{ki}$. Because, in most cases, we have only ordinal information about the weights (and no trade-offs), our interest is in the sign of the net concordance index μ_{ik}. If the sign is positive, this will indicate that alternative *i* is more attractive than alternative *k*; otherwise, the opposite holds.

For detailed studies about the application of MCDA in the field of transport policy evaluation, we refer to Vreeker et al. (2001), Nijkamp and Vreeker (2005), and Yeo and Song (2006).

Overview of Effects

In the evaluation of the different alternatives, several criteria have been incorporated which allow for a simultaneous assessment of various aspects. In the current evaluation not only the economic costs and gains are investigated but also a number of criteria have been added that assess the alternatives on safety aspects, and environmental and nature effects. The underlying thought is the principle that the optimization of functional requirements cannot result in unnecessary pressure on society, for example, relating to nature and the environment.

In this case study, the following nine main categories of criteria have been used: Coast and Sea, Nature, Environment, Landscape, Recreational activities, Transport, Logistics, Integral safety, and Economics and Space.

The scores on the different criteria have been determined in various measuring units (for example, euros, ha and decibels). Scores of economic criteria are predominantly described in terms of money, whereas nature and environmental criteria are predominantly described in a qualitative way. Some criteria are cost indicators, whereas others are benefit indicators. In order to still be able to make use of the effect overview, all scores have been standardized. Ultimately, all criteria were given values by way of a seven-point scale (see Table 15.5):

- a very large positive effect equals 7;
- a positive effect equals 6;
- a neutral effect equals 5;
- a slightly negative effect equals 4;
- a negative effect equals 3;
- a large negative effect equals 2;
- a very large negative effect equals 1.

Table 15.5 Summary of used criteria and weights

	A1	A2	A3	B1	B2	B3a	B3b	C1	C2
Coast and Sea									
Coastline land reclamation	1	3	3	5	5	5	5	3	1
Coastline other	5	5	5	5	5	5	5	5	5
Coastal safety	5	5	5	5	5	5	5	5	5
Northern delta	5	5	5	5	5	5	5	5	5
Tidal movement	3	3	5	5	5	5	5	3	5
Coastal maintenance	1	1	6	5	5	5	5	6	1
Dispersement of substances	5	5	5	5	5	5	5	5	5
Nature									
Diversity ecosystems	6	6	5	5	5	5	5	5	6
Diversity species	7	6	5	5	5	5	5	5	7
Soundness	3	3	3	5	5	5	5	5	3
Ecologically functioning water system	6	5	5	5	5	5	5	5	6
Environment									
Water	5	5	5	5	5	5	5	5	5
Air quality	5	5	5	5	5	5	5	5	5
Noise	3	3	3	5	5	5	5	3	3
Landscape									
Perceived openness	1	3	4	5	5	5	5	4	1
Variation in spatial size	7	6	6	5	5	5	5	6	7
Structure and cohesion	7	6	3	5	5	5	5	3	7
Identity and image	7	6	5	5	5	5	5	5	7
Recreational activities									
Beach activities	3	5	5	6	6	6	6	6	7
Watersports	5	6	5	5	5	5	5	5	5
Nature-related activities	6	6	5	5	6	5	6	5	7
Overnight recreational stay	6	6	5	5	6	5	6	5	7
Active outdoor sports	7	7	6	3	3	3	3	7	7
Harbour-related recreational activities	6	6	6	5	5	5	5	6	6
Visiting attractions	5	7	5	5	5	5	5	5	5
Transport and logistics	5	5	5	5	5	5	5	5	5
Integral safety									
Safety from flooding	7	7	7	5	5	5	5	5	5
External safety	5	5	5	5	5	5	5	5	5
Nautical safety	5	3	1	5	5	5	5	1	5
Tackling incidents and providing aid	6	5	6	5	5	5	5	6	6
Risks for environment and nature	5	5	5	5	5	5	5	5	5
Economy									
Direct employment	6	6	6	6	6	6	6	6	6
Investments costs	2	3	4	3	2	3	2	3	1
Net added value	6	6	6	6	6	6	6	6	6
Space									
Available harbour and industrial area until 2020	7	7	7	3	3	3	1	7	7
Expansion harbour and industrial area after 2020	7	6	6	6	6	6	6	6	7

Assigning Weights to the Criteria

In every decision process, and particularly in a discussion relating to spatial planning, there is always a certain tension between the interests of different stakeholders. These stakeholders all view the problem of spatial shortage in Rotterdam harbour from a different perspective. Environmental organizations, for example, will put more emphasis on the presence of nature and the preservation of the environment. The business community is more interested in the availability of actual space and when this will become available. Against this background, the Ministry of Economic Affairs will focus on the costs of the project and the contribution of the harbour to the Dutch economy.

It should be noted that this research was not commissioned by any of the stakeholders. As a consequence we could not address the choice problem from the perspective of one stakeholder and using his or her weight set in the final evaluation of alternatives. To overcome this problem and to delimit the outer boundaries of the policy field, we elaborated relatively extreme alternatives of the starting points: one alternative in which the economic effects of the different options were emphasized; and one in which emphasis was laid on environmental effects and the consequences for the living environment. At the same time, an alternative with a more neutral connotation (a compromise alternative) was worked out.

Weights set emphasizing economic effects. From an economic perspective, most value is given to the cost effects of investments, and the net added value. Therefore, in the weights set, more weight is granted to the costs of a certain alternative. The question whether the different alternatives offer enough space to accommodate the calculated spatial demand in 2020 and afterwards is also deemed important from an economic perspective. This aspect is also granted a higher weight. On the other hand, the environment, nature, integral safety and effects on the coast and the sea do deserve some attention. Although they offer less than the costs and gains, the different environmental alternatives will deliver some benefits. The first column of Table 15.6 gives a detailed overview of the weights given to the various criteria when the economic effects are emphasized. From this table it becomes clear that 70 per cent of the 100 points are given to criteria that fall in the economic domain.

Weights set to emphasize nature and environment. From an environmental perspective the following groups of criteria deserve the most attention: nature, environment and landscape, closely followed by coast and sea, which both have a direct relation to the quality of the living environment. Relatively little emphasis is put on economic and spatial effects. This also goes for the aspect of integral safety. The second column of Table 15.6 gives a detailed overview of the weights given to the various criteria when emphasis is placed on the interests of nature and environment. Within this scenario, in total, 75 per cent of the weight points are given to the criteria belonging to nature, environment and landscape.

Weights set for the compromise scenario. To determine how the results of the compromise scenario would relate to the results of the above sets of weights, a third weight set is added to the effect overview.

Table 15.6 Alternatives and criteria scores

	Economy	Spatial quality	Compromise
Coast and Sea	*2.5 %*	*5.0 %*	*5.0 %*
Coastline land reclamation			
Coastline other			
Coastal safety			
Northern delta			
Tidal movement			
Coastal maintenance			
Dispersement of substances			
Nature	*2.5 %*	*25.0 %*	*11.67 %*
Diversity ecosystems			
Diversity species			
Soundness			
Ecologically functioning water system			
Environment	*2.5 %*	*25.0 %*	*11.67 %*
Water			
Air quality			
Noise			
Landscape	*2.5 %*	*25.0 %*	*11.67 %*
Perceived openness			
Variation in spatial size			
Structure and cohesion			
Identity and image			
Recreational activities	*5.0 %*	*5.0 %*	*10.0 %*
Beach activities			
Watersports			
Nature-related activities			
Overnight recreational stay			
Active outdoor sports			
Harbour-related recreational activities			
Visiting attractions			
Transport and logistics	*2.5 %*	*2.5 %*	*5.0 %*
Integral safety	*2.5 %*	*2.5 %*	*5.0 %*
Safety from flooding			
External safety			
Nautical safety			
Tackling incidents and providing aid			
Risks for environment and nature			
Economy	*70.0 %*	*5.0 %*	*35.0 %*
Direct employment	10.0 %	1.0 %	11.67
Investments costs	35.0 %	2.0 %	11.67
Net added value	25.0 %	2.0 %	11.67
Space	*10.0 %*	*5.0 %*	*5.0 %*
Available harbour and industrial area until 2020	7.0 %	2.5 %	2.5 %
Expansion harbour and industrial area after 2020	3.0 %	2.5 %	2.5 %
Total	*100.0 %*	*100.0 %*	*100.0 %*

MCA Results

The evaluation of alternatives by way of Regime Analysis in this study took place in two steps. First, an analysis was performed per group (Coast and Sea, Nature, Environment, Landscape, Recreational Activities, Transport, Integral Safety, Economy and Space) on the subsequent scores of the alternatives on the sub-criteria (see Table 15.7). These results (probabilities) served as the input for the ultimate evaluation of the alternatives. The final evaluation was done using the three weight sets explained above. The first weight set emphasizes the economic effects ignored by the different alternatives. The second weight set focuses on the living environment and the accompanying effects, and the third is a more or less neutral weight set which forms a compromise between the economic and environmental perspective.

Table 15.8 shows the ranking of alternatives including the accompanying probabilities for the three different weight vectors.

From the table above it may be concluded that the discussion should predominantly focus on the question which land reclamation alternative best suits the wishes and requirements of the various stakeholders. The variants of intensive land use would not solve the spatial shortage that will appear in 2020. Moreover, the variants are relatively expensive compared with the land reclamation variants.

Table 15.7 MCA results

Criteria		Result								
		A1	A2	A3	B1	B2	B3a	B3b	C1	C2
Coast and Sea	P	4	3	1	1	1	1	1	2	3
	S	0	0.19	0.75	0.75	0.75	0.75	0.75	0.33	0.19
Nature	P	1	2	4	3	3	3	3	4	1
	S	0.94	0.75	0.06	0.44	0.44	0.44	0.44	0.06	0.94
Environment	P	2	2	2	1	1	1	1	2	2
	S	0.25	0.25	0.25	0.81	0.81	0.81	0.81	0.25	0.25
Landscape	P	1	2	4	3	3	3	3	4	1
	S	0.94	0.75	0.06	0.44	0.44	0.44	0.44	0.06	0.94
Recreational activities	P	2	1	4	5	4	5	4	3	1
	S	0.75	0.94	0.37	0.06	0.37	0.06	0.37	0.62	0.94
Transport	P	1	1	1	1	1	1	1	1	1
	S	0.5	0.5	0.5	0.5	0.5	0.5	0.5	0.5	0.5
Safety	P	1	4	3	5	5	5	5	6	2
	S	1	0.44	0.75	0.31	0.31	0.31	0.31	0.25	0.81
Economy	P	3	2	1	2	3	2	3	2	4
	S	0.25	0.69	1	0.69	0.25	0.69	0.25	0.69	0
Space	P	1	2	2	3	3	3	4	2	1
	S	0.94	0.63	0.63	0.25	0.25	0.25	0	0.63	0.94

P = Position in the ranking.
S = Score.

Table 15.8 Ranking of alternatives according to three perspectives

	Economy	Spatial quality	Compromise
1	Land reclamation Variant A1 (0.84)	Combination Variant C2 (1.00)	Land reclamation Variant A1 (1.00)
2	Land reclamation Variant A3 (0.78)	Land reclamation Variant A1 (0.88)	Combination Variant C2 (0.88)
3	Land reclamation Variant A2 (0.75)	Land reclamation Variant A2 (0.75)	Land reclamation Variant A2 (0.75)
4	Combination Variant C2 (0.73)	Intensive land use Variant B2 (0.50)	Intensive land use Variant B1 (0.54)
5	Intensive land use Variant B1 (0.41)	Intensive land use Variant B1 (0.47)	Intensive land use Variant B3a (0.54)
6	Intensive land use Variant B3a (0.41)	Intensive land use Variant B3a (0.47)	Intensive land use Variant B2 (0.38)
7	Combination Variant C1 (0.38)	Intensive land use Variant B3b (0.25)	Intensive land use Variant B3b (0.25)
8	Intensive land use Variant B2 (0.16)	Land reclamation Variant A3 (0.18)	Land reclamation Variant A3 (0.17)
9	Intensive land use Variant B3b (0.04)	Combination Variant C1 (0.00)	Combination Variant C1 (0.00)

Looking at the results, it can be concluded that Variant C2 and Variant A1 obtain a high ranking, irrespective of the weight set used. Both alternatives provide for the development of a nature area of 750 ha, thereby adding positive effects to the nature and landscape indicators.

Additionally, a sufficient number of ha to accommodate the spatial demand will become available and expansion possibilities can be described as good. With regard to the investment costs, both alternatives score badly. However, the economic cost-effectiveness of the land reclamation will be more than ensured. It should be noted that the alternative Variant C2 is based on Variant A1. The phasing of the development of the land reclamation is what distinguishes the two alternatives from each other.

The MCA clearly shows that intensive land use is no option to solve the spatial shortage. The related alternatives are expensive, do not provide new nature areas, and will increase the pressure between city and harbour. In contrast, the alternatives of land reclamation offer a proper solution for the spatial shortage, new nature areas and future expansion possibilities. It is possible to consider combining a land reclamation alternative with urban infill and intensifying the use of the existing harbour area. Research shows that it is relatively easy and cheap to free up a large amount of land within the existing harbour area, in order to postpone the development of the reclaimed land. This way it will be easier to estimate what the actual spatial shortage will be.

Conclusions

This chapter has clearly shown that port expansion nowadays is characterized by conflicting objectives. This is especially highlighted by the wish of the Dutch government to improve the quality of the economy and the social environment. Multicriteria evaluation methods are particularly suitable to take heterogeneous and conflicting dimensions of decision problems into account. These methods do so by incorporating differing evaluation criteria in the decision-making process that reflect the perspective of various stakeholders. Furthermore, various sets of weights can also be applied to reflect these different perspectives. In our analysis, we applied weight sets that reflect the following two perspectives: *economy* and *spatial quality*. The analysis showed that Alternative A1 (land reclamation) is the most preferred alternative from an economic perspective and Combination Variant C2 from an environmental perspective.

In May 2001 the Dutch government decided to develop a nature area separately from the port expansion. To achieve the strengthening of the local economy and social environment, three types of measures have been designed:

- group of projects that are aimed to intensify the current land use in the port area. In addition, these measures are aimed at improving the living environment in the surrounding areas;
- the expansion of the port with land reclamation;
- the development 750 hectares of nature outside the project area to compensate for damage caused by the land reclamation.

The designed policy package thus forms a combination of the alternatives *land reclamation* and *urban infill*. The compensation of environmental damage will not take place on the areas of land reclamation themselves but is to be achieved by means of the development of a nature area elsewhere in the Rotterdam area.

However, in January 2005 the Dutch Council of State did not approve the proposal for the reclamation of land. It decided that the current information available is not sufficient to judge the environmental consequences. Further environmental research will be conducted to obtain additional information.

In December 2006 the adjusted plans for the land reclamation were passed by the Dutch House of Representatives and the Senate. These adjusted plans overcome all criticism as expressed by the Dutch Council of State. It has been concluded that the environmental impacts are less than expected.

In April 2007 the City of Rotterdam presented a draft version of the legally binding development plan for the harbour expansion. Stakeholders could comment on it until 31 May. The definitive development plan will be established in the months after this date.

It is expected that the actual works for the expansion plan will start in 2008 and that ships can use the new harbour facilities in 2013.

It should be emphasized that MCDA, like cost benefit analysis, is a valuable rational decision-support tool. In actual decision-making situations, other factors and (policy) powers may influence the decision-making process. The case study shows, however, that the application of an MCDA method in the field of transport policy making provides decision makers with important information about the feasibility of alternatives. Furthermore, MCDA methods can be used, by applying extreme visions to the choice problem, in order to delimit the edges of the policy field concerned. This information can inform relevant stakeholders about possible conflicts that may occur in the decision-making arena.

Conflicts may arise about the selection of the judgement criteria and the associated scores (effects of the alternatives). Regime Analysis does not require a standardization of effect scores. As a consequence, stakeholders can easily understand the input used by the method, and hence an important source for conflicts is eliminated. The outcomes of the method may, however, be less clear to the stakeholders involved. This does not, however, apply to the produced ranking. But what does the calculated overall performance exactly imply?

As there is no single evaluation method that can satisfactorily and unequivocally evaluate all of the complex aspects of choice possibilities and is also intuitively understandable, the role of the decision analyst is very important in such situations.

He or she needs to guide the decision maker in choosing an appropriate method and to assist in communicating the many aspects (objectives, procedures, inputs, outcomes and so on) of the decision-making process.

References

van Ansem, I. (2000), 'Het Dreigend Ruimtetekort in de Rotterdamse Haven; Een Multicriteria Analyse van de Verschillende Oplossingsalternatieven' (in Dutch) [Imminent shortage of space in the Rotterdam Port area; A Multicriteria Analysis of a Possible Solution], Wageningen University.

Ballestero, E. and Romero, C. (1998), 'Uncertainty and the Evaluation of Public Investment Decisions', *American Economic Review* 60, 364–78.

CPB (1997), 'Economische en Ruimtelijke Versterking van Mainport Rotterdam' (in Dutch) ['Economic and Spatial Enhancement of the Mainport Rotterdam'], Werkdocument No. 92, Den Haag.

—— (1998), '2020: Integrale Verkenningen voor Haven en Industrie, een Beoordeling van Onderzoek van het Gemeentelijk Havenbedrijf Rotterdam' (in Dutch) ['2020: An explorative study of development options for Port activities, an assessment of results findings of the Rotterdam Port Authority'], Werkdocument No. 108, Den Haag.

GHR (1998), 'Integrale Verkenningen voor Haven en Industrie 2020' (in Dutch) ['An explorative study of development options for Port activities 2020'], Rotterdam.

—— (1999), 'Ruimtelijke Dynamiek in de Rotterdamse Haven' (in Dutch) ['Spatial Dynamics in the Rotterdam Port area'], Rotterdam.

Hinloopen, E., Nijkamp, P. and Rietveld, P. (1983), 'Qualitative Discrete Multiple Criteria Choice Models in Regional Planning', *Regional Science and Urban Economics* 13:1, 77–102.

Korhonen, P. (2005), 'Interactive Methods', in J. Figueira, S. Greco and M. Ehrgott (eds), *Multiple Criteria Decision Analysis: State of the Art Surveys* (New York: Springer), pp. 641–65.

Munda, G., Nijkamp, P. and Rietveld, P. (1995), 'Monetary and Non-Monetary Evaluation Methods in Sustainable Development Planning', *Economie Appliquee* 18:2, 143–60.

Nijkamp, P. and Vreeker, R. (2005), 'Multicriteria Evaluation of Transport Policies', in D.A. Hensher and K.J. Button (eds), *Handbook of Transport Strategy, Policy and Institutions* (Amsterdam New York: Elsevier), pp. 507–25.

Nijkamp, P., Rietveld, P. and Voogd, H. (1990), *Multicriteria Analysis for Physical Planning* (Amsterdam: Elsevier).

Nijkamp, P., Ubbels, B.J. and Verhoef, E.T. (2005), 'Transport Investment Appraisal and the Environment', in D.A. Hensher and K.J. Button (eds), *Handbook of Transport Strategy, Policy and Institutions* (Amsterdam New York: Elsevier), pp. 333–55.

PMR (1999), 'Interimrapportage Op Koers' (in Dutch), ['Interim Report, Project Mainport Development Rotterdam'], Project Mainportontwikkeling Rotterdam, Den Haag.

—— (2001), 'Mer Hoofdrapport', Project Mainportontwikkeling Rotterdam (in Dutch), [Environmental Effect Study', Project Mainport Development],Rotterdam.

Roy, B. (1968), 'Classement et Choix en Presence de Points de Vue Multiple (La Methode Electre)', *R.I.R.O.* 2, 57–75.

—— (1972), 'Decision avec Criteres Multiple', *Metra* 11, 121–51.

Roy, B. and Jacquet-Lagreze, E. (1977), 'Concepts and Methods Used in Multicriterion Decision Models: Their Applications to Transportation Problems', in H. Strobel, R. Genser and M.M. Etschmaier (eds), *Optimization Applied to Transportation Systems: Proceedings of the IFAC-IIASA Workshop Held in Vienna, Austria, on 17-19 February, 1976* (Laxenburg: International Institute for Applied Systems Analysis), pp. 9–26.

SBRG (1999), 'Integrale Projectnota Bestaand Rotterdams Gebied. Ruimte voor de Haven, Kwaliteit voor de Leefomgeving', Samenwerkingsverband Bestaand Rotterdams Gebied, Den Haag (in Dutch), ['Integral Development Report for the existing Rotterdam Port Area. Space for Port and Environmental Quality' Coalition existing in the Rotterdam Port Area, The Hague].

SM2V (S.T.M.) (1999), 'Projectnota Landaanwinning', Projectgroep Projectnota Landaanwinning, Den Haag (in Dutch), ['Study for Land Reclamation', Commission for the study of land reclamation, The Hague].

Vincke, P. (1992), *Multicriteria Decision-Aid* (Chichester New York: Wiley).

Vreeker, R., Nijkamp, P. and Ter Welle, C. (2001), 'A Multicriteria Decision Support Methodology for Evaluation Airport Expansion Plans', *Transportation Research Part D: Transport and Environment* 7, 27–47.

Yeo, G.-T. and Song, D.-W. (2006), 'An Application of the Hierarchical Fuzzy Process to Container Port Competition: Policy and Strategic Implications', *Transportation* 33:4, 409–22.

Chapter 16

Integrated Restructuring of Urban Transport Networks: A Case Study on the Athens Tramway System

George Patris, Maria Giaoutzi and John Mourmouris

Development of the Athens Tramway System

Over the years Athens has undergone significant infrastructure developments leading to a remarkable restructuring, especially in the field of transportation. New peripheral roads have been created or significantly improved, which bypass the city centre and offer alternative routes, while the modernised modes of public transport have improved both the quality of services in transportation and living standards.

Amongst other things, a Light Railway Network (Tramway system) was designed connecting the northern suburbs with the coastal part of the city. In this connection, a number of studies were conducted, proposing various scenarios for the alignment of the Tramway system.

The spreading of the Greater Athens Area in recent decades and the rapid increase of private car ownership have caused serious problems in traffic management, especially for the city centre. It was decided by the late 1980s that an efficient and reliable system of public transport would best meet the transport demand created by the compact form of the city centre. The reason to develop such a system was to divert passengers from private cars so that heavy traffic, bottlenecks, congestion and environmental pollution could be diminished.

More precisely, the development of a tramway network was based on the assumption that such a system would best serve the objectives of environmental protection, cost effectiveness and best integration with the rest of the system in order to maximize the share of public transport in everyday movements and improve the traffic conditions of the city.

This particular analysis is part of a number of studies commissioned by the State in order to identify the best alternative for such an important infrastructure investment. The analysis focuses independently on specific issues, as described in the following article, without making any comparison with other studies or other perspectives (for example, other stakeholders such as the local authorities).

The study was conducted in 1999, and therefore the analysis and comments presented below reflect the rationale of the time. However, the focus of the chapter is on the approach that was implemented rather than on the analysis of the transport problem as such.

Short Description of the Proposed Alternatives

The idea of creating a tramway network in the area of Athens had been developed in the 1990s. A number of studies were conducted, resulting in various alternative alignments of the network.

The first task of the present study was to gather all these proposals and evaluate their results with regard to the expected benefits.

The first screening of the proposals took into consideration the level of depth reached by each study, concerning the availability of technical and socio-economic studies, as well as the impact assessment of each proposed alternative on traffic, urban development and the environment. As a result 11 alternatives qualified to be selected for further evaluation, as can be seen in Table 16.1 above (see also the map of Athens in Annex 16.A).

Evaluation Framework

The focus of the present chapter is on the evaluation framework developed for the selection of the best-case scenario for the development of a light railway network (Tramway system) in the city of Athens.

As an input to this process a set of alternatives, developed in the context of a number of preparatory studies, were used. The outcome of the study had to comply with the objectives of the project: namely, the improvement of the transportation services and environmental protection.

The outline of the evaluation process can be seen in Figure 16.1.

The specific problem was to evaluate a number of alternative options, each of them accomplishing similar but different objectives. In this context, several factors had to be taken into consideration of both a qualitative and a quantitative nature. It was decided that the most suitable approach for the specific nature of the problem was that of multicriteria analysis (MCA).

Table 16.1 Alternatives selected for evaluation

	Alternative No.	Description
1.	1a	Patissia – Delta Falirou – Glyfada
2.	1b	Aigyptou Sq. – Delta Falirou – Glyfada
3.	2a	Patissia – Delta Falirou – Chatzikiriakio
4.	2b	Aigyptou Sq.– Delta Falirou – Chatzikiriakio
5.	3	Patissia – Delta Falirou
6.	4a	Patissia – Delta Falirou – Glyfada – Chatzikiriakio
7.	4b	Patissia – Delta Falirou – Glyfada – Neo Faliro
8.	5	Patissia – Delta Falirou – Circle
9.	6	Patissia – Delta Falirou – Glyfada – loop
10.	7	Patissia – Delta Falirou – Chatzikiriakio – loop
11.	8	Patissia – Delta Falirou – Glyfada – Neo Faliro – loop

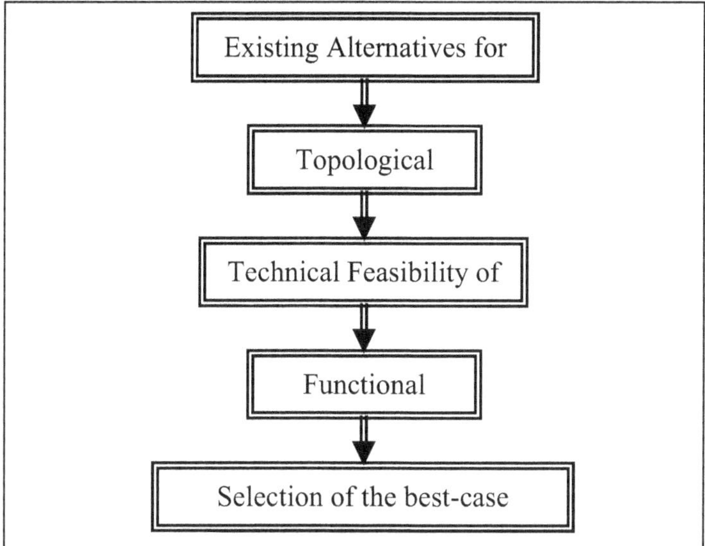

Figure 16.1 Chart of evaluation stages

Multicriteria analysis is designed to help decision makers to integrate the different options, reflecting the opinions of the actors concerned, into a prospective or retrospective framework (Tavistock Institute 2003). MCA describes any structured approach used to determine overall preferences for alternative options, where the options meet several objectives. In MCA, desirable objectives are specified and corresponding attributes or indicators are identified.

The actual measurement of the indicators is not necessarily in monetary terms, but is often based on the quantitative analysis (by means of scoring, ranking and weighting) of a wide range of qualitative indicators and criteria. Moreover, various environmental and functional indicators can be developed together with the economic costs and benefits identified. In addition, MCA provides techniques for comparing and ranking different outcomes, even though a variety of indicators are used (Joerin et al. 2001; Vreeker et al. 2001).

After the selection of the evaluation method, special attention must be paid to its appropriate application in order to cover the specific needs of the problem. The alternative to be selected as the best should accomplish two different objectives: first, it should be integrated into the existing urban environment; and, second, it should operate in a way that would satisfy the transportation needs.

The focus in the present chapter is on the process of selecting alternatives, in the first part of the study (see Figure 16.1), where the evaluation process follows four successive stages.

More precisely, in the first round of the selection process – called the topological evaluation – the alternative scenarios are evaluated on the basis of their topological characteristics. The prevailing alternatives that meet the thresholds set at this stage will then be further elaborated as to their technical feasibility, in order to specify the technical details/specifications of the proposed alignment of the Tramway system.

In the second round of the process – called the functional evaluation – the prevailing set of alternative scenarios will then be evaluated on the basis of their functional characteristics in serving the various transportation needs.

Both stages of the process involve a number of quantitative and qualitative indicators/criteria especially prepared for this purpose.

For the topological evaluation, the criteria selected took into consideration the spatial characteristics of the study area at the time.

At that time, Athens was already a congested city with an urban transport network comprising bus (and trolley) lines and a Metro line, and two additional Metro lines were under construction. In this context each alternative for the Tramway development had to be evaluated against its potential to integrate the existing (and future) urban transport system and to contribute to the improvement of the transport services in the urban context.

Factors taken into consideration were:

- the impacts of the proposed alternatives for the Public Transport System;
- the impacts of the proposed alternatives on transport services and the urban environment; and
- the technical feasibility of each alternative.

The assessment of the technical feasibility of the proposed alternative scenarios included the study of the technical parameters also involved in the construction phase. This step of the process falls outside the scope of this chapter.

The functional evaluation focus is on the assessment of the impacts of the Tramway system on both the transportation services and the environment, during the operation phase, taking into account certain characteristics of the areas crossed by the specific Tramway alignments.

More specifically the attributes taken into consideration were:

- transportation patterns in the study areas;
- spatial and socio economic characteristics;
- likely impacts on the environment; and
- economic/financial impacts.

A number of indicators were developed to assess the characteristics of each alternative scenario and their impacts on both the urban environment and the transport network. Those indicators were used as criteria in the context of multicriteria analysis in order to rank the alternatives and support the decision process,

Evaluation Methodology

The methodology used for the stages of both the 'topological' and the 'functional' evaluation is multicriteria analysis, and more specifically the Regime method. This approach was selected because of its ability to combine criteria of both a quantitative and a qualitative nature (van Herwijnen 1999).

The multicriteria analysis follows the stages presented below:

- *Description of the evaluation criteria*: At this stage the evaluation criteria are developed, and grouped on the basis of their impacts. In addition, the mathematical calculation of the score of each criterion is presented.
- *Development of the Impacts Matrix*: The Impacts Matrix includes the scores of the criteria described for each of the selected alternatives.
- *Criteria characteristics*: At this stage the characteristics of each criterion are determined. These characteristics include the type (quantitative/ qualitative), the scale (ratio/binary/interval/ordinal) and the trend (benefit/cost) of each criterion.
- *Attaching weights to the criteria*: In order to determine the proper weight for each criterion, two techniques were applied: the rank order and the pairwise. The calculations include both techniques in order to assess the reliability of the results.
- *Calculation of results.*

Topological Evaluation

The topological evaluation involves ranking the alternative scenarios on the basis of their topological characteristics, in order to enable the first screening of the proposed alternatives for further examination. According to the results of the analysis, certain thresholds are set to define which alternatives could be considered for further evaluation. The objective of the thresholds set is to identify all those alternatives which have comparable results in the topological evaluation and to allow the next stages of evaluation to identify the most attractive one. The prevailing alternatives that met the thresholds were then further elaborated as to their technical feasibility, in order to set the technical specifications of the proposed alignments of the Tramway system. The evaluation followed the stages of the multicriteria analysis described above (see Figure 16.1).

Description of criteria The topological evaluation was performed on the basis of three different groups of criteria describing the spatial characteristics of each alternative as can be seen below:

- criteria of *Group A* that refer to the likely impacts on the Public Transport System by each alternative;
- criteria of *Group B* that refer to the impacts on both urban utilities, such as level and quality of the transport services, and quality of the urban environment;
- criteria of *Group C* that refer to the feasibility of each alternative, taking into account the technical works required and their impacts on traffic.

More specifically, the three groups consist of the following criteria:

Group A: Public Transport System (PTS)

- *Cohesion of PTS*: This criterion describes the contribution of each alternative to the cohesion of the Athens Public Transport System. The relevant index 'σ_i' represents the ratio of the number of transfer stations included in alternative *i*, over the number of transfer stations of the whole network:

$$\sigma_i = \frac{\sum M_i}{\sum_{i=1}^{n} M_{network}} \tag{16.1}$$

Variations of this index may include only the stations that facilitate the transfer of passengers from Tram to Metro (and vice versa) or from Tram to bus and trolley lines.

- *Potential for expansion*: This criterion describes the potential of each alternative for the future expansion of the Tram network. The relevant qualitative index comprises three variations:
  - expansion from a single terminal of the line;
  - expansion from all terminals of the line;
  - expansion from intermediate stations of the line.
- *Restructuring of bus and trolley lines*: This criterion describes the level of 'inconvenience' that each alternative will create and the range of changes that have to be made in the other means of public transport (bus and trolley lines) for the integration of the specific alternative in the Athens PTS. This level A_i can be determined by the number of bus and trolley lines that are affected by the specific alternative, divided by the total number of bus and trolley lines of the whole network:

$$A_i = \frac{\text{Nr of lines affected by alternative } i}{\text{Total nr of lines affected}} \tag{16.2}$$

Group B: Urban Utilities

- *Service of major transport 'poles'*:[1] This criterion describes the level of services provided by each alternative to existing areas and new areas under development that may increase transport demand within the urban area. The relevant index E_i is calculated as the ratio of the number of areas served by solution i over the number of areas for the whole Tram network:

1 The term 'poles' is used to describe urban areas that attract or generate traffic movements and are therefore important for the public transport network.

$$E_i = \frac{\sum \Pi_i}{\sum_{i=1}^{n} \Pi_{i,\ network}} \tag{16.3}$$

- *Quality of transportation*: This criterion describes improvements in transportation services stemming from the introduction of the Tramway system into the Athens PTS. This improvement refers to the following three characteristics:
  - reduction of transfers for travelling to the city centre;
  - increase in the frequency of the service;
  - traffic level of services along the Tramline.

  The above-mentioned effects are described by a relevant qualitative index.

- *Environmental impacts*: The criterion describes the positive impacts on the urban environment following the various interventions that will accompany the construction of the Tramline. These include improvement of the urban environment, construction of pedestrian facilities, creation of green areas and zones, and so on. This criterion is presented by a qualitative index.

Group C: Technical Feasibility

- *Engineering works*: This criterion refers to the number and type of technical works and engineering constructions that have to be constructed along each alternative Tramline. For this purpose three different scales (large, medium and small) were used that characterize the nature of the works. The corresponding index T_i for each scale is calculated as the ratio of the number of works (for the specific alternative and of the particular scale) over the number of works for the whole Tramway network (of the same scale)

$$T_{ij} = \frac{\text{Nr of technical works of scale } j \text{ for alternative } i}{\text{Nr of technical works of scale } j \text{ for the whole Tram network}} \tag{16.4}$$

- *Impacts on traffic*: This criterion describes the impacts that each alternative has on traffic. The integration of the Tramline into the existing traffic required certain traffic interventions such as modification of junctions, installation of traffic lights and other traffic management measures. Those traffic interventions were considered separately from the other engineering works, to avoid double counting. The relevant index K_i is the number of traffic interventions required for the alternative i divided by the number of interventions required for the whole Tramway network.

Development of the Impacts Matrix The scores of the criteria, presented in Table 16.A1 (see Annex 16.A), are demonstrated on the vertical axis of the table, while the 11 alternatives are displayed on the horizontal axis.

Definition of the type, scale and trend of the selected criteria The characteristics of the selected criteria are presented in Table 16.2.

Table 16.2 Characteristics of the criteria – Topological evaluation

No.	Criterion	Type	Scale	Trend
1	*Cohesion*			
1.1	No. of Tram/Metro transfer stations	Quantitative	Ratio	benefit
1.2	No. of Tram/bus transfer stations	Quantitative	Ratio	benefit
2	*Ability to extend*			
2.1	Not from one end	Qualitative	Binary	cost
2.2	From all ends	Qualitative	Binary	benefit
2.3	From intermediate stations	Quantitative	Interval	benefit
3	*Restructuring of bus and trolley lines*	*Quantitative*	*Ratio*	*cost*
4	*Level of Service*			
4.1	Category A	Quantitative	Ratio	benefit
4.2	Category B	Quantitative	Ratio	benefit
4.3	Category C	Quantitative	Ratio	benefit
5	*Quality of transportation*			
5.1	No. of transfers	Qualitative	Ordinal	1 is best
5.2	Frequency	Qualitative	Ordinal	1 is best
5.3	Traffic flow	Qualitative	Ordinal	1 is best
6	*Environmental effects*	*Qualitative*	*Ordinal*	*1 is best*
7	*Engineering works/projects*			
7.1	small scale	Quantitative	Ratio	cost
7.2	medium scale	Quantitative	Ratio	cost
7.3	large scale	Quantitative	Ratio	cost
8	*Effects on traffic*	Qualitative	Ordinal	1 is best

Definition of weights As already mentioned, the weights for each criterion were determined by using both the rank order and the pairwise methods. The ranking of criteria is presented in Table 16.3.

Calculation of results By applying the Regime method, the topological evaluation was carried out. The outcome of the process is presented in Table 16.3. The results show that the alternatives can be grouped into three categories based on their scores by using both rank order and pairwise comparison, as shown in Table 16.4.

Table 16.3 Results of the topological evaluation

Rank order weights		Ranking			Pairwise		
Ranking		*Rank*		*Scores*	*Rank*		*Scores*
1:	C1.1	1:	7	0.95 ←	1:	7	0.95 ←
2:	C4.1	2:	8	0.87 ←	2:	8	0.86 ←
3:	C1.2	3:	6	0.76 ←	3:	6	0.76 ←
4:	C4.2	4:	2a	0.65	4:	2a	0.66
5:	C7.1	5:	4a	0.65		4a	0.63
	C7.2	6:	5	0.45	6:	5	0.45
	C7.3	7:	la	0.39	7:	la	0.40
8:	C3	8:	4b	0.35	8:	4b	0.35
9:	C2.2	9:	1b	0.22	9:	1b	0.22
	C2.3	10:	2b	0.14	10:	2b	0.14
11:	C8	11:	3	0.08	11:	3	0.08
12:	C4.3						
13:	C2.1						
14:	C5.1						
	C5.2						
	C5.3						
17:	C6						

Table 16.4 Grouped results of the topological evaluation

	Alternatives	Scoring
Group 1	6, 7, 8	0.95–0.76
Group 2	2A, 4A, 5	0.66–0.45
Group 3	1A, 1B, 2B, 3, 4B	0.49–0.08

According to the topological evaluation three dominant alternatives were identified (6, 7 and 8). These were then further evaluated on the basis of their functional characteristics. The prevailing alternatives are presented in Table 16.5.

Table 16.5 Prevailing scenarios

Alternative No.	Description
6:	Patissia – Delta Falirou – Glyfada – loop
7:	Patissia – Delta Falirou – Chatzikiriakio – loop
8:	Patissia – Delta Falirou – Glyfada – Neo Faliro – loop

The topological evaluation was followed by a thorough analysis of the technical feasibility of each alternative in order to define the technical details of each prevailing alignment, for example, the allocation of tracks in relation to the road axis, the effects on junctions and pavements.

After having completed the above stage, where all the technical aspects were clearly defined, the process could then proceed to the stage of the functional evaluation of the alternative scenarios.

Functional Evaluation

The three prevailing alternatives with the highest scores in the topological evaluation proceeded to the next stage of the methodological framework, a *functional evaluation*, in order to determine the alternative that best serves the objectives of the project.

After the agreement of the relevant authority for the development of Tram, the functional analysis also includes the versions of these three alternatives that did not include the circular/loop at the centre (that is, alternatives 1a, 2a and 4b). In the following, the A-Series results will be those results of the evaluation that refer only to the three alternatives of the topological evaluation, while the B-Series results will be the results that refer to all six alternatives

This process follows the stages of the multicriteria analysis.

Description of the criteria Four groups of criteria have been used at this stage as described below:

- Criteria of *Group A* refer to the existing transport patterns describing the modal split and the trip purpose.
- Criteria of *Group B* refer to the spatial patterns describing the effects on economic activities and spatial developments.
- Criteria of *Group C* refer to the effects on the environment due to the construction of the Tramlines and the restructuring of the bus and trolley lines.
- Criteria of *Group D* refer to the financial dimension of each alternative taking into account the construction costs.

More specifically, the four groups consist of the following criteria:

Group A: Transportation Patterns

- *Modal split*: This criterion describes the modal split amongst the various modes of transport. The modes taken into consideration are: private cars (PC), trucks, buses and motorcycles. The relevant index B_i is the ratio of the number of traffic movements M using the mode i along the road section k over the length D of the section.

$$B_i = \frac{\sum_{i=1}^{n} M_{ik}}{D_k} \tag{16.5}$$

For the requirements of this index, five road sections were identified that correspond to the main road sections that compose the selected alternatives.

- *Trips to work*: This criterion provides information on the movements with trip purpose '*work*', in the various areas affected by the Tram. The information used is based on the Origin-Destination Matrix developed in the context of the Metro network development study. The index L_{ij} used is the number of trip movements T, from each area of origin i to all other destination areas j.

$$L_{ij} = \sum_{i=1, j=1}^{k} T_{ij} \tag{16.6}$$

- *Trips for other purposes*: The criterion provides information on the trips for other purposes, within the range served by the Tramway network. The index E_{ij} includes the trips for other purposes.

$$E_{ij} = \sum_{i=1, j=1}^{k} T_{ij} \tag{16.7}$$

- *Bus and trolley Routes*: The criterion describes the transport services (bus and trolley lines) provided in each alternative scenario. It is assumed that the Tram offers a better quality of service and will therefore attract passengers from bus and trolley services. For this purpose, two sub-criteria were developed: the first refers to bus and trolley lines with their origin and destination within the areas crossed by the Tramline (intra-region), and the second refers to a wider area, in which a bus or a trolley line crosses or passes close to the Tramline. In both cases the relevant index T_i is calculated as the number of lines affected by alternative i, multiplied by the frequency of routes of each line.

$$T_{i\text{-intra}} = (\text{No. of lines}) \times (\text{frequency of routes});$$
$$T_{i\text{-inter}} = (\text{No. of lines}) \times (\text{frequency of routes}) \tag{16.8}$$

- *Performance of the existing transport modes*: This criterion provides information on the performance of the existing transport modes in the intra-region study area. The relevant index A_i is the number of passengers per vehicle and per kilometre in the intra-region lines of each alternative i:

$$A_i = \sum_{j=1}^{m} \frac{T_j}{L_j \times f_j} \tag{16.9}$$

where:
$j = 1, \ldots, m$ = bus and trolley lines in the intra-region of alternative I;
T_j = no. of passengers transported by line j;
L_j = length of line j /Km;
F_j = no. of routes of line j.

Group B: Spatial Patterns

- *Economic activities*: This particular criterion refers to the structure of the economic activities in the areas directly affected by the Tramway operation. The related index LQ_i describes the spatial concentration of the economic activities, and is calculated as the ratio of the employment in each sector of the economy in the area affected by the alternative i, over the ratio of the employment in the same sector in the Greater Athens Area.

$$LQ_i = \sum \frac{E_i / \sum E_i}{E / \sum E} \tag{16.10}$$

where:

E_i = employment of each sector of the economy (primary, secondary, tertiary) in the area i;
E = employment, of each sector of the economy, in the Greater Athens Area.

The above-described index is calculated for the three sectors of the economy, as well as for the basic employment (four indices in total).

- *Spatial development*: This criterion includes three sub-criteria that define the impacts caused by the development of the Tramway system:
  - technical interventions;
  - access to the existing Olympic Complexes;
  - access to the new Olympic Complexes.

The Olympic Complexes are deemed to be important poles generating movements.
The first index PE_i refers to the total length of the streets to be pedestrianized:

$$PE_i = \sum_{i=1}^{m} E_i \tag{16.11}$$

where:
$i = 1, \ldots, m$ = no. of technical interventions;
E_i = pedestrianized length of the technical intervention i.

The next two indices *Oi* and *Oj* refer to the transport services of the existing and new Olympic Complexes (OC), respectively:

$$O_i = \frac{\text{No. of new OC of alternative } i}{\text{No. of new OC for the whole Tram network}};$$

$$O_j = \frac{\text{No. of existing OC of alternative } j}{\text{No. of existing OC for the whole Tram network}}; \tag{16.12}$$

Group C: Environmental impacts

- *Environmental benefit*: The development of the Tramline facilitates the removal of some of the existing bus lines that fall within the range of service of the Tramway. The goal is the improvement of the environmental quality in the areas by reducing the gas emissions from traffic.

  The related index Y_i is calculated as follows:

$$Y_i = \sum_{i=1}^{n} K_i - \sum_{j=1}^{k} \Delta_j \tag{16.13}$$

  where:
  $i = 1, \ldots, n$ = no. of bus/trolley lines crossing the area;
  $j = 1, \ldots, m$ = no. of bus/trolley lines that are being abolished;
  K_i = no. of bus/trolley routes crossing the area (intra and inter);
  Δ_j = no. of bus/trolley routes that are being abolished.

- *Physical structure*: The morphology and physical structure of the area greatly affect the environmental quality of the areas crossed by the Tramline. In an area with tall buildings and narrow streets the emissions from buses is more likely to remain in the area and aggravate the environmental conditions. The replacement of buses by the Tram in areas with such a physical structure creates significant environmental benefits. Therefore it is assumed that the more environmentally decayed an area was before the development of the Tramway system, the more it benefits after its construction. The three qualitative indices below refer to the:
  - existence of buildings on the one side of the street;
  - existence of buildings on both sides of the street;
  - ability of emissions to spread (relation between street width and length of the buildings).

Group D: Financial dimension

- *Construction costs*: The financial cost of each alternative is calculated by taking into account the various technical aspects, such as the total length, the number of stations or the road construction costs.

Development of the Impacts Matrix The Impacts Matrix includes the scores of the criteria described for each of the selected alternatives. The criteria are shown on the vertical axis of the matrix, while the alternatives are displayed on the horizontal axis. As described earlier, the functional analysis also included the versions of the three alternatives that did not include the circular/loop at the centre (that is, alternatives 1a, 2a and 4b) on the request of the authority responsible for the development of the Tram. In this respect, the A-Series results are the evaluation results that refer only to the three alternatives of the topological evaluation, while the B-Series results are the results that refer to all six alternatives.

The Impacts Matrix for all the six alternatives is presented in Table 16.A2 in Annex 16.A.

Types, scale and trend of the selected criteria In order to implement the multicriteria analysis, it is important to determine the characteristics: namely, the type (quantitative/qualitative), scale (ratio/binary/interval/ordinal) and trend (benefit/cost) of each criterion, as presented in Table 16.6.

Table 16.6 Characteristics of the criteria – Functional evaluation

No.	Criterion	Type	Scale	Direction
1.	*Transportation Patterns*			
1.1	Modal split	Quantitative	interval	Benefit
1.2	Transportation for work	Quantitative	ratio	Benefit
1.3	Transportation not for work	Quantitative	ratio	Benefit
1.4	intra – bus/trolley routes	Quantitative	interval	Benefit
1.5	inter – bus/trolley routes	Quantitative	interval	Benefit
1.6	Performance of existing transportation modes	Quantitative	ratio	Cost
2.	*Spatial patterns*			
2.1	Economic activities in primary sector	Quantitative	ratio	Benefit
2.2	Economic activities in secondary sector	Quantitative	ratio	Benefit
2.3	Economic activities in tertiary sector	Quantitative	ratio	Benefit
2.4	Economic activities in basic employment	Quantitative	ratio	Benefit
2.5	Technical interventions	Quantitative	interval	Benefit
2.6	Access to existing Olympic Complexes	Qualitative	ratio	Benefit
2.7	Access to new Olympic Complexes	Qualitative	ratio	Benefit
3.	*Environmental effects*			
3.1	Environmental compensation	Quantitative	interval	Benefit
3.2	Existence of buildings on the one side of the street	Qualitative	binary	Yes is best
3.3	Existence of buildings on both sides of the street	Qualitative	binary	No is best
3.4	Ability of emissions to spread	Qualitative	—/+++	—/+++
4.	*Financial dimension*			
4.1	Construction costs	Quantitative	monetary	Cost

Attaching weights – Implementation of the algorithm In order to test the sensitivity of the method, the process of attaching weights took place in five different ways (see Cases 1–5). In Case 1 the same weights were attached to all criteria, while in the remaining Cases 2–5 priority was given to each of the four groups of criteria

The overall results of all five cases are presented in Table 16.7.

Table 16.7 Results of the functional evaluation – Cases 1–5/A-Series

Alternative	Case 1	Case 2	Case 3	Case 4	Case 5
8	*1.00*	*1.00*	*1.00*	*1.00*	*1.00*
6	0.50	0.48	0.50	0.45	0.49
7	0.00	0.02	0.00	0.05	0.01

The score of each alternative represents the average probability that this alternative prevails over any other alternative.

The high and stable scores of *alternative 8 in all five cases* proves the high stability of the system and its low sensitivity to the alternative weights. It is clearly evident that this alternative is the most suitable for the development of the Tram network in the city of Athens.

The same remarks can also be made for the B-Series, the results of which are shown in Table 16.8.

Table 16.8 Results of the functional evaluation – Cases 1–5/B-Series

Alternative	Case 1	Case 2	Case 3	Case 4	Case 5
8	*1.00*	*1.00*	*1.00*	*1.00*	*1.00*
4b	0.80	0.80	0.79	0.77	0.77
6	0.60	0.59	0.61	0.56	0.61
1a	0.32	0.21	0.24	0.15	0.22
7	0.28	0.39	0.36	0.47	0.39
2a	0.00	0.01	0.00	0.06	0.01

In addition, it should be noted that alternative 4b is the second-best alternative in all five cases. The fact that alternative 4b is the same as 8 without the loop strengthens the accuracy of the results and the reliability of the methodology implemented.

Conclusions

The present chapter has presented the first part of the evaluation framework developed for the selection of the best-case alternative for the Athens Tramway system. The purpose was to define the most suitable alignment of the Tramline that would best serve the needs of the city and of the (at that time) forthcoming Olympic Games. A number of pre-selected alternatives were evaluated aiming at the selection of the best-case scenario for the integration of the Tram into Athens's public transport system and urban environment.

The analysis was performed in successive stages. The first stage of the analysis evaluated the integration of each alternative into the existing urban and public transport system, while the functional evaluation that followed assessed the operational characteristics of each alternative. In between the two stages a technical feasibility analysis confirmed that the selected alternatives could be constructed.

A number of criteria, both qualitative and quantitative, were selected for the needs of each stage of the evaluation procedure but also for the main goal of the total endeavour.

The implementation of the methodology was performed in a way that guarantees the reliability of the results. In this context several scenarios were studied in order to test the sensitivity of the method.

The methodology clearly identified the alternative that best meets the criteria set. The selected alternative prevailed in both the topological and the functional evaluation, a fact which strengthens the accuracy of the results and the reliability of the methodology implemented.

The evaluation described above provides a solid basis for the selection of the most suitable alternative. It is recognized, however, that such an analysis has certain limits, and it is difficult to address every possible issue. In the next stage of the research study, further examination concerning the transport demand and the costs of construction and operation is necessary for the in-depth analysis of such a serious infrastructure investment.

In addition, experience has shown that, in the course of action, other factors arise that could not be predicted or correctly assessed in advance. In the particular case under study, the public acceptance of the solution selected has proven to be such a crucial factor: the reaction of the local authorities and the citizens have sometimes been the most important difficulty to overcome

Despite the difficulties faced, the construction of the Tramline in Athens is now a fact for the capital of Greece. The alignment that was finally constructed was not identical to the alternative 8 described above, purely for practical reasons, that is, because of the short period of time between the beginning of the construction and the Olympic Games that took place in Athens in August 2004. The alignment constructed reached the two south ends (Glyfada and Neo Faliro) as described in alternative 8 but, instead of reaching the northern end (Patissia), it only reached the centre of Athens (Syntagma Square) in order to cause minimum side effects to other works in progress in that area.

Three years after its construction there are still certain aspects that need to be clarified concerning the effectiveness of the mode and its integration into the public transport system. The alignment chosen is, however, regarded to be a successful one. The whole construction did not meet any serious technical problems and it was possible to be concluded in time to serve the increased transportation needs during the Athens Olympic Games. After facing some early-stage problems that were related to the integration of a totally new mode in the city, the Tram is currently progressing well with a satisfactory volume of passengers and a high level in the quality of services provided. In this context, the first extension of the line is currently in progress, reaching along the coast to the south-east areas (Voula). On the other hand, the extension towards the west (also along the coast) is expected to start in the coming months, taking the Tram to the heart of Pireaus, the main port of Athens. These extensions will certainly enhance the role of the Tram in the public transport system and prove the successful implementation of the mode.

References

van Herwijnen, M. (1999), 'Spatial Decision Support for Environmental Management', PhD Thesis, VU University Amsterdam.

Joerin, F., Thériault, M. and Musy, A. (2001), 'Using GIS and Outranking Multicriteria Analysis for Land-use Suitability Assessment', *International Journal of Geographical Information Science* 15:2, 153–74.

Tavistock Institute (2003), *The Evaluation of Socio-Economic Development – The Guide* (London: Tavistock Institute).

Vreeker, R., Nijkamp, P. and Ter Welle, C. (2001), 'A Multicriteria Decision Support Methodology for Evaluating Airport Expansion Plans', *Transportation Research Part D* 7:1, 27–47.

Annex 16.A

Table 16.A1 Impacts Matrix – Topological evaluation

No.	Criterion	1a	1b	2a	2b	3	4a	4b	5	6	7	8
1	*Cohesion*											
1.1	No. of Metro transfer stations	0.50	0.50	0.67	0.67	0.50	0.67	0.67	0.83	0.83	1.00	1.00
1.2	No. of bus transfer stations	0.83	0.83	0.72	0.72	0.67	0.89	0.83	1.00	0.83	0.72	0.83
2	*Ability to extend*											
2.1	Not from one end	No	No	Yes	Yes	No	No	Yes	No	No	Yes	No
2.2	From all ends	Yes	Yes	No	No	Yes	No	Yes	Yes	Yes	No	Yes
2.3	From intermediate stations	1	1	2	2	1	2	1	1	1	2	1
3	*Restructuring of bus/ trolley lines*	*0.84*	*0.84*	*0.91*	*0.91*	*0.75*	*1.00*	*0.97*	*0.75*	*0.84*	*0.91*	*0.97*
4	*Level of Service*											
4.1	Category A	0.89	0.89	0.78	0.78	0.67	1.00	0.89	0.67	0.89	0.78	0.89
4.2	Category B	0.67	0.67	0.83	0.83	0.50	1.00	1.00	0.50	0.67	0.83	1.00
4.3	Category C	0.92	0.77	0.77	0.62	0.69	1.00	1.00	0.69	0.92	0.77	1.00
5	*Quality of transportation*											
5.1	No. of transfers	5	5	2	1	5	3	4	2	2	1	2
5.2	Frequency	4	1	4	1	5	3	4	2	1	1	2
5.3	Traffic flow	5	5	5	1	3	4	1	4	5	5	4
6	*Environmental effects*	*3*	*5*	*3*	*5*	*4*	*2*	*3*	*3*	*1*	*2*	*1*
7	*Engineering works*											
7.1	small scale	0.61	0.52	0.59	0.49	0.35	0.85	0.71	0.51	0.76	0.74	0.86
7.2	medium scale	0.53	0.53	0.41	0.41	0.12	0.82	0.65	0.29	0.71	0.59	0.82
7.3	large scale	Yes	Yes	Yes	Yes	Yes	Yes	Yes	Yes	Yes	Yes	Yes
8	*Effects on traffic*	*6*	*7*	*3*	*3*	*8*	*1*	*2*	*8*	*6*	*3*	*2*

Table 16.A2 Impacts Matrix – Functional evaluation

No.	Criterion	6	7	8	1a	2a	4b
1.	*Transportation Patterns*						
1.1	Modal split	113,182	115,514	125,042	113,182	115,514	125,042
1.2	Transportation for work	196,318	216,898	236,142	139,030	159,610	178,854
1.3	Transportation not for work	229,853	261,302	284,678	163,479	194,928	218,304
1.4	intra – bus/trolley routes	2,692	2,862	2,692	2,692	2,862	2,692
1.5	inter – bus/trolley routes	5,149	5,078	5,345	4,001	3,930	4,197
1.6	Performance of existing transportation means	206.17	221.54	206.17	206.17	221.54	206.17
2.	*Spatial patterns*						
2.1	Economic activities in the primary sector	5.49	3.51	7.05	5.49	3.51	7.05
2.2	Economic activities in the secondary sector	6.35	4.90	8.49	6.35	4.90	8.49
2.3	Economic activities in the tertiary sector	7.19	5.04	9.16	7.19	5.04	9.16
2.4	Economic activities in basic employment	6.17	4.55	8.22	6.17	4.55	8.22
2.5	Technical interventions	12,800	8,903	12,800	9,200	5,303	9,200
2.6	Access to existing Olympic Complexes	1	1	1	1	1	1
2.7	Access to new Olympic Complexes	0.33	0.67	1	0.33	0.67	1
3.	*Environmental effects*						
3.1	Environmental benefit	5,093	5,192	4,907	4,109	4,208	4,305
3.2	Existence of buildings on the one side of the street	Yes	Yes	Yes	Yes	Yes	Yes
3.3	Existence of buildings on both sides of the street	Yes	Yes	Yes	Yes	Yes	Yes
3.4	Ability of emissions to spread	++	+	++	++	+	++
4.	*Financial dimension*						
4.1	Construction costs	151,000	144,000	171,500	129,000	121,500	148,500

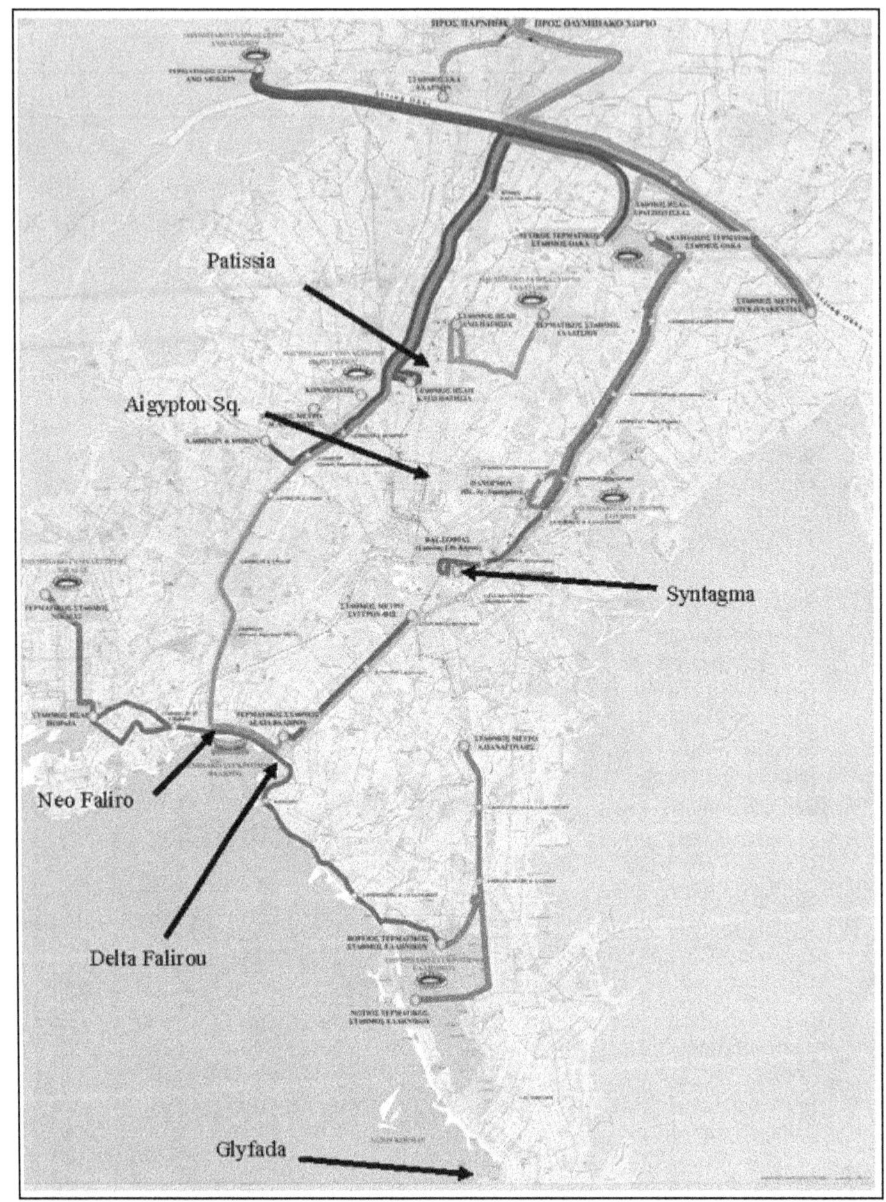

Figure 16.A1 Athens map

Chapter 17

Experiments on Fiscal Federalism in the Netherlands: Institutional Coordination of Parking Policy in Urban Networks

Caroline Rodenburg, Mark Koetse and Piet Rietveld

Introduction[1]

Parking is an issue that is increasingly calling for attention in policy making and in scientific research. The main reason for this is the existence of parking externalities, mainly in urban areas. One can think, for instance, of the time that is spent on searching for a parking space, especially in cities and during the weekend. There are various market imperfections in the market for parking, such as congestion costs and information problems, leading to substantial inefficiencies in parking behaviour (that is, search and waiting costs). Many local authorities have therefore introduced parking tariffs or put restrictions on parking duration and supply in order to internalize some of the externalities.

Besides efficiency arguments, other objectives of a parking policy may exist, such as the generation of revenues. To what extent economic efficiency and equity play a role in actual policy making depends on, amongst other things, the goals of decision makers at various levels of government. It is therefore not surprising that substantial variations may exist between municipalities, and that the interests of local, regional and central government may diverge. In this respect the governmental structure underlying the parking policies may have a considerable impact on the outcomes of the policies implemented.

From an institutional standpoint, there has always been an interest in the influence of governmental structures on the effectiveness and outcomes of policy making. One of the reasons for this is the potential heterogeneity of the relationship. For instance, the impact of federal systems may depend on specific federal structures and on the specific policy area. In general, arguments in favour of a (national) federal system are that it would: give incentives for an efficient allocation of resources; stimulate political participation and democratic representation; and protect basic liberties and freedoms. On the other hand, the control of central government over an economy

1 This contribution is based on a document prepared for the EU project 'Transport Institutions in the Policy Process' (TIPP). The project was funded by the European Commission and was part of the 5th Framework Programme of the European Union.

(Wibbels 2000), and over the actual impact of policy on a society (Castles 1999) is limited.

In this contribution we look at the institutional background of Dutch parking policies in order to analyse the influence of federal structures on parking policy and the functioning of the Dutch parking market. The structure of the paper is as follows. In the next section we discuss the concept of fiscal federalism. We use this concept as our framework for analysis. Then the working of the parking market is discussed. Special attention is paid to market failures and their associated externalities. This is followed by a section which considers parking policy and government structure in the Netherlands. We then present three case studies on different forms of Dutch parking policies. Finally, there is a conclusion.

Institutional Framework: Fiscal Federalism

Public choice theory has led to the insight that governments do not operate as benevolent and omniscient entities, but rather are run by politicians and bureaucrats with their own agendas, and with cognitive limitations. The consequence is that the public sector operates in a world of second-best, that perfect outcomes cannot be expected, and that trade-offs (for example, between efficiency and equity, or between the well-being of opposing groups) are inevitable. An important aspect of public choice theory is (fiscal) federalism. In assessing the role of different governmental tiers and analysing the impact of different federal structures on (the effects of) parking policies, this concept provides a useful framework.

Inman and Rubinfeld (1997) mention three distinctive characteristics of federal structures that can be manipulated and that influence the political decision making and its results. These characteristics are: the number of lower-tier local authorities; their representation in the central government; and the assignment of policy responsibilities between the vertical tiers of government. On the basis of these three characteristics, Inman and Rubinfeld (1997) identify three types of archetypal federal structures, that is, *economic federalism*, *cooperative federalism* and *democratic federalism*. What is interesting in this distinction is that each type implements the most decentralized government capable of internalizing all economic externalities. As such, each structure embraces economic efficiency as its main objective. Moreover, each structure adheres to the Tiebout model (see Bewley 1981), in that it assumes that public goods can be efficiently provided by small communities that are in competition with each other. The number of lower-tier local authorities is such that all economies of scale in providing public services are just exhausted, or, in the case of contestable public services, when the average cost per person of providing a given level of the public service equals the marginal cost of adding another user (Inman and Rubinfeld 1997). The difference between the structures lies in the constitutional constraint that each type imposes on central government policy decisions. More specifically, the structures differ in their vision of how to provide pure public goods and regulate inefficiencies due to competition between lower-tier local authorities. The representation of these lower-tier authorities in central government and the assignment of policy responsibilities between the vertical tiers are therefore key issues.

Economic federalism aims at implementing the most decentralized structure of government capable of internalizing all economic externalities, subject to the constraint that all central government policies be decided by an elected or an appointed central planner (Inman and Rubinfeld 1997). In essence, the structure of a government under economic federalism is that the central government is responsible for those public activities that entail significant externalities on a large geographical scale, while lower governmental tiers get assigned public tasks with little or no externalities (Oates 1972). Inefficiencies due to competition between lower-tier governments (for example, tax benefits, welfare assistance) are regulated by the central planner in this structure.

Cooperative federalism is more pessimistic about the ability of a central planner to regulate inefficiencies caused by lower-tier government policies. It therefore has the same objective as economic federalism, but the constitutional constraint is more restrictive in that all central government policies must be agreed to unanimously by the elected representatives from each of the lower-tier governments. Thus, the main objective of a central government under cooperative federalism is enforcing contracts among lower-tier governments to ensure the provision of pure public goods, and correcting the inefficiencies caused by lower-tier fiscal competition. Some of the important points of the critique on the efficiency of such Coasian bargaining between lower-tier governments are the failure to agree on the division of surplus, asymmetric information (and rent-seeking behaviour), and problems of enforceability. Moreover, largely because of these problems, bargaining processes will be expensive and time consuming (that is, transaction costs are high). Related to the latter is the problem that systems in which (near) unanimity is required for policy implementation might suffer from joint decision traps (see Scharpf 1988). Because unanimity is required, each player in the decision process has a veto. Hence, as long as existing policy is preferred to new policy proposals by even one player, existing policies will not change (Blom-Hansen 1999) and will be locked in the institutional setting.

Democratic federalism takes a position between economic and cooperative federalism in that all central government policies must be agreed to by a majority of elected representatives from lower-tier governments (instead of unanimous agreement). It seeks to balance the efficiency (transaction cost) gains from centrality with the efficiency gains of cooperation (regulating inefficiencies from competition).

A more pragmatic distinction that probably suits our purposes best is between power sharing and power separating federal structures (see Braun et al. 2002).[2] In the *power sharing* model most of the policy decisions are made at the federal level where both lower-tier authorities and the central government are represented. However, the implementation of policies is almost completely in the hands of lower-tier governments. Hence, this model resembles cooperative federalism or democratic

2 A somewhat related distinction, which is based mostly on the behaviour of governmental layers, is between cooperative and competitive models (see Braun 2000).

federalism, depending on the degree to which parties included in decision making have to agree in order to get a policy proposal implemented (unanimity versus majority). Good examples of such a model are the federal structures in Germany and Switzerland. In the *power separation* model, there is a separation of decision making and implementation between sub-national actors. Both the federal government and lower-tier governments have guaranteed authority in a large number of policy issues. This model therefore resembles most economic federalism, in which competition between lower-tier governments predominates. A good example of such a model is the federal structure in Canada. Inefficiencies might arise in this model when policy externalities are not resolved by collective bargaining. On the other hand, some inefficiencies due to the deadlocks inherent in the power sharing model, such as joint decision traps, are prevented (see Painter 1991).

Although the above-mentioned rather archetypal federal structures seem to exist in reality (*power sharing* in Germany and Switzerland and *power separation* in Canada), federal structures in most countries will be a mix of different forms. Moreover, federal structures may vary between different policy areas. It is important for our purposes to analyse the federal structure regarding parking policies in the Netherlands. In the next section, we therefore give a description of issues related to the parking market in general, after which we describe Dutch parking policies at the national, regional and local level.

Parking Market and Parking Market Externalities

Theoretical Framework

In the transport research literature, parking has received relatively little attention (Young 2000). This is rather surprising, since parking is a major policy issue in many cities. As with many other markets, the market for parking is characterized by various distortions. Examples of distortions are externalities, lack of information, monopoly power of the owner of private parking places, and high transaction costs. We will discuss some of these in more detail here.

Parking and market distortions Parking is an activity that may easily give rise to *externalities*. A well-known example is a lorry parked on a street, to be loaded or unloaded in a city centre, which blocks the way for other road users. A less extreme example is that parking a car on a parking place may have an effect on the time other road users need to find a parking place. Also non-road users may be affected by parked cars: parked cars reduce the scenic quality of the city centres for both visitors and residents, and they reduce the opportunity to use the road for other purposes, for instance, as a playground for children. Parking is also related indirectly to the externalities of car traffic such as noise, emissions and accident risks. Another market imperfection associated with parking is the *lack of information* on demand and supply. This may lead car drivers to look for a parking place, implying search and waiting costs that could be avoided had adequate information been available. The *monopoly pricing* issue is relevant because parking facilities usually serve a very local market, and within many cities there are clear barriers to entry for new entrants on the parking area market. *Transaction costs* relate to the costs of letting car users pay for the use of parking places: for example, by introducing metering for on-street parking. These costs may be so high that the authority responsible for providing the parking places may decide to offer them for free. The problem with free parking places is illustrated in Figure 17.1.

In the short run the volume of parking places is given. When the downward sloping demand curve intersects the supply curve, a positive parking charge (p) would be needed to arrive at the market equilibrium, but with a price equal to zero there will be excess demand leading to queues and waiting costs. Another possible distortion is that private suppliers of parking space provide the parking for free: for example, in the case of parking provided by employers to employees (see, for example, Shoup 1997, and Rietveld and van Ommeren 2002). This would imply that commuters by car do not pay the total costs of their trip, that is, the price they pay for a car trip is too low.

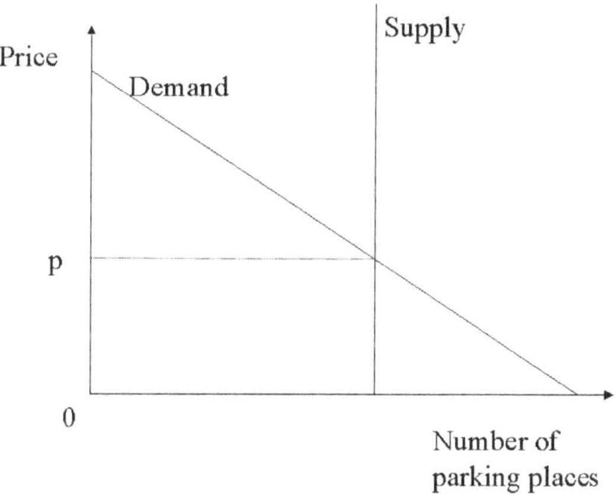

Figure 17.1 Demand and supply on the parking market

Approaches to deal with market distortions Several approaches exist to address these market imperfections. Some of the problems can be overcome by using *new technologies*. For example, ICT can be used to develop dynamic parking information systems so that an unnecessary search for parking places can be avoided. Also the transaction costs incurred in parking can be reduced by using ICT. *Pricing* measures via the introduction of paid parking are introduced as a direct means to address parking externalities, or as a second-best instrument to correct for other externalities. Also various kinds of *rationing* are adopted. For example, restrictions on the duration of parking may help to keep some of the parking problems manageable.

Or giving parking rights to only some categories of users (for example, residents) may help to solve the mismatch between demand and supply. The *supply of parking infrastructure* is also a policy often used by local authorities to reduce some of the related problems. For example, in order to overcome the negative external effects of car traffic and of parking in historical cities, parking may be concentrated at the fringe of the cities, after which public transport services are offered to bring visitors to their final destination. In this policy, the location of parking facilities, as well as the function of parking places in multi-modal transport chains, play a central role.

Parking in urban areas leads to various problems in various contexts. From the Dutch experience the following distinctions can be made:

- parking in residential areas: new city quarters versus existing areas;
- parking at workplaces: new versus existing.

As explained in Rietveld (2006), capacity planning and parking norms play an important role in the case of new areas, whereas prices are more important in the situation of existing areas. The background is that, at the stage of designing land use in a new area, it is quite cheap to take into account the parking needs of the car users in a quantitative sense. When land use is already fixed, however, expansion of parking space is very expensive, so in this case capacities are assumed as given and price measures become pertinent.

Parking Policy and Government Structure in the Netherlands

In the Netherlands, a three-level government structure can be found with a central level, an intermediate level, and usually a complex set of local level authorities. The wide variety of multi-level governments is evident, with most local units administering a population of less than 4,000 people, and most intermediate level governments administering a population of 50,000 to 200,000 people (Bennett 1980). In such a wide variety of government structures, the division of responsibilities between central, intermediate, and local government is often very complex.

In general, Dutch parking policy has always mainly been an issue at the municipal level. Some policy directions have been dictated by the central government, but the real design of parking policies is a matter of decision making by local authorities. Three periods of parking policy can be distinguished. Until the end of the 1970s, parking policy in the Netherlands was generally steered by the demand for parking places (demand following policy). If demand increased, the number of parking places would simply be increased as well. The demand for parking places was determined autonomously and was not influenced by policy measures that discourage parking.

This policy was curbed when the growth in car traffic threatened the accessibility of inner cities. As a reaction to these negative effects, the national government chose a more steering type of parking policy aimed at the preservation of the quality of the inner city, while at the same time safeguarding an easily accessible inner city. Crucial factors for such a successful steering parking policy are (Reisen and Hus 2003):

- Good alternatives for parking places in the inner city.
- The policy needs to be balanced: target groups need to show the desired behaviour, while unwanted effects need to be prevented.
- Stakeholders should understand the measures and have confidence in the future accessibility, liveability and vitality of the city.
- The city council needs to be convinced of the necessity for a steering policy and stick to the policy path chosen; it is not easy to introduce parking licences or to increase parking tariffs.

From that moment onwards, satisfying the demand for parking was no longer the main aim of parking policy. Parking was being used as an instrument to influence car mobility, and commuter traffic in particular. Nowadays, the term used for this policy is that of 'integrated parking policy', in which parking is seen as one aspect of overall urban policy making.

In the decision process of integrated parking policy, three mechanisms play a role (Martens et al. 2003):

1. the people (Who are involved in the process, and how do they co-operate?);
2. the markets (Which market responses will occur?);
3. the power (Which authorities and formal rules need to be involved in order to reach specific outcomes?).

Concerning the second element, it is tempting to have the market define parking policy. Possibilities for this exist. The construction of parking places costs money, but it can also yield revenue (paid parking). If the government concerned defines the conditions for parking facilities, and in particular the price charged to the user, effective demand may be derived. However, it may be difficult to predict the demand for parking in the long run on account of, for example, office construction.

Mobility management, including parking policy, is seen as the coordinating policy field for influencing the demand of urban traffic. Specific instruments that have been used by the various levels of government to reach this goal are parking norms, that is, restrictions in the number of parking places per location type, parking tariffs, and a regional approach to parking policy. The current approach for public parking is based on consumer preferences. Quality is an important aspect in this respect. In recent years, many park & ride facilities have been built around cities, and inner cities have been made partly car-free. Furthermore, starting up commercial activities related to parking becomes more and more attractive for private parties since money can be earned from providing parking facilities.

Some expected future trends that could direct and support the development of parking policy in the Netherlands in the coming years are (Bootsma 2003):

- technological development: for example, parking with the help of mobile phone communication, automatic parking, and use of debit cards;
- decreasing influence of central government on parking policy and increasing responsibilities of regional and local governments: some municipalities already

cooperate with private parties and have tendered the parking management in the region;
- on-street parking vs. underground parking: the importance of multiple land use and spatial quality claims will become more relevant.

As a summary some illustrative data about Dutch parking policies are presented in Table 17.1. They deal with different ways to regulate parking, objectives, type of parking, and policy focus. One of the interesting things to observe is that the bulk of the municipalities with more than 30,000 inhabitants have introduced some form of paid parking. Moreover, the majority of these municipalities also make use of price differentiation.

Table 17.1 Dutch parking policy figures

Policies implemented	By x % of the municipalities	Comment:
Use of paid parking	80 %	(municipalities with more than 30,000 inhabitants)
Use of price differentiation	86 %	(of all municipalities that introduced paid parking)
Use of parking permits	65 %	
Use of fines to prevent parking violations	almost all	(municipalities with more than 50,000 inhabitants)
General objectives:		
Accessibility and liveability	19 %	
Influencing parking pressure	10 %	
Type of parking:		
On-street	69 %	
Parking lot	50 %	
Garages	23 %	
Focus of parking policy on:		
Inhabitants	25 %	
Visitors	13 %	

Source: Zwartjes et al. (2001).

Case Studies

Introduction

Municipalities have several opportunities to regulate parking. They control the supply of parking space, the spatial dispersion of parking places, and parking facilities on private territory by means of spatial planning instruments and building Acts. As the administrator of municipal roads, municipalities can implement parking bans or reserve parking places for specific target groups. Moreover, they can prevent long-term parking by means of parking zones. Other possibilities to regulate parking are the introduction of paid parking by using parking meters and parking licences.

By using the above-mentioned instruments, controlling the mix of parking facilities, and regulating the relative price ratio between different parking facilities, municipalities have a major influence on the demand for, and the supply of, parking facilities. The following sections illustrate different considerations that have to be made in the process of introducing and executing parking policy. We focus on different forms of parking policy and their implications for policy processes, as well as on the coordination of national and local policies and coordination problems that exist between local authorities.

Case Study I: ABC-Policy

For almost 15 years, location policy has been a very important element of parking policy in the Netherlands. During the period of steering parking policy the national government introduced strict rules for local authorities. Municipalities were mainly responsible for carrying out parking policies for which the national government created the legal basis. In this process, the provinces acquired a substantial role in parking policies by means of translating national norms into local and regional policy (Bootsma 2003), thereby making the provinces responsible for ensuring the local effectiveness of parking policy.

Provinces were also responsible for the categorization and positioning of what is called the 'ABC policy' in 1988. The focus of this rather stringent policy was on firms at new locations, with the aim of having 'the right firm in the right place' and to reduce car use by tuning the accessibility profiles of locations with the mobility profiles of companies. For each company, a location profile (A, B, or C) was determined and companies were encouraged to settle at a location with a matching profile. The national Ministry of Transport defined three accessibility profiles:

1. A-locations: Locations that are easily accessible by public transport at national, regional, and local level. Accessibility by car is considered less important. Stringent parking norms are set and good conditions for the use of bicycles should be provided. The presence of services contributes to an attractive environment (for example, locations close to intercity railway stations in city centres).
2. B-locations: Locations that are easily accessible via public transport at regional and local level and reasonably accessible by car at regional and local level. Limitations on parking facilities are implemented, especially for long-term parking, and the locations are easily accessible by bike (for example, hospitals, and offices).
3. C-locations: Optimal accessibility by car at regional and local level. No requirements are in place concerning the accessibility by public transport. Parking facilities are adjusted to the type of company, and there should be congestion-free connections to a main transport artery. In other words, these locations are suitable for activities with few visitors/employees that are dependent on good accessibility by car/truck: for example, the production and distribution sector.

A distinction has been made between A-, B-, and C-locations in the Randstad area (the urbanized area in the western part of the Netherlands) and elsewhere. Table 17.2 shows the number of parking places allowed at different types of locations. The municipal governments were responsible for carrying out this policy. They had to check each new company on the principles of the ABC policy and determine accompanying parking norms in spatial plans. The introduction of the ABC policy meant a revolution in parking policy. Before 1988 firms usually had to obey minimum parking norms, but then they suddenly had to face maximum parking norms.

Table 17.2 Number of parking places allowed at different types of locations

Location	Number of parking places per 100 employees
A-location Randstad	10
A-location elsewhere	20
B-location Randstad	20
B-location elsewhere	40
C-location	No restrictions applied

Source: Bootsma (2003).

If we summarize the results of the ABC policy at the *national* level, we can say that the general goal (reducing the growth of car traffic) has not been reached, but that at many places a change in the modal split of transport to and from companies has taken place. However, the growth in the number of companies at B- and C-locations was relatively larger than at A-locations (Bootsma 2003). About one-third of the new companies settled down at the 'right' location, about one-third at a 'wrong' location, and for about one-third the results were neutral. Furthermore, many B-locations were insufficiently accessible by public transport, whereas the accessibility by car was of C-location quality. A policy evaluation observes three reasons for the failure of the ABC policy (Kamer van Koophandel 2002):

1. Competition between municipalities weakens the willingness to put stringent restrictions on the location of new companies.
2. The specific market demand for A- and B-locations has never been calculated, so that a shortage of these locations occurred. These shortages remained mainly unobserved. The result was that companies that should ideally have settled at A- or B-locations could only find space at C-locations.
3. The maximum possible effect of location policy on transport is limited because of the limited possibilities for substitution of car by public transport and bicycle.

At the level of the *municipalities*, this policy has not been very successful either. Very often, local authorities attach more importance to employment growth that is correlated with the location of a company within the municipal borders, than to the consequences it has for mobility in the area. Other problems that arise are that public transport possibilities are often not of sufficient quality; that congestion is often not severe enough to reduce car mobility; and that there is policy competition between municipalities in terms of how flexible they are with parking norms.

Because of these implementation problems, the national government dropped the ABC parking policy in 2001 and abolished the national norms mentioned in Table 17.3 (in Case Study III below). It is interesting to observe that the ABC policy will have speeded up the introduction of paid parking in the public sphere because maximum parking standards obviously imply a transfer of the parking burden from the private areas of firms to the public domain. The next section will discuss this issue in more detail.

Case Study II: Paid on-Street Parking

As well as negative externalities in the parking market in general (see earlier section called 'Parking Market and Parking Market Externalities'), there are also specific direct and indirect negative externalities to be defined for on-street parking (Feitelson and Rotem 2004). The direct externalities include the effects of impervious parking surfaces on water flows and water quality. It leads to greater surface flows, often bearing pollutants. In addition, the land used for parking is not available as open space, and parking lots often have negative visual effects. Surface parking also affects development density, development cost and urban design, which indirectly affects urban travel patterns and transport modes.

On the other hand, taxing of surface parking may have several positive side effects. In particular, taxation has the potential to provide an incentive for municipalities to intensify the use of parking. So one of the starting points of an active parking policy is that parking should have its own social-economic price. Scarcity and price are the most important instruments to change the competitive position of the car compared with alternative transport modes. Since the beginning of the 1990s, price instruments, such as parking fees, have been used by an increasing number of municipalities in the Netherlands and led to the encouragement of the use of a regulating parking policy (see Table 17.2). Since paid parking is a local tax and parking fees are set by Dutch local authorities, this might give rise to inefficient pricing policies, insofar as either the costs or the benefits of a policy change fall onto groups from outside the local area (see also de Borger and Proost 2004). To prevent causing welfare effects for groups outside the local area, it is important that local authorities set efficient parking prices. There are, however, a number of different reasons why a local transport authority may fail to set efficient parking prices.

A first example is the introduction of parking fines. Apart from the municipal authority collecting taxes, in a typical urban transport market central government collects tax revenue from the consumption of fuel, or from vehicles themselves, or collects revenue from fine payments (speeding, illegal parking and so on). The fact that fines are collected at the national level leads to an unambiguous incentive for the municipal authority to set the meter fee at too low a level. This keeps revenues within the local community since low fees encourage drivers to pay at the meter (rather than risk the fine).

Local authorities will probably perceive the transfer of revenue towards the national level as a loss of local welfare. As a result, they may use the tax instruments that are under their control to both correct for local externalities and reduce transfers to the central government. The presence of such transfers to the central government may distort local government decision making, and it is usually referred to as *vertical tax competition* (Calthrop 2002).

Since 1 January 1991, municipalities in the Netherlands no longer have to transfer their parking benefits to the national government, which has resulted in the large-scale introduction of paid parking. In the last 15 years, the benefits from parking fees have increased considerably. In 1985, the benefits were estimated to be around €83 million, whereas in 1990 estimations were around €135 million. Estimates for 1995 and 2000 amounted to as much as €344 million and €629 million, respectively (CBS 2000). Not only in absolute figures, but also in relative amounts, the increase in benefits is considerable. The share of parking fees in total municipal benefits was 2.7 per cent in 1990, and increased to 5.8 per cent in 2000 (CBS 2000). This increase is even larger within the three big cities of the Netherlands: Amsterdam, Rotterdam and The Hague. For these cities, the share of parking fees in total municipal benefits was 2.7 per cent in 1990, 6.6 per cent in 1995, and even 14.2 per cent in 2000 (CBS 2000). The increase in benefits is mainly the result of increasing parking tariffs and an increase in the number of parking places.

A second example of inefficient pricing policy is that downtown parking places are not only used by local residents but also by tourists and commuters. The municipal authorities may rationally decide to ignore the consequences of policy changes on the welfare of non-residents, and instead attempt to extract rent from this group of people who live outside the municipality. This is an example of *tax-exporting (or horizontal tax competition)* (Calthrop 2002) and is a well-known phenomenon in the tax competition literature. Revenue-raising concerns, rather than social-welfare concerns, may play an important role in determining parking prices on these markets, that is, the actual price may exceed the efficient price and social welfare is not optimized.

Besides the examples of existing inadequacies of local government decision making discussed above, there may be good reasons why decentralized decision making is beneficial, perhaps because of better information on local traffic conditions, or values of time. The examples of institutional failure given here should be seen as just that. If institutions may need to be reformed, we should account for several factors omitted from this case study.

In the Netherlands, the introduction of pricing has proceeded further with parking than with driving (toll collection). This is probably because parking charges can easily be implemented in a gradual way so that citizens get used to them, whereas road charging programmes were conceptualized as large high-tech initiatives that cannot easily be tested on a smaller scale. Furthermore, the low-tech character of parking charges and the availability of alternatives, that is, parking facilities somewhat further away, will have made its acceptance easier. Involvement of all actors in the development process of parking policy is important in order to obtain support for the complex issue of the distribution of scarce space. The involvement will improve if the municipal parking policy is transparent in its goals, target groups, instruments, tariffs, tariff differentiation and expected effects.

The variation in tariffs within and between municipalities is substantial. Tariffs for on-street parking vary from €0.57 per hour up to €3.20 per hour, while there are also municipalities that have not yet introduced paid on-street parking (see Table 17.2). With regard to differentiation in parking tariffs, there is hardly any exchange of information between municipalities, so there is no possibility of systematic learning, which leads to negative effects (van Dijken 2002). The decision process for parking tariffs is often a matter of 'trial and error' in a complex administrative and political force field. Since there is a lack of *ex-ante* insight into the behaviour of inhabitants, employees and visitors in the short and long term, the search for the optimal tariff structure will remain a challenge. Concrete monitoring of behaviour, behavioural experiments and pilots, but also the development and exchange of knowledge and experience in this field between municipalities could improve the process. Besides that, in order to come to an optimal tariff structure, it is important to have insight into the actual (social) costs of parking in municipalities.

A final issue is that many urban areas are covered by multiple jurisdictions (Feitelson and Rotem 2004). In such a case, if only a single jurisdiction imposes an on-street parking tax, some spillover effects to neighbouring jurisdictions may be expected. For example, when municipalities have a financial interest in maintaining their parking policy, the number of regulated parking places increases, as well as the tariffs, in order to diminish the pressure on parking places. This, however, leads to increasing pressure on parking places in the surrounding areas, which in turn causes inconvenience for the inhabitants and companies in this area. For example, in the short term drivers may park their car in neighbouring jurisdictions where parking taxes have not yet been introduced. In the long term, firms may choose to locate in jurisdictions where no on-street parking tax is imposed. This will, however, be dependent upon the degree to which the neighbourhoods are perfect substitutes. The only way to curb such developments is to introduce a similar parking regime in the neighbouring areas as well (see also the next section).

Case Study III: Coordination of Parking Policy between Various City Quarters in the City of Amsterdam

In order to decrease parking externalities and increase local welfare, one important way for local authorities to achieve this is by the introduction of paid parking. A related objective when parking policy is introduced at the district level is to avoid negative spillovers to other city districts. If, for instance, paid parking is introduced in a specific city quarter, residents living in that city quarter may move their cars to neighbouring quarters where parking for residents is still free. This increases the external effects of traffic and parking, such as noise, nuisance, emissions and congestion in parking spaces, at the location with free parking. Clearly, a local authority would benefit from coordinating parking policies between city quarters.

Parking policy in the city of Amsterdam certainly shows similarities with the issues described above. The objectives of introducing paid parking in the city centre were closely related to reducing parking externalities (congestion), but also more generally to reducing traffic externalities (noise). Specifically, the central aims of parking policy in Amsterdam were to improve the liveability and accessibility of the city. In practice, this meant reducing commuter traffic and increasing the number of users (the flow) per parking space. Concrete parking policies were implemented in 1991 and consisted of introducing parking tariffs solely in the centre of Amsterdam (city district '*Binnenstad*'). Parking in other city districts was still free at that time and, compared with current standards, the tariff in the city centre was still modest at €1.13 per hour.[3]

In 1993 the parking tariff increased steeply, and paid parking was introduced in the other city districts within the A10 orbital motorway as well, although the tariffs in these districts were half as much as in the city centre. In 1996 paid parking was even expanded to include some city districts outside the A10 orbital motorway. We thus see a gradual diffusion of paid parking from the centre of the city outwards towards the periphery – a pattern consistent with the initial introduction of paid parking at locations where externalities are most severe, and the subsequent introduction and diffusion of paid parking because of negative traffic and parking spillovers of the initial policy in neighbouring city districts (see Table 17.3).

The current tariff structure is almost identical to that for 2001. The differences between the period before 2001 and the period after 2001 pertain to whether or not

Table 17.3 Development of parking tariffs for on-street parking in Amsterdam for various city districts from Monday to Saturday from 9.00 to 19.00 hours (hourly tariffs)

Year	Tariff A: City centre (+ museum quarter since 2002)	Tariff B: City districts inside the A10 motorway (+ the quarter near the World Trade Center (WTC)	Tariff C: City districts outside the A10 motorway
1991	€1.13	–	–
1992	€1.13	–	–
1993	€1.82	€0.91	–
1994	€1.82	€0.91	–
1995	€1.82	€0.91	–
1996	€1.93	€1.02	€0.68
1997	€2.16	€1.25	€0.79
1998	€2.16	€1.25	€0.79
1999	€2.27	€1.36	€0.79
2000	€2.38	€1.47	€0.79
2001	€2.61	€1.59	€0.79

3 Prices and monetary figures before the official introduction of the euro were in guilders. They are converted to euros using the current exchange rate (1 euro = 2.20171 guilders).

paid parking is introduced between 19.00 and 24.00 hours, and to whether or not paid parking is introduced on Sundays. Furthermore, tariffs are higher, equalling €3.20 in the city centre, €1.90 in city districts inside the A10 motorway and €1.00 in city districts outside the A10 motorway.[4] Besides on-street paid parking there are other parking possibilities as well. In Table 17.4 the prices of parking permits for residents are given for the various city districts in Amsterdam.

Again, prices in the inner city are highest by far, but leaving aside the inner city there is not much price differentiation. The existing differences do, however, again point to decreasing prices towards the outer limits of Amsterdam. The same can be said for hourly parking prices in (private and public) parking garages; the cheapest parking facilities are all on the outside of the city. However, it should be noted that most parking garages are near the city centre (basically from *Byzanthium* to *Heinekenplein*), so prices differ very little between these locations. It is interesting to see that, when comparing prices in parking garages with the A-tariff in 2004, only three parking garages are more expensive than the A-tariff.

Although the central aims of parking policy remain unchanged, the Amsterdam City Council has decided to implement a higher degree of flexibility in its parking policy from 2005 onwards – mainly because of complaints from the city districts about the lack of such a policy. An important part of this higher degree of flexibility is that city districts will have more influence on the specific design of the parking policy to be implemented within their own jurisdictions. In other words, as explicitly stated in local government documents, the principle of subsidiarity is introduced (see Gemeente Amsterdam 2004).

Table 17.4 Prices of parking permits in various city districts in Amsterdam in 2004

	Resident permit	Company permit
Centrum (Binnenstad)	€107.28	€171.60
Oud-West	€63.90	€102.24
Westerpark	€63.90	€102.24
Zeeburg	€63.90	€102.24
Bos en Lommer	€63.90	€102.24
De Baarsjes	€63.90	€102.24
Oost/Watergraafsmeer	€63.90	€102.24
Zuideramstel	€63.90	€102.24
Oostelijke Eilanden	€60.90	€97.50
Oud-Zuid	€60.90	€97.50
Osdorp	€46.98	€75.12
Westpoort	€46.98	€75.12
Geuzenveld/Slotermeer	€46.98	€75.12

4 Other options to pay for on-street parking are, amongst others, parking permits for a day, a week and a month.

Until now prices for on-street parking and parking permits have been set and allocated to specific locations by the local authority. Within the new policy proposals, the local authority still sets the tariffs for on-street parking, and has also introduced two new tariffs. These tariffs are the A+ tariff, which is substantially higher than the A- tariff, and what is called the 'Blue zone' tariff, which comes down to a mere €0.10 per hour and is implemented mainly to facilitate parking for shopping and at cemeteries and sports facilities. As Table 17.5 shows, a gradual increase in on-street parking tariffs is planned up to 2009.

One of the main differences with the old parking policy is that, with exception of the 'Blue zone' tariff, city districts are now free to implement the tariff of their own choice at each location. 'Blue zone' tariffs, as mentioned above, can only be implemented at specifically defined shopping centres or streets and at cemeteries and sports facilities. Moreover, in the 'Blue zone' there are parking duration restrictions.

Another aspect of the new policy is that city districts can themselves set the prices of parking permits, as long as they are not outside the scope of the central aims of parking policy in Amsterdam (see earlier in this subsection). The latter restriction is enforced by the local authority which has the power to turn down a tariff proposal by a city district.

A final important issue is that city districts have the option to reintroduce free parking on Saturdays (Sundays have always been excluded from paid parking). This way city districts can shield attractive and important economic and cultural areas from the possible harmful consequences of paid parking on Saturdays.[5]

Although city districts are probably pleased with their increased degree of influence and control over parking policies within their own jurisdictions, there are possible problems associated with such a decrease in the level of centralization:

- pressure from residents and the business community driving down the price of parking;

Table 17.5 Tariff structure for on-street parking in Amsterdam from 2004 onwards (hourly tariffs)

Year	Tariff A+	Tariff A	Tariff B	Tariff C	'Blue zone'
2004	€4.20	€3.20	€1.90	€1.00	€0.10
2005	€4.40	€3.40	€2.00	€1.10	€0.10
2006	€4.60	€3.60	€2.10	€1.20	€0.10
2007	€4.80	€3.80	€2.20	€1.30	€0.10
2008	€5.00	€4.00	€2.30	€1.40	€0.10
2009	€5.20	€4.20	€2.40	€1.50	€0.10

5 Other differences with the previous parking policy are the increased number of parking permits for people working in care institutions, the police and schoolteachers. Furthermore, city districts have the option to decide when a parking permit for a resident remains valid when he or she moves into the city district.

- competition between city districts, driving down the price of parking;
- legality of large differences in parking tariffs between comparable city districts.

The first two problems are consequences of the fact that city districts may have different interests than the local authority, such as maintaining the competitive position of their own shops vis-à-vis those in other city districts, and may be more susceptible to influences from interest groups such as residents and the business community. These issues may drive down the price of parking, which is harmful for welfare because of the various externalities in the parking market (see section called 'Parking Market and Parking Market Externalities'), assuming that parking tariffs are not above the optimum in the first place, that is, above marginal external costs.

Moreover, the local authority has built in a couple of restrictions on the decision-making power of city districts. As already discussed, the local authority still sets the parking tariffs and a zero tariff is not one of them. Also the prices of parking permits are set by city districts but must be agreed upon by the local authority since the proposed prices may contradict the central aims of the Amsterdam City Council parking policy.

Furthermore, an interesting feature of the new policy is the introduction of what is referred to as 'incentive de courage'. The incentive for on-street parking lies in the fact that a large part of the extra revenues from introducing the A+ tariff instead of the A- tariff accrues to the city district instead of to the local authority. Since 2002 the structure of distributing paid parking revenues has been as follows.[6] City districts have to remit 16 per cent of gross revenues from the B- and C-tariff and 50 per cent of the part of the A-tariff that is above the B-tariff (note that the costs of maintenance and operation of on-street parking have to be paid by the city districts). For 2005 this means that city districts would have to remit 16 per cent of €2.00 and 50 per cent of the remaining €1.40. Under the new regime, city districts would only have to remit 16 per cent of the extra revenues from introducing the A+ tariff (instead of 50 per cent as is the case with the extra revenues from the A-tariff).[7]

For parking permits something similar holds. For each city district the local authority sets a standard price for a residential parking permit. A large part of the extra revenues from introducing a price that is higher than the standard price again goes to the city district. This incentive has been created to give city districts the chance to generate revenues to build parking garages, thereby increasing parking capacity and decreasing negative parking externalities. However, since city districts now have a direct incentive to introduce the A+ tariff everywhere, the danger of such an incentive is that it may work too well. This may drive the price of parking to a level that is welfare inefficient, that is, above the socially optimal price that exactly internalizes all marginal external costs. Moreover, extra revenues from the A+ tariff

---

6 Before 2002 the structure was principally the same but different in the exact details of distribution.

7 Introducing the A+ tariff thus requires three stages of remittance, that is, 16 per cent over the B-tariff, 50 per cent over the part of the A-tariff that is above the B-tariff, and 16 per cent of the part of the A+ tariff that is above the A-tariff.

are not earmarked by the local authority, and can thus be used for purposes other than building parking garages. The central point is, therefore, that the city district may become a 'selfish leviathan' that aims to maximize parking revenues rather than a benevolent authority that aims to promote the overall benefits of the district.

Finally, let us leave aside interest group influence and competition between city districts. Note, furthermore, that city districts may be more able than a local authority to assess the extent of the negative externalities in their own local parking market, and that they may actually be more able to judge the extent to which paid parking is harmful for the local economy (apart from the influence of interest groups on such a judgement). In these situations, and assuming that the local authority and city district both have social welfare maximization as their main policy goal, introducing the proposed flexibility in Amsterdam's parking policy may have positive welfare consequences.

Considering the variety of possible effects set out above, the question of whether a further decentralization of parking policy from the local authority to city districts will have positive or negative consequences is an empirical one. Unfortunately, however, from the moment of its introduction, monitoring activities to measure the impact of paid parking have been rather inefficient. It will therefore be difficult to assess the consequences of the suggested changes in parking policy.

Conclusions

Parking is a theme in transport that is relatively under-researched. This is an unsatisfactory situation, because parking problems are prominent in urban areas. These problems are caused by several market imperfections in the field of urban parking, such as information costs, search and waiting costs, and visual hindrance. These imperfections may be remedied by the public sector. However, government involvement in parking does not necessarily solve the problem, since it may also give rise to various additional problems (externalities) that may partially or even completely offset a policy's initial welfare-enhancing effects. In the present study we have focused on some examples, with special emphasis on the concept of fiscal federalism. The presented case studies show the relevance of this concept by illustrating that parking policies may lead to welfare losses due to spatial spillovers; policies by one district may well imply negative effects on neighbouring districts. Moreover, potential conflicts exist between government bodies at various spatial levels (for instance, national versus local objectives). Although we address different parking policies at different levels of aggregation in the three case studies, some similarities can de identified.

Two issues that stand out are that the 'ABC policy' did not work well at all, and that the parking policy in Amsterdam only started to be successful after the change in policy regime in 1991. Arguably a common cause in both cases is that a conflict arose between the policy goals of the central government (which formulated the policy) and the local authority (which implemented the policy). This is underlined by the fact that parking policy became successful only after the local authorities were given complete control over policy making.

A conclusion that can be drawn from these findings is that, when policy making has a top-down character in which policy is made by central government and implemented by lower level authorities, in order to avoid conflicting goals and undesirable outcomes, it is necessary to take into account the relevance of (financial) incentives for the lower levels of government. In the light of this, it is interesting to observe that, in the third case study, the regime shift in Amsterdam's parking policy largely consisted of shifting parking policy responsibility from the local authority to the city districts. In addition, the implemented policies are monitored on their compatibility with initially defined policy goals. Thus greater flexibility in policy making is created (in which incentives are almost automatic) without causing conflict in policy goals, in this case between the local authority and the city districts. Although the outcome of the regime shift is still uncertain, it serves as an interesting example of the balance between control and incentive.

References

Bennett, R.J. (1980), *The Geography of Public Finance; Welfare under Fiscal Federalism and Local Government Finance* (London: Methuen).

Bewley, T.F. (1981), 'A Critique of Tiebout's Theory of Local Public Expenditures', *Econometrica* 49:3, 713–40.

Blom-Hansen, J. (1999), 'Avoiding the 'Joint-decision Trap': Lessons from Intergovernmental Relations in Scandinavia', *European Journal of Political Research* 35:1, 35–67.

Bootsma, G. (2003), 'Parkeerbeleid in NL; Trends door de Tijd Heen', (in Dutch), ['Parking Policy in the Netherlands: Trends through Time'] in Congresbundel Colloquium Vervoersplanologisch Speurwerk 2003, *No Pay, No Queue? Oplossingen voor Bereikbaarheidsproblemen in Steden* (Delft: CVS), pp. 1249–69.

de Borger, B. and Proost, S. (2004), 'Vertical and Horizontal Competition in Transport'. Paper prepared for the STELLA meeting of Athens, June, <http://www.stellaproject.org/FocusGroup5/Athens2004/Papers/DeBorger-Proost.doc>.

Braun, D. (2000), 'Territorial Division of Power and Public Policy-Making: An Overview', in D. Braun (ed.), *Public Policy and Fiscal Federalism* (Aldershot Burlington: Ashgate).

Braun, D., Bullinger, A. and Wälti, S. (2002), 'The Influence of Federalism on Fiscal Policy Making', *European Journal of Political Research* 41:1, 115–45.

Castles, F.G. (1999), 'Decentralization and the Post-war Political Economy', *European Journal of Political Research* 36:5, 27–53.

Calthrop, E. (2002), 'Institutional Issues in On-Street Parking'. Paper prepared for the STELLA Transatlantic Thematic Network (Focus Group 5), Brussels, April.

CBS (Statistics Netherlands) (2000), 'Parkeerbelastingen steeds Belangrijker voor Gemeenten; De Auto voor de Gemeente' (in Dutch) ['Parking Taxes Ever More Important for Municipalities'], <http://www.cbs.nl/nl/publicaties/ /publicaties/algemeen/index/index1021.pdf>.

van Dijken, K. (2002), 'Parkeren in Nederland: Omvang, Kosten, Opbrengsten, Beleid'. (in Dutch), ['Parking in the Netherlands: Magnitude, Costs, Revenue,

Policy'], Economisch Onderzoek voor de Publieke Sector, IOO/AVV, Zoetermeer/ Rotterdam.

Feitelson, E., and Rotem, O. (2004), 'The Case for Taxing Surface Parking', *Transportation Research Part D* 9:4, 319–33.

Gemeente Amsterdam (2004), 'Raadsvoordracht Menukaart Parkeerregelingen, Amsterdam' (in Dutch) ['City Council Meeting Menu Parking Measures'], <http://www.ivv.amsterdam.nl/nieuws/parkeren_ 06052004.htm>.

Inman, R.P. and Rubinfeld, D.L. (1997), 'Rethinking Federalism', *Journal of Economic Perspectives* 11:4, 43–64.

Kamer van Koophandel (2002), 'Visie KvK Noordwest-Holland; Locatiebeleid voor Noordwest-Holland' (in Dutch) ['Vision Netherlands Chamber of Commerce North West Holland: Location Policy for North West Holland'], Alkmaar, <http://assets.kvk.nl/assets/ /NoordwestHolland/artikelassets/Kamerbulletin/ Locatiebeleid.pdf>.

Martens, M., Luipen, B. and de Wit, T. (2003), 'Geintegreerd Parkeerbeleid Gaat niet Vanzelf; Ervaringen met het Proces', (in Dutch), ['Integrated Parking Policy: Experiences with Policy Processes'], in Congresbundel Colloquium Vervoersplanologisch Speurwerk 2003, *No Pay, No Queue? Oplossingen voor Bereikbaarheidsproblemen in Steden* (Delft: CVS), pp. 1271–9.

Oates, W.E. (1972), *Fiscal Federalism* (New York: Harcourt Brace Jovanovich).

Painter, M. (1991), 'Intergovernmental Relations in Canada: An Institutional Analysis', *Canadian Journal of Political Science* 24:2, 269–89.

Reisen, A.A.J. and Hus, J. (2003), 'To Pay or Not to Pay; Keuzemogelijkheden in Stedelijk Gebied' (in Dutch), ['To Pay or Not to Pay: Choice Options in Urban Areas'] in Congresbundel Colloquium Vervoersplanologisch Speurwerk 2003, *No Pay, No Queue? Oplossingen voor Bereikbaarheidsproblemen in Steden* (Delft: CVS), pp. 1281–300.

Rietveld, P. (2006), *Urban Transport Policy; The Dutch Case*, in R.J Arnott and D.P. McMillen (eds), *A Companion to Urban Economics* (Oxford: Blackwell), pp. 292–310.

Rietveld, P. and van Ommeren, J. (2002), 'Company Cars and Company Paid Parking', in W.R. Black and P. Nijkamp (eds), *Social Change and Sustainable Transport* (Bloomington: Indiana University Press), pp. 201–08.

Scharpf, F.W. (1988), 'The Joint-decision Trap: Lessons from German Federalism and European Integration', *Public Administration* 66:3, 239–78.

Shoup, D. (1997), 'Evaluating the Effects of Cashing Out Employer Paid Parking; Eight Case Studies', *Transport Policy* 4:4, 201–16.

Wibbels, E. (2000), 'Federalism and the Politics of Macroeconomic Policy and Performance', *American Journal of Political Science* 44:4, 687–702.

Young, W. (2000), 'Modeling Parking', in D.A. Hensher and K.J. Button (eds), *Handbook of Transport Modelling* (Oxford: Pergamon), pp. 409–20.

Zwartjes, S.J., van Dijken, K. and Harteveld, C. (2001), 'De Publieke Kosten van Parkeren: Wie Betaalt Wat?' (in Dutch), [The Public Costs of Parking: Who Pays for What?], in Congresbundel Colloquium Vervoersplanologisch Speurwerk 2001, *Wie Doet Wat?* (Delft: CVS), pp. 571–85.

Index

For Product Safety Concerns and Information please contact our EU representative GPSR@taylorandfrancis.com
Taylor & Francis Verlag GmbH, Kaufingerstraße 24, 80331 München, Germany

www.ingramcontent.com/pod-product-compliance
Lightning Source LLC
Chambersburg PA
CBHW060329220326
41598CB00023B/2657

* 9 7 8 1 1 3 8 2 6 7 5 9 6 *